Interpreting Quantum Mechanics

This novel text directly addresses common claims and misconceptions around quantum mechanics and presents a fresh and modern understanding of this fundamental and essential physical theory. It begins with a non-mathematical introduction to some of the more controversial topics in the foundations of quantum mechanics. For those more familiar with the theoretical framework of quantum mechanics, the text moves on to a general introduction to quantum field theory, followed by a detailed discussion of cutting-edge topics in this area such as decoherence and spontaneous coherence. Several important philosophical problems in quantum mechanics are considered, and their interpretations are compared, notably the Copenhagen and many-worlds interpretations. The inclusion of frequent real-world examples, such as superconductors and superfluids, ensures the book remains grounded in modern research. This book will be a valuable resource for students and researchers in both physics and the philosophy of science interested in the foundations of quantum mechanics.

David W. Snoke is a Distinguished Professor of Physics at the University of Pittsburgh and leads a laboratory studying fundamental optical effects. In 2006 he was elected a Fellow of the American Physical Society "for his pioneering work on the experimental and theoretical understanding of dynamical optical processes." He has published over 180 articles in science and philosophy journals, and five books, including *Solid State Physics* (2nd edition published by Cambridge University Press, 2020), *Universal Themes of Bose–Einstein Condensation* (Cambridge University Press, 2017), and the well-known "green book," *Bose–Einstein Condensation* (Cambridge University Press, 1996).

Interpreting Quantum Mechanics
Modern Foundations

David W. Snoke

University of Pittsburgh

CAMBRIDGE
UNIVERSITY PRESS

Shaftesbury Road, Cambridge CB2 8EA, United Kingdom

One Liberty Plaza, 20th Floor, New York, NY 10006, USA

477 Williamstown Road, Port Melbourne, VIC 3207, Australia

314–321, 3rd Floor, Plot 3, Splendor Forum, Jasola District Centre, New Delhi – 110025, India

103 Penang Road, #05–06/07, Visioncrest Commercial, Singapore 238467

Cambridge University Press is part of Cambridge University Press & Assessment, a department of the University of Cambridge.

We share the University's mission to contribute to society through the pursuit of education, learning and research at the highest international levels of excellence.

www.cambridge.org
Information on this title: www.cambridge.org/9781009261555

DOI: 10.1017/9781009261562

First published 2024

A catalogue record for this publication is available from the British Library

A Cataloging-in-Publication data record for this book is available from the Library of Congress

ISBN 978-1-009-26155-5 Hardback

Cambridge University Press & Assessment has no responsibility for the persistence or accuracy of URLs for external or third-party internet websites referred to in this publication and does not guarantee that any content on such websites is, or will remain, accurate or appropriate.

Oh, the depth of the riches and wisdom and knowledge of God!
How unsearchable are his judgments and how inscrutable his ways!
—Romans 11:33

Contents

Preface

The overall premise of this book is that, while quantum mechanics is strange in some ways, it is far less strange than you probably think. Most of the intuition you need to understand quantum mechanics can be drawn from your intuition about water waves.

Much of this book is aimed at bringing the discussion of quantum mechanics up to the present day in regard to what quantum physicists know. This will include a fair amount of "debunking" of claims made in the past about quantum mechanics that even many physicists today assume are true. For example, the Planck radiation spectrum and the photo-electric effect have nothing to do with proving that particles exist as localized little objects, as we will see. This book will also include critiques of some widely held interpretations of quantum mechanics.

When the modern understanding is taken into account, much of the strangeness of quantum mechanics disappears, but not all of it. There are some truly strange results, and most of these involve *nonlocality*, the apparent effect that things in one place can affect things far away without any signal (that we know of) traveling from one place to the other. There are also open questions, such as whether the randomness we see is the result of tiny, but real, fluctuations not presently accounted for by the equations of quantum mechanics.

This book can be read at several levels. Part I requires no mathematical knowledge, and gives the overall perspective of this book. I strongly encourage advanced readers not to skip this part, because it includes many new perspectives on what we think we know about quantum mechanics. Part II requires only introductory college math, and works out some basic examples relevant to Part I.

A major contention of this book is that the proper way to start thinking about all philosophy of quantum mechanics is with quantum field theory. But many philosophers and even many physicists never study quantum field theory, because it is assumed to be a very high-level theory understandable only to a few experts. Part III of this book gives an introduction to all of the essential elements of quantum field theory needed to think about the philosophy properly. This section starts at the beginning but will be most suited for people who have already taken at least one upper-level course on quantum mechanics.

While most of the philosophy of quantum mechanics can be discussed without math, there are some arguments that require math. Part IV is a supplement to Part I that gives specific mathematical arguments relevant to the philosophical interpretations under debate.

Finally, Part V presents advanced theory of decoherence, to which I and my coworkers have made original contributions in the literature. This material is appropriate for students who have taken graduate quantum mechanics and quantum field theory or quantum optics classes. Some of the results, however, are accessible to people with less training, if they are willing to skip over the proofs.

I have talked with too many people over the years about quantum philosophy to properly thank them all. Particular discussions that come to mind are those with Harvey Brown, Časlav Brukner, Erica Carlson, Andrew Daley, Steve Girvin, Bob Griffiths, Richard Jones, Andrew Jordan, Ruth Kastner, Tony Leggett, Roger Mong, John Norton, John Sipe, Fernando Sols, David Wallace, Peter Zoller, and Wojciech Zurek. I also thank all of the graduate students at the University of Pittsburgh who endured the philosophical tangents in my classes, and my wife Sandra for her constant support.

Soli Deo Gloria

Part I

A Nonmathematical Exposition of Quantum Mechanics and Quantum Field Theory

It's All Fields and Waves

If you have read anything about quantum mechanics before, you probably expect that this book will begin by talking about particles. Most popular books on quantum mechanics talk about spooky behavior of particles that are here and not here, jump from one place to another without any cause, and so on.

There *is* something mysterious in quantum mechanics, but it actually has very little to do with the existence of particles. In talking of the philosophy of quantum mechanics, many people have focused on the topic of "wave-particle duality." In this book, we will see that modern theory does not treat these two things on an equal footing. Waves are fundamental in quantum theory; particles are not. To use a big word, particles are "epiphenomena," that is, theoretical constructs that are useful in some circumstances but that can be completely derived from other, more fundamental elements of the theory.

Saying something is not fundamental does not mean that it is a useless concept. For example, I could explain the sliding of different solids by invoking the concept of a friction force; the friction force is a very useful concept. But if we wanted, we could derive it from microscopic forces between atoms. Friction force is not a fundamental force of nature; it is a handy concept in many cases, such as when many atoms in two solids are sliding against each other. In other cases, if we insisted on the existence of friction force, we would run into strange conundrums. For example, if we wanted to talk about the friction force between single atoms, we would end up speaking nonsense. This often happens when a concept is pushed beyond the limits of its applicability.

Many scientists learned the particle picture of quantum mechanics in introductory physics in college and never questioned it afterwards. But anyone who has gone on to study the advanced theory of quantum mechanics taught in graduate school, known as *quantum field theory*, knows that particles arise as oscillations of the underlying fields in nature. This field theory is the most accurate theory we have for quantum mechanics; the Schrödinger single-particle equation model presented in introductory classes is known to be a simplified case of the more general field theory. Instead of discussing quantum mechanics in terms of that simplified model, let us go right to the heart of the more basic theory, the theory of fields.

1.1 Fields

Talking about fields sounds like science fiction, but it is not really so strange. The best way to imagine a field is to think about the physical system that was the historical basis

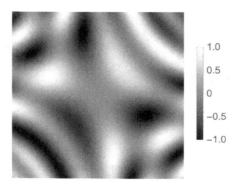

Figure 1.1   An example of a scalar field. The gray scale represents the numerical values.

of field theory: a fluid, like water. All of our mathematics for fields was developed in the early 1800s to describe the flow of fluids. This mathematical framework was then adopted by James Maxwell in the 1860s to describe the fields of electricity and magnetism, and then his concepts were eventually adapted by Paul Dirac and others in the 1920s and 1930s to give us our modern formulation of quantum mechanics. Along the way, most physics departments stopped teaching the theory of fluids ("hydrodynamics") and now start with just Maxwell's field theory. But thinking about water flow helps us to visualize fields.

Imagine a container filled with a fluid like water everywhere inside its volume. We can describe this in mathematical language by assigning numbers to every point inside the fluid. One example would be to write down a number for the density of the fluid at each point. The density of the fluid at all locations is an example of what is called a *scalar field* – at each point in space, there is just one number to write down, namely the density of the fluid at that point. Figure 1.1 illustrates this idea.

At every point in the fluid, we could also identify the direction and speed of its flow. To do this, we would need several numbers. One number could give us the speed, in miles per hour or meters per second, and another number or two could give us the direction. For example, we could give the angle of the flow relative to due east. If we allow up-and-down motion, we could also define the angle of ascent relative to the ground.

In this case, instead of just one number, we would have two or three numbers to assign to every location inside the fluid, to describe the flow speed and direction at each point. This set of numbers is collectively called a *vector*. But just like the case of the density, we could assign a numerical measurement to every point in space. The distribution of velocities of the fluid is what we call a *vector field*.

Figure 1.2 shows two ways of representing a vector field using pictures. In the first case, the direction of the flow is indicated by little arrows at each point in space. In the second, the direction is given by *field lines*. In both of these ways of illustrating fields, there is a lot of empty space between the arrows or lines, which is not labeled. One must imagine that, for a real field, every single point in space could have an arrow or line assigned to it. The illustrations draw the arrows or lines more sparsely just to make the picture easier to understand.

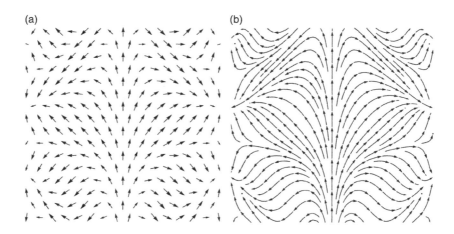

Figure 1.2 Two ways to draw a vector field. (a) Arrows giving the strength and direction of the flow speed at many points in space. (b) The same field represented by field lines.

Thinking about a fluid like water helps us to understand what we think of as "real." We have seen that there are several different ways of describing a water field. One way is as a set of numbers. (In most cases, we don't actually have to write down all the numbers individually; instead, we can write down a mathematical formula that tells us how to calculate the numbers.) For the density field of the water, another way to describe the field is with a grayscale image like that shown in Figure 1.1. For the vector field of the water flow, other ways of describing it are with little arrows or with field lines, as in Figure 1.2. In each case, what is real is the water itself and its properties, specifically density and velocity. The sets of numbers or vectors, or the pictures, are not the fundamental reality; rather, these are ways that we describe the underlying reality. We could describe the field with different numbers. For example, instead of writing the speed of the fluid in miles per hour, we could write the speed in kilometers per second, or furlongs per fortnight.

It is a main contention of this book that this distinction doesn't change when we switch to talking about quantum fields. Just as water is a "thing" that can be described by a mathematical formalism, so the fields of quantum mechanics are "things" that exist, whether or not we choose to write down mathematical descriptions or pictures for them. This is not always how physicists and philosophers talk, but, as we will see, if we accept the notion of the reality of water flow, then it is artificial to drop the notion of reality for other types of fields.

Electric and magnetic fields. In the early 1800s, the famous experimental physicist Michael Faraday made the intellectual leap that electricity and magnetism could be described by fields very similar to the flow of fluids; he actually drew pictures of electric and magnetic fields circulating around in space like water. Figure 1.3 shows how a magnetic field can be seen directly in the alignment of small iron filings. The somewhat disturbing implication of Faraday's work is that electric and magnetic fields flow through all space, even through matter such as our bodies.

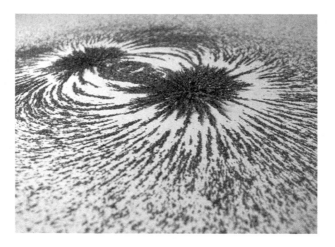

Magnetic field lines from a bar magnet (placed under the board) seen in the alignment of small iron filings. (Windell H. Oskay (www.evilmadscientist.com).)

The Scottish physicist James Clerk Maxwell later succeeded in writing down mathematical equations that accurately described this flow. Figure 1.4 gives these equations, but there is no need to understand these equations to see their main feature: they are very short and compact, written in just four lines. Maxwell's equations remain to this day one of the most impactful and elegant successes of physics. Although these equations are short, they fully describe a huge range of effects of electricity, magnetism, light and optics, X-ray radiation and gamma rays, infrared heat radiation, microwaves, and radio waves. They led directly to the technology of radio and television which revolutionized communications. They also needed no corrections when Einstein's theory of relativity came along. In fact, the accuracy of Maxwell's equations played a major role in inspiring Albert Einstein to come up with his theory of relativity; Einstein wanted to keep Maxwell's equations the same for all observers in the universe. In a way, Einstein was just mopping up the implications of what Maxwell wrote down. Not only that, Maxwell's field theory is not changed significantly by

$$\nabla \cdot \vec{E} = \frac{\rho}{\epsilon_0}$$

$$\nabla \cdot \vec{B} = 0$$

$$\nabla \times \vec{E} = -\frac{\partial \vec{B}}{\partial t}$$

$$\nabla \times \vec{B} = \mu_0 \epsilon_0 \frac{\partial \vec{E}}{\partial t} + \mu_0 \vec{J}$$

Maxwell's equations for electric field $\vec{E}$ and magnetic field $\vec{B}$, for a charge density ρ and current density $\vec{J}$. The values of ε_0 and μ_0 are universal constants that together give the speed of light, according to $c = 1/\sqrt{\varepsilon_0 \mu_0}$.

quantum mechanics. In quantum theory, we learn that Maxwell's equations are correct for a special class of fields known as *coherent* fields.

Every physicist today learns Maxwell's equations as an example of a field theory. Just as we can assign a direction or density to water flow, these equations assign a direction and strengths of the electric and magnetic fields at every point in space. The electric and magnetic fields at every point in space (including inside your body) make up together what is called the *electromagnetic field*.

Understanding the nature of the electromagnetic field is one of the first major conceptual leaps that modern physics requires. Are electromagnetic fields "real"? In the case of water flow, the field is obviously "real" by any normal definition: we can look at the surface of a river and see the flow of the water. That is, we can see its velocity field directly. In the case of the electromagnetic field, the concept is a little more difficult. What we measure are forces; for example, the electric force needed to deflect a pointer in a meter – the same electric force you feel when you rub a balloon on your hair and stick it to a wall. These forces always act on objects. One could therefore argue that only the objects and forces are real, and the field in the empty space between them is just a mathematical trick for keeping track of the complicated forces between the objects. Some people do argue this way (e.g., Mead 2000). But the vast majority of scientists today view electromagnetic field as "real." In fact, their sense of the reality of forces and fields is inverted: the field is the real thing, which flows throughout all space, and the forces we measure are just the specific effects of that field in particular times and places.

One reason for seeing the electromagnetic field as real is that electric and magnetic fields can carry energy and pressure through the vacuum of empty space. We all have a very immediate experience of this. The sun sends light radiation through the vacuum of outer space to us every day, and this is responsible for the warmth of the earth, and our life. A laser beam traveling through outer space is also an example of the field carrying energy and pressure. You might not think that light exerts pressure, but it does – a laser beam, and even a light bulb, pushes any object it hits, with a tiny pressure. Some scientists have proposed using vast sails in outer space to use the pressure from sunlight to sail around the solar system.

In the 1800s, it bothered people to think about how the vacuum of outer space could carry energy and pressure. Some argued that the particle picture is needed to understand this. In this view, light particles (photons) travel through outer space like little bullets. It is easy to imagine energy and momentum being carried by these little particles, and pressure coming from them when they hit something.

But the particle picture is not necessary for understanding how light carries energy and momentum through space. All the theory we need is in Maxwell's equations, which tell us exactly how much energy and pressure is carried by light waves through vacuum. The electromagnetic field carries these just as a water wave does, and Maxwell used the mathematics of fluid pressure to describe how electromagnetic waves do this.

Quantum fields. The next conceptual leap of modern physics is to envision the field that underlies everything we think of as normal matter. Just as there is an electromagnetic field that fills all of space everywhere, the fundamental theory of quantum mechanics says that there is also a "matter field" that extends everywhere in space. This field is described by equations, based largely on the work of Paul Dirac, that are very similar in form to Maxwell's equations.

In classical electromagnetic field theory, the field at each point in space is determined by the electric and magnetic forces there. This has the advantage that we can equate the numbers of the field to physical measurements that we could have made, at least in principle. In quantum field theory, there is an additional conceptual hurdle, which is that the number value given to the field at each point doesn't even correspond to a force. We can only deduce the values of these numbers indirectly. For this reason, some physicists who accept that the electromagnetic field is real get off the boat at this point and say that the matter field is not real.

But in quantum field theory, the equations that describe the electromagnetic field and the equations that describe the matter field *have the same fundamental nature.*[1] We will look into this in more detail in the coming chapters; for those who are mathematically inclined, the basic math is given in Part III of this book. The general point is this, however: we have an ascending ladder of fields introduced in physics, from water currents and air flow, to electric and magnetic forces, to matter fields, which have exactly the same type of math. We are comfortable with calling the water currents "real," and most physicists are comfortable with calling electromagnetic fields "real." Why then make a cutoff and refuse to call quantum matter fields real, when the basic structure of the mathematics is the same, and the way the equations predict the results of experiments are the same? It is true that, just because some equations are formally similar, we need not take them as describing the same thing; science has many examples of equations that are similar but describe different things. But in the case of matter fields and classical fields, not only are the mathematical structures the same; the elements in the different fields are interchangeable – matter fields generate light and sound fields, and vice versa. (We will come back to this in Section 4.5.) And all classical fields can be deduced as special cases from quantum fields. There is simply no fundamental reason to draw a sharp line between the nature of physical reality of different types of fields.

---

[1] The only significant difference is that what we typically think of as matter fields are *fermion* fields, while electromagnetic fields are *boson* fields. The distinction between these two will be discussed in Section 2.4; in the math, the only difference is a change of a single sign from positive to negative. Classical fields such as water velocity fields (described by the *Navier–Stokes* equations) and electromagnetic fields (described by Maxwell's equations) are now understood to be special cases of the more general formalism for quantum boson fields, known as *coherent states*, discussed mathematically in Section 12.5.

A commonly repeated statement is that the mathematical structure of quantum fields involves *operators*, while classical fields do not, and that this gives an essential difference between the two (the mathematics of operators are discussed in Section 11.1). This is actually a misunderstanding based on different uses of the term "field" in the literature. Classical and quantum fields can both be acted on by mathematical operators, and in both cases, the operators by themselves do not carry any information about actual physical states. When the operators have a spatial dependence, one can talk of an "operator field," but this is not a physical entity for either classical or quantum fields. For the mathematical details of this, see the discussion of spatial field operators in Section 12.2.3.

Others have argued (e.g., Griffiths 2003) that the fact that matter waves involve complex numbers means that they are fundamentally different from electromagnetic waves. But as shown mathematically in Section 13.1.2, complex numbers are just a useful bookkeeping device to keep track of two degrees of freedom of a fermion field. Section 9.1 also addresses this view.

The perspective shared by most physicists familiar with quantum theory is that matter fields really do exist. All the classical fields we are familiar with, such as the electric fields that raise our hair in the presence of static electricity, and currents in water, can be derived from underlying quantum fields. The quantum fields are more fundamental, or one might say "more real," than electric forces and water currents.

Quantum field theory is not some alternative version of quantum mechanics. It is the most generally accepted, basic theory of quantum mechanics as understood by experts. And it makes no sharp distinction between light waves, electron waves, and water waves.

1.2 Waves

The previous section introduced the concept of a *field*. The next important concept is a *wave*. Just as we defined a field with the concrete example of a fluid, we can use a very tangible example to define a wave, which is just what its name sounds like: a water wave. If you go to the ocean and see waves, or if you make waves in a swimming pool, you know what a wave is. The important thing to keep in mind about waves is that energy and pressure are carried by waves from one place to another even though the water itself goes nowhere. Think about it: if someone were sitting on the opposite side of a swimming pool, you could push the water, creating a wave, and the wave would move to that person and hit them. Yet the water level in the pool would stay the same. You put energy and pressure into the water, and the water carried that energy and momentum somewhere else, even while the water itself stayed in the same place.

The water in this case is the fluid field we discussed in the previous section. The wave is an oscillation of that field. In the same way, the electromagnetic field and the quantum matter field can have wave oscillations.

Let us define some generic terms used for all waves. Table 1.1 gives a summary of these. The first important term is the *wavelength* of a wave, which is the distance between

Table 1.1 Standard wave terms.			
Name	Usual symbol	Definition	Measurement
Wavelength	λ	Distance between the crests of a wave	Measured as a length
Frequency	f	Number of wave crests passing a point per second	"Hz" = number per second
Phase	ϕ	Degree of completion of the oscillation cycle of a wave	Ranges from 0 to 360°
Amplitude	ψ	Range of variation of whatever is oscillating	Different for each field

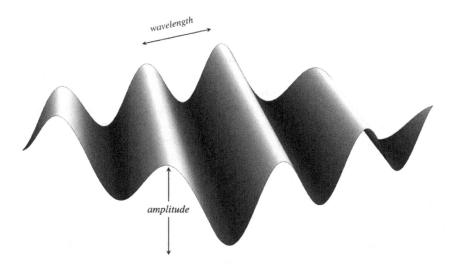

wavelength

amplitude

Figure 1.5 A typical wave, with the definition of two terms for describing the wave.

its crests, as illustrated in Figure 1.5. The next is the *frequency* of a wave, which is the number of wave crests moving past some location within a period of time. The frequency and wavelength of a wave are not independent of each other: if we pick one, then we can deduce the other from the properties of the field. A fast oscillation has high frequency (many wave crests per second) and short wavelength; a slow oscillation has low frequency and long wavelength. Frequency is measured in crests per second; instead of saying "crests per second," scientists and engineers use the term *Hertz*, after the famous scientist of that name, and abbreviate this as *Hz*. The oscillating electric wave in your home electric circuits has a frequency of 60 Hz (60 crests per second), while radio waves have frequencies of millions of crests per second, that is, mega-Hertz (written in the metric system as MHz). Sound waves in the audible range have frequencies of tens of Hz up to a few thousand Hz, and ultrasound waves, used for seeing inside a human body, have frequencies around 1 MHz.

The electromagnetic field can oscillate just like the surface of a swimming pool. When the electromagnetic field oscillates, we detect it as light waves, radio waves, or microwaves, and so on. The only difference between these types of waves is the *frequency* of the waves. Radio waves have much longer wavelength than light waves but are otherwise just the same.

The quantum matter field can also oscillate. Matter waves can have short wavelength or long wavelength. In general, it takes more energy to create a wave with high frequency than it does to create a slow oscillation with low frequency.

In each type of wave, we call the range of oscillation the *amplitude* of the wave, as illustrated in Figure 1.5. In the case of water waves, it is easy to see the amplitude: as the water goes up and down, the amplitude is the height of the waves. It is not so easy to visualize the oscillation of electromagnetic waves or matter waves. But something is really oscillating. In the case of an electromagnetic wave, the strength and direction of the electric

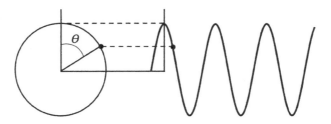

Figure 1.6   The phase of a wave equated to the position of a clock hand. The angle θ is the *phase angle* corresponding to a point in the cycle.

force oscillates. If you had a small charged object, it would feel an oscillating force if an electromagnetic wave passed by. The force would change direction from left to right, or up to down, and then back the other way. The amplitude in this case would be measured as the maximum force felt by the charged object. In the case of a matter wave, nothing is moving left to right or up to down. But something is still oscillating.

One last general wave term is the *phase*. This concept, like many other scientific concepts, suffers from degradation in the popular media: "set phasors to stun." Phase is actually very simple. Consider the left side of Figure 1.6, with a clock hand that can go around a circle. The phase of the clock hand is its position in the circle. This can be measured as an angle. For example, we could define 12:00 PM as angle zero, 3:00 PM as the 90° angle, 6:00 PM as 180°, and so on. The phase goes all the way around to 360° and then starts over again.

We can extend this definition of phase to any cyclic oscillation. (This is why we talk of the phases of the moon, in its cycle from new to full and back.) With a water wave, the wave goes up and down. We can define the highest point as phase of zero, and the lowest point, when it is halfway through its cycle, as phase of 180°; when it gets back to the highest point, the phase reaches 360° and starts over. Figure 1.6 illustrates how the position in an oscillation can be equated to the position of a clock hand. The water is not really moving in a circle; it is moving up and down, but since it is moving through a periodic cycle, we can equate the different stages in its cycle as different phase angles around a circle.

In a matter wave, the phase of the matter field also oscillates in a cycle, so that its phase goes from 0 to 180°, 360°, and so on. The phase of a matter wave is not just a fictional concept. It can be measured in experiments, for example, in the current of superconductors and in small ("mesoscopic") electronic circuits. One way this is done is with interference and tunneling experiments, which are described in Section 1.3.

1.3 Basic Wave Effects

All of these waves in fields, whether we are talking about water waves, sound waves (oscillations of the air), electromagnetic waves (light, radio, etc.), or matter waves, have certain common properties. These properties all seem normal when we are thinking about typical "energy" waves like sound and light but can seem strange when we are talking about matter.

The first basic wave property is called *superposition*. This property means that two waves can pass through each other, and, while they are passing through each other, the behavior of the whole system is just the sum of the two waves.

This property is not absolute. It breaks down when there are "nonlinear" terms in the wave equations, and, in almost all real systems, there are some nonlinear terms that crop up in some cases. Many modern physics experiments study these nonlinear effects. But for the most part, we can assume that the fields are linear and therefore the principle of superposition is true. As discussed in Section 2.3, nonlinear effects are one reason why the notion of indivisible quantum particles breaks down in some cases.

Waves passing through each other seem normal for waves that we think of as pure energy waves. For example, it doesn't seem strange that a beam of light from one flashlight can pass through the beam of light from another flashlight, or that a water wave created by a person splashing on one side of a pool can pass though the wave created by another person on a different side of the pool. Our normal experience, however, is that solid matter does not pass through other matter. Matter waves passing through other matter waves seem very strange. It can happen, however. The reason why it does not happen more often has to do with the specific property of matter waves called *Pauli exclusion*, which we will discuss in Chapter 2.

Given the property of superposition of waves, a direct consequence is the possibility of *interference* of waves. Interference occurs when one wave cancels out another wave. Suppose that two water waves are passing through each other, as shown in Figure 1.7. At some points, the wave crest from one wave is moving up while the crest from the other wave is moving down. The two waves will cancel at those points, leaving the water motionless there. At other points, the two waves will add up, to give a wave higher than either. You may not have seen interference of water waves (though it is easy to demonstrate), but you have certainly seen it with light waves, and may have heard it with sound waves. If you have seen rings of different colors on an oil slick, as shown in Figure 1.8, you have seen interference

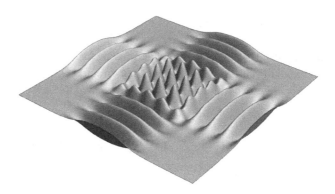

Figure 1.7   Computed interference pattern of two waves passing through each other.

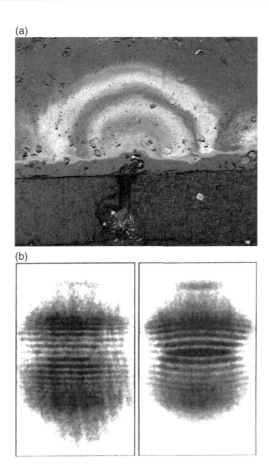

(a)

(b)

Figure 1.8 Pictures of interference in the real world. (a) Fringes of light reflected from oil, caused by the interference of light from the top surface and the bottom surface of the oil, with slight thickness variations of the oil (Creative Commons Attribution – Share Alike 2.5 Generic). (b) Interference patterns of matter waves composed of whole atoms, measured at very low temperature. From Andrews 1997.

of light. What is happening is that the light waves bouncing off the top surface of the oil are interfering with light waves bouncing off the bottom surface, so that in some places the light of a certain color cancels out. Sound can do the same thing. There is now an industry selling active sound-reduction headphones. These generate sound exactly opposite of the sound coming in, canceling it out and creating silence.

Again, this doesn't seem so strange with light and sound waves, but very strange for matter. Imagine taking a solid object and canceling it out with another object, so that you have nothing. Yet experiments have been done that do essentially that: electronic current can be canceled out through interference. When electric current sent down one wire is sent into the same place as electric current sent down another wire, for certain experimental conditions, they can cancel out, giving no current. This is not science fiction; it is reality, and the

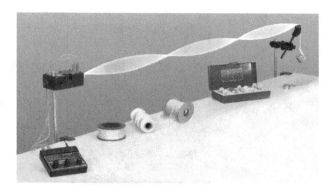

Figure 1.9 A vibrational resonance of a string produced by an oscillator at one end. The frequency is such that three half-wavelengths fit exactly into the available space. (Greg Severn, University of San Diego.)

basis of useful devices for measuring magnetic field, as discussed further in Sections 2.6 and 8.5. More recently, whole atoms have been canceled out, as shown in Figure 1.8(b), to make fringes similar to the ones you see on an oil slick.

Another wave effect that will be extremely important in all of this book is *resonance*. Resonance means that some things have "natural" frequencies that they vibrate at. This is familiar to all musicians. For example, a tuning fork resonates at a certain frequency if you hit it, as does a stretched piano wire on a piano. Figure 1.9 shows an example of a resonant mode of a taut string.

An object can have more than one resonance. For example, if you blow into a bugle, you can get different notes by blowing differently; the different notes correspond to different resonances of the sound wave in the air in the bugle. In the same way, a tuning fork or a bell can have "overtones" (higher-frequency resonances) if you hit it hard enough.

A crucial property of wave resonances is that they can be sharply defined, with jumps in between. Blowing into a bugle, you get distinct notes, not a continuous smear of sound. Because resonances can be sharply defined, and the system can stay in a resonance for a long time, they are often called stable *states* of the system. There is always one lowest-frequency resonance, which corresponds to the longest wavelength; this is called the *ground state* of the system. Higher-energy states, which have shorter wavelength and higher frequency, are called *excited states*.

In quantum mechanics, not only sound waves and electromagnetic waves but also matter waves can have resonances. The famous "jumps" between electronic states of the atom, which we learn in chemistry class, correspond to transitions between resonances of the electron matter wave. These jumps are not instantaneous, however; we will return to discuss this in Chapter 3.

Resonances are not always sharply defined. If there is energy loss in the system, the resonances may be smeared out. For example, when you tune a radio, you are tuning it to resonate at different frequencies of radio waves. Each radio station sends out its signal at a different frequency; this is what the call numbers stand for (e.g., 104.5 MHz is 104.5 million wave crests per second). In a cheap radio, the resonances of the radio are not very

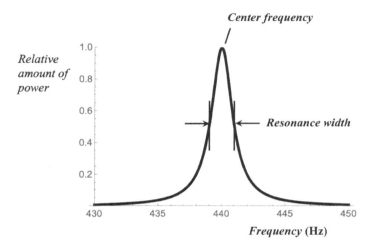

Figure 1.10 A typical resonance curve. The ratio of the center frequency to the width of the resonance is called its Q-factor.

distinct, and you can pick up more than one radio station at the same time. Figure 1.10 shows some of the characteristics of a resonance.

One more basic wave effect is *partial transmission*. You are quite familiar with this effect of waves: a person may be speaking in another room, and you hear the sound coming through the wall. The sound is quieter than it would be if you were in the same room as that person, because only part of the sound gets through the wall and the rest is reflected. The same thing happens with light: if light hits a glass surface, part of it goes through, and part of it is reflected, as shown in Figure 1.11. The same thing happens with the windows in your house.

Think how strange this would be if normal matter acted this way in daily life. Imagine a person hitting a wall, and one copy of the person passing through the wall while another copy of the person bounces off! We are used to matter being indivisible, going one way or

Figure 1.11 A laser beam being split into two by a glass surface. (C.-M. Zetterling, Royal Institute of Technology, Sweden.)

another but not both. We can break an object into two parts, but we never see one object that does two things simultaneously.

Yet this effect does happen with matter, also. When matter is partially transmitted through a barrier, it is known as *tunneling*. This has been seen in all kinds of experiments; for example, electrons in a superconductor can tunnel through barriers, and nuclear matter inside an atom can tunnel out, leading to the well-known effect of nuclear radiation. We will return to discuss this in Chapter 8.

1.4 The Return of the Ether

In the case of water waves, it is clear what is oscillating, or waving: the water. What is waving in the case of electromagnetic waves or matter waves? This question bothered many people in the 1800s. The answer, as understood in modern physics, is clear: the field is oscillating. Both the electromagnetic field and the quantum matter field, which exist everywhere in space, can oscillate in a measurable way.

This sounds a lot like saying that the "ether" oscillates. The concept of the *ether* is a very ancient concept, which scholars envisioned as a fluid filling all space. (From this term, we get the word "ethereal," referring to something invisible and immaterial.) Many people who have read about philosophy of science have heard that the concept of the ether was put to death at the end of the 1800s, and now nobody believes in it. But actually, scientists have just substituted another word for the ether, and called it the "field"!

Let us take a short excursion into basic logic. A longstanding conclusion of pure logic is the principle that "nothing cannot exist." This is a tautology: by using the verb "exist," we imply that something exists which is doing the existing. Therefore, to exist is to *be something*. But if that something exists, then it is not "nothing." Thus, to say that "nothing" exists is to engage in self-contradiction, to say that something exists, but it doesn't.

When we talk about the vacuum of outer space, we mean that there is no *mass* there, but it is not "nothing." There is always a *field* there; specifically, both the electromagnetic field and quantum matter field. These fields may have no waves happening there, but the fields still exist, just as in a pool with no waves, the water still exists.

So throughout all space, "something" exists, which we call a "field," and that field can oscillate. We can call this the ether, if we want. Why then did the old concept of the ether get rejected?

Essentially, the old ether concept that got rejected was one that was pinned to a static concept of space and time. Einstein showed that our sense of time passing and our sense of distance depend on our point of view, and people could watch the same events and disagree about when and where they took place, even while agreeing that the same events happened.

The old concept of the ether was, in effect, nailed down to space, like the water in a lake. If you are in a boat and you go fast enough, you can measure your speed relative to the water, and you can even go faster than the waves you create. Every boater is familiar with this effect: as the boat speeds up, the waves stretch out to the rear. Waves only advance out in front of the boat if it is going slower than the waves do.

If light waves were like that, then if you were on a rocket going near the speed of light, and you shone a flashlight forward, you would see the light piling up at the front of the rocket, like water waves piling up at the front of a speeding boat. Einstein was specifically concerned to make sure that his theory did not allow this to happen, because he was certain that Maxwell's equations are universally true from every perspective in the universe. He based this on the intuition that in the vacuum of outer space, far away from any matter, there would be no way to know whether you were moving or not.

When the proper math of Einstein's relativity is taken into account, the result is that no matter how fast you go, light waves travel away from you at the same speed in all directions, and therefore you can't measure your speed relative to the field by looking at the light waves. The same is true for the quantum matter field: both the electromagnetic and quantum matter fields are *relativistic* fields. That means that the field equations are the same no matter how fast you go. (This incidentally implies that the speed of light which goes into the equations must always be the same, a premise that is often used in calculations of relativity.)

It is important to understand what relativity says and what it does not say. It does not say that anyone can believe anything he or she wants. It says that every observer everywhere can compute the numerical values of the fields using the same set of equations. If they are traveling at different speeds, they will come up with different numbers for the field values. But the laws of relativity give very definite rules for how to transform the field values for an observer at one time and place to the field values measured by an observer at another location, traveling at a different speed.

Thus, the field is always still there – it does not vanish for any observer. Every observer can detect a field filling all of space, and every observer can perceive that the field can oscillate, leading to waves. So it is very much like the ether of old. It is just not a nailed-down ether. It looks different from different perspectives, but it is always there. Physicists agree that the field is everywhere present, but some do not like to call the field a "medium," because, by their definition, a medium is something pinned down, like the old version of the ether. But if we define a medium as simply whatever it is that is oscillating, then it is perfectly appropriate to say that the electromagnetic and quantum matter fields are the mediums for waves.

Not only is the vacuum field not "nothing," but, in the modern view, the vacuum field is quite active. The remnants of the light emitted at the Big Bang still fill all of outer space, which means that there are electromagnetic waves bouncing around everywhere. The vacuum is filled with electromagnetic and matter fields, and these fields are as real and active as any physical entity we know.

References

M. R. Andrews, C. G. Townsend, H.-J. Miesner, et al., "Observation of interference between two Bose condensates," *Science* **275**, 637 (1997).

R. Griffiths, *Consistent Quantum Theory*, (Cambridge University Press, 2003).

C. Mead, *Collective Electrodynamics*, (MIT Press, 2000).

How Fields Generate Particles

In Chapter 1, I presented the noncontroversial core teaching of modern quantum mechanics, namely that fields exist throughout all space and that there are waves in these, described by equations that have many similarities for all types of fields. We haven't yet talked about particles. As mentioned at the beginning of Chapter 1, particles are "epiphenomena," a secondary effect of waves. We are now ready to talk about how they are related to waves.

2.1 Field Resonances

In Section 1.3, we briefly looked at the standard wave effect known as *resonance*. In general, any physical system has natural modes of oscillation, and most systems have more than one such natural oscillation. A typical example is a string tied at both ends, like the one shown in Figure 1.9. Any wave tied down at both ends must have a wave amplitude of nearly zero at those points.

As shown in Figure 2.1, there is more than one wavelength which can have that property. Any wave that goes through an integer number of half-waves along the length of the string will have no motion at the ends. Recall that the wavelength is defined as the distance from one crest of the wave to the next. The wave with longest wavelength, called the *fundamental* mode, is the one with just one half-wave on the string. At each end, the wave is pinned down, while in the middle the wave oscillates back and forth with large amplitude. The wave with the next-longest wavelength is the one that has a single full wave on the string. Both ends are pinned down and have zero amplitude, but there is also a point of zero amplitude in the middle of the string, even though the string is not pinned down there. This point is called a *node*. Good guitar players are familiar with the existence of nodes. If you put your finger exactly at the midpoint of a guitar string and pluck the string at a different spot, you will hear a higher note, which is equal to twice the normal frequency of the string. A wave with half the wavelength of another wave, and therefore twice its frequency, is defined as one *octave* higher, in musical terms. As shown in Figure 2.1, there are many higher wave modes on the string, each with shorter wavelength and higher frequency. These are the resonant frequencies of the guitar string.

Since any number of half-waves fit in the space available, one can have resonances of the string at any integer number times the fundamental mode. This principle underlies the whole western music system. Table 2.1 gives the standard "well-tempered" musical scale. Going to a resonance with twice the frequency of another corresponds to going one octave

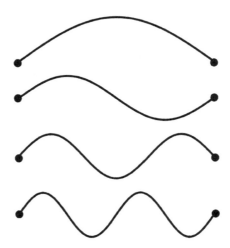

Figure 2.1   Modes of vibration, or resonances, of a guitar string pinned down at both ends.

higher. Going to a resonance frequency three times higher corresponds to one octave plus one-*fifth*. One can then go an octave below, or two octaves below, to get the same note on the musical scale in a lower range. By successively tripling and then halving, one gets all the notes of the western musical scale, with some small adjustments made so that, after twelve times tripling and then halving, one comes back to the same starting note.

As discussed in Section 1.3, if the resonances are sharp (i.e., they have high "Q," as defined in Figure 1.10), then in between the resonant modes, there will be gaps. The system

Table 2.1 The "well-tempered scale" of western music, also called the "circle of fifths," which arises from octaves being a factor-of-two ratio of frequency, and an octave plus a fifth being a factor-of-three ratio of frequency.

Note	f (Hz)	×3	$\times \frac{3}{2}$	$\times \frac{3}{4}$
A	440	1,320	660	
E	659	1,977	989	494
B	494	1,482	741	
F# (G♭)	740	2,220	1,110	555
C# (D♭)	554	1,662	831	
G# (A♭)	831	2,493	1,246	623
D# (E♭)	622	1,866	933	467
A# (B♭)	466	1,398	699	
F	699	2,097	1,048	524
C	523	1,569	785	
G	784	2,352	1,176	588
D	587	1,761	881	
A	880			

will not oscillate very well at frequencies between the resonances. Thus, there are jumps of frequency from one resonance to another.

What I have just described, a sequence of resonances of a wave, is all you need to know about where particles and quantum jumps come from. The word "quantum" in quantum mechanics refers to the jumps between different resonances of a wave. These jumps are not some new law of physics added to the theory of waves in fields discussed in Chapter 1. They are a natural implication of any field theory.

2.2 Two Types of Quantization

I have argued in Section 2.1 that waves in the electromagnetic field are just like waves on a guitar string, and so are matter waves. In the case of waves on the guitar string, the quantum jumps were caused by pinning down the string at both ends. Only certain wavelengths could fit on the string then. What is equivalent to pinning down the ends of the string, for electromagnetic waves and matter waves in vacuum?

It turns out that there are two different answers to this question. One possibility is that we do something which forces the electric field to be zero at certain places. For electromagnetic waves, for example light waves, this could be done by putting two mirrors opposite each other. The light cannot go through the mirror, so right at the surface of the mirror, the amplitude of the light wave is zero. If we put two mirrors opposite each other, we will have a situation almost exactly the same as a guitar string tied down at two ends. Only certain wavelengths of light wave can fit in the space between the two mirrors (this is called an *optical cavity*, and it is often used in making lasers). Just as in the case of the guitar string, there will be a series of different resonant frequencies of the light wave like that shown in Figure 2.1.

It is also possible to make a circular optical cavity, using a set of mirrors to send the light in one direction continuously. In this case, the constraint on the wave is that the wave must come back to where it started by the time it goes all the way around the circle. The longest wavelength that can fit is equal to the circumference of the circle. Shorter wavelengths can also fit, as in the case of a guitar string, equal to one-half, one-third, and so on, of the longest allowed wavelength.

In the case of a matter wave, for example an electron wave, the same thing can happen. This is the case for an electron wave going in a circular path around an atom. Figure 2.2 illustrates how only certain wavelengths of the electron matter wave can fit in a given orbit. This is the cause of the jumps between different states of an electron going on a circular path around an atom. The electron wave around an atom has various resonances just like a guitar string.

Second quantization. But there is a second type of resonance that occurs in field theory. The resonances we have discussed so far, for both electromagnetic waves and matter waves, as well as guitar strings (sound waves), are all known as "classical" wave resonances. That type of wave resonance was already well known in the 1800s.

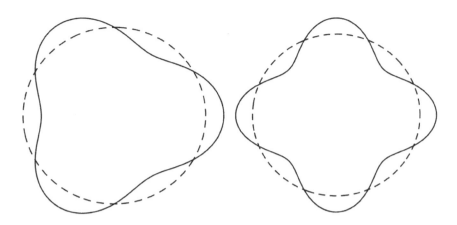

Illustration of two different natural modes of oscillation, that is, resonances, of an electron wave around an atom.

Quantum field theory says that there is another, deeper, type of resonance. To see this type of resonance, we must focus on a single point of a wave. Think of looking at a single spot on the guitar string. If you pluck the string, that part of the string will vibrate. If you pluck the string harder, you will get a high amplitude of the wave at that point (i.e., that point on the string will move a lot), and if you pluck the string more weakly, the amplitude will be lower (that point will not move very much). What is the range of all possible amplitudes you could generate?

In classical field theory, the wave amplitude can be any amount; there is no limit on how much or how little a part of the string can move during its oscillation. In quantum field theory, there are resonances in the amplitudes. The amplitude of the wave at any given point in space cannot have just any value – only certain values will work. The mathematical argument for this is given in Chapter 12. This is a standard result of the fundamental theories of modern physics and is true for sound waves in water or on guitar strings, for electromagnetic waves like light, and for matter waves.

We can see the basic reason for this new type of resonance with the following argument. Figure 2.3 shows a plot of the "stiffness" of a string as a function of how far it is pushed from its resting point. This is characterized by the amount of energy needed to push it that far. As seen in this figure, there is a valley, so to speak, of low stiffness near to the center position, and increasing stiffness as the wire is pushed away from the center in either direction. This is a typical behavior for all kinds of springy things, including real springs, but also guitar strings, water waves, and sound waves in air, and all the fields we discussed in Chapter 1.

Looking at this valley, one can see that it is a type of confinement like a string pinned down at both ends, but instead of the motion being completely stopped at the ends, there is gradual suppression of the motion away from the center, as the wire gets stiffer and stiffer. The relatively simple math of Section 9.4 shows that only certain waves will work because there is a *boundary condition* that the wave is confined to fit in a certain region.

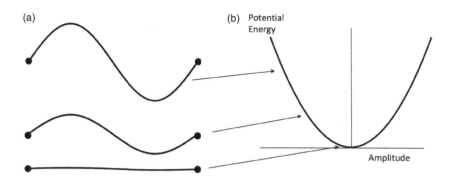

Figure 2.3 (a) Various positions of a string during its oscillation, with different amplitudes. (b) The potential energy of the string (that is, its "stiffness") as a function of its amplitude.

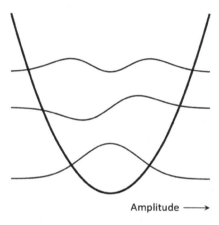

Figure 2.4 Various resonant modes of the *amplitude* function of a wave on a string.

Figure 2.4 shows resonant waves confined in this type of valley. Instead of being pinned to zero amplitude at the sides, the waves fade away to zero gradually.

Keep in mind, though, the difference of the two types of resonance. In the first type, classical quantization (called *first* quantization), what we had was resonances of the *wavelength* of the wave going along the string. Now we are talking about resonances of the *amplitude*, that is, in the distances that the string can move in the perpendicular direction, side to side. This is known as *second quantization*. Importantly, *each* possible wavelength in the first quantization has a whole set of resonances of its possible amplitudes in its second quantization.

Although this is mostly a nonmathematical discussion, let us look at just a little math here. We write the symbol ψ (the Greek letter "psi") as the general symbol for the amplitude of any wave, whether it is a sound wave, light wave, or matter wave. The *intensity* of a wave, that is, the amount of energy it carries, is proportional to the square of the amplitude, ψ^2. This agrees with our experience that a wave with more motion carries more energy.

The result of second quantization from the math of Section 12.2 is that ψ^2 has resonances proportional to $(N + \frac{1}{2})$, where N is any nonzero integer. In other words, the jumps in the energy of the wave are proportional to an integer N. The amount of energy change in each jump in amplitude is given by

$$E = hf, \tag{2.2.1}$$

which is known as the *DeBroglie relation*, where f is the frequency of the wave, defined in Section 1.2, and h is a universal constant of nature, known as *Planck's constant*.

Planck's constant is very small (6.6×10^{-34} Joule-seconds). Without worrying about the exact value of this number, we can say that, for typical fields, this means that the jumps in the resonant amplitudes are extremely small – these resonances are so close together that, in normal life, we can set ψ^2 to have any value we want, just as in classical field theory. To detect the jumps in the value of ψ^2 we must have extremely high accuracy of our measurement. But these jumps in amplitude are measurable and real, and have important consequences, as we will see.

This rule for quantized jumps in amplitude is a general result of quantum field theory that applies very generally to water waves, sound waves, and electromagnetic waves. As we will discuss in Section 2.4, a similar rule applies to quantum matter waves, namely that the wave energy can only be increased in jumps of energy hf. For electromagnetic waves and quantum matter waves, it is a little more abstract to imagine what is oscillating that is analogous to the side-to-side motion of the guitar string. But the same rule applies to all of these types of waves – amplitudes have quantized resonances.

2.3 Resonances as Particles

We have already stated that waves with greater amplitude carry more energy, which stands to reason from our experience. The rule we gave in Section 2.2 for the quantization of amplitudes says that, to make a wave go to higher amplitude, we must add energy in lumps, or quanta; we cannot just add any arbitrary amount of energy. The extra amount of energy we must add is equal to Planck's constant times the frequency of the wave.

In quantum field theory, this amount of energy is defined as the addition of a "particle." Each field has an associated particle: a sound wave (or water wave) has energy quanta called *phonons*, the electromagnetic field (whether light waves, radio waves, etc.) has the associated particles called *photons*, and there are various matter fields with associated particles known as *electrons*, *quarks*, and so on.

All of these particles are *excitations* of the field. "Excitation" is just the physics term for putting energy in. The physical picture is the following: the *vacuum* is equated with a field sitting in its quietest state, with no extra energy. As discussed at the end of Chapter 1, that means that, even in the vacuum of outer space, there is still "something" there, namely the field. When a field is caused to oscillate at a higher amplitude by putting in a quantum of energy, we can call this "adding a particle." In physics terminology, the vacuum state of the field has been "excited" to a higher-energy state.

Nothing in any of this has given any footing to thinking of particles as little billiard balls flying around.[1] For the same reason, there is no reason to ponder whether particles are "point" particles or whether they have some spatial size. As we will discuss in Section 2.8, it is actually problematic to try to assign a spatial size to many types of particles, and doing so leads to mathematical nonsense in some cases. There is no reason to try. The basic notion of particles in quantum field theory is nothing more or less than a specific amount of energy put into a field to cause it to vibrate at a higher amplitude. That vibration might be extended over space (as in the case of a guitar string or a laser beam) or it might be localized to a small region like an atom.

Note the progression we have gone through. We started with a field, for which our basic example was a guitar string. We found two types of resonances of the string: a series of resonances in the wavelengths of the wave, and another series of resonances in the amplitude of the waves of each wavelength. The series of resonances in the amplitude turned out to be equally spaced in energy. Therefore, we adopted the convenient picture of adding energy to the wave as adding particles, each with a fixed amount of energy.

What if a field existed in which the energy was not proportional to an integer N but instead, for example, was proportional to N^2, or, say, $N + 0.001N^2$, that is, with a small *nonlinear* term? Then there would still be amplitude resonances, but it would not be so easy to interpret the amplitude resonances as corresponding to integer numbers of particles. As you went from one resonance to the next, the amount of energy you added each time would not be a constant; it would depend on where you were in the series of resonances.

In fact, in every real field, the energies of the resonances are *not* exactly proportional to N. Treating the jumps in energy as exactly proportional to N is an approximation; a very good approximation for most fields but still an approximation. The energy in all real fields must be corrected by extra nonlinear terms, which are usually ignored because they are small, but they are very real, and lead to very important physical effects. For example, heat flow is largely controlled by nonlinear terms in sound fields; light of one color can be turned into other colors in laser laboratories using nonlinear optical effects. Even in the vacuum of outer space, the amplitude resonances of light waves do not have energies exactly proportional to an integer number; at very high intensities, nonlinear terms occur even in vacuum.

It is often a useful approximation to treat each field as having states with an exact number of particles, and then to treat the nonlinear terms as small perturbations, which cause the field to jump between states with different numbers of particles. But that approach is just an approximation, and there is no fundamental reason to always stick with it. One could instead find the exact resonances, in which case every jump in energy would be by a

---

[1] David Baker (2009) has argued that because all quantum field states can be written in terms of Fock states (defined mathematically in Sections 12.1 and 13.1), which he treats as states with a definite number of particles, quantum field theory cannot escape the need for a definition of particles. However, Fock states are simply states with definite amplitudes in all of the allowed wave states; as we have seen, there is no reason to treat these states of well-defined amplitude as corresponding to localized, compact objects. Paul Teller (1995) makes a distinction between *quanta*, as jumps in the energy/amplitude of Fock states, and *particles*, as localized objects with a definite history, and presents several arguments as to why the quanta of Fock states cannot be thought of as traditional "particles."

different amount. Then it would not be as appealing to talk of particles because each jump in amplitude would have a different energy from all the others.

2.4 Bosons and Fermions

So far, everything we have discussed has been quite general for all types of fields. But there are actually two different classes of fields, with a crucial difference between them. The first type of field is called *bosonic* (named after the Indian physicist Satyendra Nath Bose), and the second type is called *fermionic* (named after the Italian physicist Enrico Fermi). Sound waves (including water waves and guitar strings), with their associated phonon particles, and electromagnetic waves (including light, radio, microwaves, and X-rays), with their associated photon particles, are examples of bosonic fields, and all of the associated particles of these are collectively called *bosons*. We often think of these as "energy" fields. On the other hand, most of the things we think of as "normal matter" are fermionic fields, with the associated particles of electrons, protons, and neutrons, as well as other more exotic particles; all of these particles collectively are called *fermions*. All the fields in the universe that we know of, whether for electrons, quarks, or obscure subatomic particles, fall into one of these two classes, and there is reason to believe that there cannot be any other classes.[2]

The primary difference between the two classes is in the number of amplitude resonances they can have. Bosonic fields have an infinite number of amplitude resonances; that is, the integer N in Eq. (2.2.1) can be equal to 0, 1, 2, and so on, with no upper limit. Fermionic fields have only two resonances: N can be equal to only 0 or 1. Figure 2.5 illustrates this difference.

Bosons   Fermions

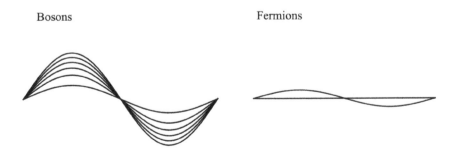

Figure 2.5 Steps in amplitude for waves in a bosonic field (no upper limit on the amplitude) and a fermionic field (only two allowed states). Note that the boson field cannot have zero amplitude; the lowest amplitude allowed has an amplitude associated with the *vacuum energy* of the state, corresponding to the $N = 0$ in the term $(N + \frac{1}{2})$.

[2] It is possible to write down theoretical equations for *anyons*, which are a different class from either of these, but the only known ways to do this require underlying fields that are either bosonic or fermionic.

This simple difference leads to enormous differences in the properties of the different fields. The fermion property of only two possible resonant states is called the *Pauli exclusion principle* and means that, if a given wave state is excited from vacuum (which we call "being occupied by a particle"), then no more energy can be added (that would require $N = 2$, which is forbidden for fermionic fields). The Pauli exclusion principle is responsible for the entire periodic table of chemistry: filling of "shells" in chemistry corresponds to putting one electron in each possible wave state around an atom. The Pauli exclusion principle is also very much responsible for why solid matter seems solid, as we will see in the next section.

On the other hand, as we have already discussed for sound waves, for example, waves on guitar strings, boson waves have no upper limit on the number N. That leads to our experience that waves like sound waves and light waves have no limit on their intensity. You can put as much sound energy or light energy in a container as you like; it will not "fill up" and prevent more sound or light from coming in (unless the container melts from all the energy in it). Not only that, more bosons in a state will make it more likely that other bosons in other states will change their state to enter into the highly occupied state. This property is called *stimulated scattering*, or in the case of photons being emitted from a light source, *stimulated emission*. This property is the basis of the laser: light that is initially random can self-organize into all being one wave state with large amplitude.

The rules for amplitude resonances of fermion and boson fields can be summarized as follows: given a system in amplitude-resonance state N, the rate of transition to the next-highest resonance is proportional to

$$\begin{cases} 1 - N & \text{for a fermion field} \\ 1 + N & \text{for a boson field.} \end{cases} \tag{2.4.1}$$

These rules are derived mathematically in Section 14.1. Note the symmetry between these two cases, which are the same except for a change from a $+$ sign to a $-$ sign. This symmetry is not accidental. Boson fields and fermion fields are two sides of the same coin; as discussed in Section 13.1, the math is the same for the two types of fields at every level except for the plus and minus signs being different. Yet this difference is crucial. For fermion fields, it means that, if $N = 1$, there can be no transition to $N = 2$, while for boson fields, the greater the N is, the greater the rate of amplification to even higher N, if there is enough energy around to add.

Students of chemistry are used to the Pauli exclusion principle and often think of the wave states that get filled up as something like little boxes that can contain only one particle. The property of stimulated emission of photons in lasers is also learned by many students. The Pauli exclusion principle and the stimulated emission principle are truly strange principles if we think of particles as little billiard balls. A "state" in chemistry is not a box that contains anything. It is a wave with a particular direction and velocity. Imagine a billiard table with fermionic billiard balls. The Pauli exclusion principle would imply that, as the balls move around the table and collide with each other, if even one ball is moving to the left with a certain speed, then after a collision, no other ball will move in that direction with the same speed! It is as though the balls do an opinion poll in advance of what the other balls are doing, and if they hear of another particle doing something, they

	Boson fields	Fermion fields
Examples	Light, sound	Electric charge, nuclear matter
Transition rate	$(1 + N)$	$(1 - N)$
Spin	$0, 1, 2, \ldots$	$\frac{1}{2}, \frac{3}{2}, \frac{5}{2}, \ldots$

Table 2.2 Properties of boson and fermion fields.

refuse to do the same thing. On the other hand, if the billiard balls were bosons, then if one ball were moving to the left, then all the other billiard balls would start moving in the same direction at the same speed, as though they were sheep who wanted to follow a leader!

If we drop the localized particle picture, these properties are not so strange for waves. As discussed in Section 1.2, waves in the same place can either add together or cancel out, depending on their phase. In fermion fields, if there is already a wave with amplitude resonance $N = 1$, extra energy added to the wave would effectively be like adding a wave with the opposite phase, canceling out the initial wave. In boson fields, the amplitudes add with the same phase, giving a greater amplitude. This effect of stimulated emission is a well-known classical wave effect, which occurs, for example, when the radiation from one radio antenna increases the radiation coming from another antenna.

Table 2.2 summarizes the differences between bosonic and fermionic fields. One other important difference is in the associated *spin* of the excitations of the field. All fields have certain allowed states of angular momentum (that is, allowed rotation rates). Bosons all have an associated angular momentum equal to an integer times Planck's constant; fermions all have half-integer values. The spin properties of fermions follow from the assumption that a fermionic wave never vanishes at any moment in time; by contrast, bosonic fields have oscillations that pass through zero during every oscillation, which means that, at certain times, bosonic waves "wink out" momentarily. (This is shown mathematically in Sections 9.1 and 13.2.) That is the only important physical difference between the two types of field.

2.5 A Wave Can Be a Very Solid Thing

All this talk of particles as resonances of a field may make some people uncomfortable. Fields seem so "ethereal"! It can make us feel that nothing we see around us is really substantial. Even we ourselves are made of waves! Particles, by contrast, seem to make us feel more solid, more real.

Actually, our sense of particles being substantial and permanent is rather misguided. All types of particles can appear and disappear under various conditions. Quantum field theory says that the electrons and protons that make up our atoms can appear out of vacuum and disappear back into it. Particles are not rock solid entities that can never vanish.

On the other hand, our fear that the wave picture alone makes us "insubstantial" is also misguided. Nothing in physics theory should change what you know and see with your own eyes: that things and people are very solid indeed. Saying they are made of "nothing but" fields does not change this.[3] This solidness arises in the context of field theory as a result of Pauli exclusion, and is known as *fermion pressure*. Section 9.3 presents this mathematically; here, we can just think about it qualitatively.

Our sense of "solidness" of things comes from the experience that, when we push on them, they do not compress very much before they push back with the same force. This property is in turn a consequence of the property that we must put energy into the system to get it to compress. So, in asking why some things seem solid, what we are really asking is why it pushes back when we push on it.

In a solid, there are many electronic wave states. Pressing on a solid reduces its volume slightly. For example, in pressing on a piece of wood, you might push its surface a micron or so. That makes the electron states have higher energy, because shorter wavelength corresponds to higher frequency, and higher frequency corresponds to higher energy. Therefore, energy is needed to compress a solid. This is no small, esoteric effect. As shown mathematically in Section 9.3, even a deflection of the surface of a piece of wood by a micron can raise the fermion pressure enough to create a substantial force pushing back against your finger. Fermion pressure, which comes from Pauli exclusion, is fundamentally the reason for our experience of the stiffness of solids.

The reason why fields seem so intangible is because our normal experience with wave effects is with boson fields such as sound waves and light waves. If all matter were made of boson fields instead of fermion fields, our experience would not be that things were so solid. There is no limit to the amount of bosonic energy you can squeeze into one place. If the volume of a bosonic system is compressed, so that the wavelengths of the states are shorter, the system can lower its energy by changing the amplitude of all the wave states, so that the longer-wavelength states have higher amplitude ("greater particle occupation") and the shorter-wavelength states have lower amplitude ("less particle occupation"). This can't happen in a fermion system since there is a maximum amplitude of all the states.

It is a conventional truism to say that atoms, and we, are "mostly empty space." That is not really correct. It assumes the picture of an atom as a little solar system with the nucleus as the sun and a point-like electron orbiting around it. Quantum field theory says that the electron matter wave fills the volume around the nucleus, and this matter wave for electrons is very incompressible.

2.6 ...And a Solid Can Be a Very Wavy Thing

As discussed in Section 2.5, our intuition about waves comes from our general experience that the waves we encounter are bosonic waves, like light and sound, while solid matter is made of fermion waves. Solid matter usually doesn't seem very wavy.

---

[3] Brain scientist and author Donald Mackay coined the phrase "nothing buttery" to describe the philosophical sleight-of-hand that says that, if something is made out of something else, then it is not real, since it is "nothing but" the stuff it is made out of. In his context, he meant that saying that brains are made of atoms and molecules has no bearing on whether our experience of thinking is real or not – of course we think! See, e.g., Mackay 1980. Section 7.2 discusses this topic further.

Table 2.3 Types of composite bosons.		
System	Boson name	Underlying particles
Superfluid helium	Helium atom	Two electrons, two protons, and two neutrons
Superconductor	Cooper pair	Two electrons in a metal
Atomic condensate	Alkali atom	N protons, neutrons, and electrons, where N is even
Neutron star	Cooper pair	Two neutrons

All of this intuition is turned on its head by a class of matter, well known in physics experiments, which has bosonic properties. This type of matter is typically called "super"-something: superfluids, superconductors, and even supergases. A general term for this kind of behavior is *Bose–Einstein condensation*, named after the two physicists who laid the groundwork for the theory, Satyendra Bose and Albert Einstein. Bose–Einstein condensates will be discussed in greater depth in Section 8.5.

These systems exist because of an additional property of quantum fields, namely that you can make bosons out of fermions. As mentioned in Section 2.4, there is an associated amount of spin for each particle; that is, each of the energy-amplitude resonances of the fields also has a certain amount of angular momentum. Because bosons have integer spin and fermions have half-integer spin, this means that two fermions can be added together to make a boson; in fact, any even number of fermions bound together make a boson. Table 2.3 lists examples of bosons made from fermions. It is a good trick to understand how two electrons can bind together in a metal, to make a Cooper pair, because electrons normally repel each other, but it can happen at low temperatures when the motion of the atoms in the surrounding solid cancels out the repulsion.

When waves are made in these systems, they are not subject to the Pauli exclusion principle, which means that waves in these systems can have large amplitude, and act just like classical waves such as sound and light. This leads to all kinds of fascinating behavior. For example, large currents of electrons in superconductors can interfere and cancel out. This is the physical basis of a very sensitive magnetic sensor known as the SQUID (*s*uperconducting *qu*antum *i*nterference *d*evice). Figure 2.6 shows an example of interference of superconducting matter. The same type of interference can be seen in currents of whole atoms, in atom condensates, as seen in Figure 1.8(b). Currents in these systems also flow without viscosity. One way to understand this is to recall the rule that transitions of bosonic fields follow the $(1+N)$ amplification effect. That means that transitions *into* wave states with large amplitude state are enhanced by a large factor N (which could be billions, for electrons in a solid) while transitions *out of* that state not enhanced. Since energy dissipation corresponds to random transitions out of the large wave state, dissipation is strongly suppressed.

Even in normal solids, the wave nature of system plays a major role. In crystals with orderly rows of atoms, the electron states form *Bloch waves*. What this means physically is that the electron waves reflect off of the rows of atoms in such a way that some wavelengths

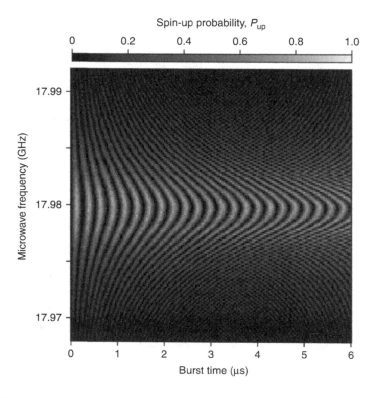

Figure 2.6 Interference fringes seen in the response of a superconducting circuit in which the electromagnetic field mixes with the wave function of the superconducting electrons. From Zhang 2018.

of the electron waves are canceled out by interference. This effect leads to forbidden ranges of energy for the electrons, known as *bandgaps*, an effect that is extremely important for all of semiconductor physics, and the basis of the entire high-tech digital industry. In this case, the electrons still act as fermions, which means that the amplitude of the Bloch waves is limited to the $N = 1$ state, but the resonances in the wavelength of the electrons are still very important.

2.7 Dirac's Beautiful Theory

All of the theory I have summarized in this chapter was already known in the 1930s, and much of the credit for organizing it into our modern system can be given to Paul Dirac. Dirac's theories are beautiful and elegant both in their reasoning and in their final form, and as important as those of Newton, Maxwell, and Einstein. As discussed in Section 13.2, Dirac took a simple intuition, namely conservation of mass at all times, and developed this into a theory of quantum mechanics that would be consistent with Einstein's theory of

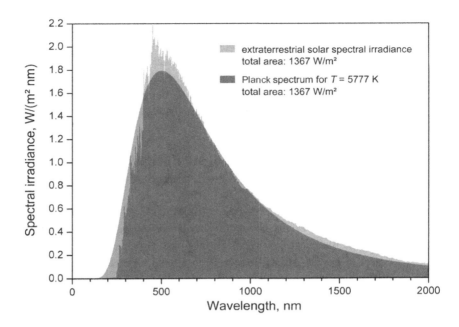

Figure 2.7 The emission spectrum of the sun as measured above the Earth's atmosphere compared to the Planck radiation formula of an object at 5777 K. From Thuillier 2003.

relativity. Along the way, he derived the spin of the electron and predicted the existence of antimatter. His work is the basis of all our modern quantum mechanical calculations.

Unfortunately, Dirac's work came late in the story of quantum mechanics. Long before Dirac completed his work, the major scientists and philosophers of the day had weighed in on the meaning of quantum mechanics.[4] Early experiments had shown that *classical* fields could not explain the behavior of light and matter. The leaders of science in the day such as Bohr and Einstein therefore jumped to the conclusion that particles are the truly fundamental entity, and the waves in the quantum theory were just theoretical constructs about our knowledge of particle motion. This view of particles as fundamental was fairly well locked in by the time Dirac completed his great work. Although his work was recognized as brilliant, the earlier philosophy of particles was grafted onto it.

An example of this can be seen in one of the major evidences used to argue for the existence of particles before Dirac, namely the Planck spectrum of light. Figure 2.7 shows the measured intensity of different wavelengths of light from the sun, compared to the Planck spectrum. This spectrum can be completely explained by Dirac's field theory without reference to particles, as shown in Section 10.1. However, long before Dirac, the same result was derived from a statistical calculation based on the assumption that electromagnetic energy must always have a definite number of particles. Although the two different

---

[4] Dirac himself did not philosophize on the ontology of the quantum mechanics, as much as on the need for aesthetic beauty in its mathematical structure (see, e.g., Kragh 1990).

methods give the same final prediction, the derivation from Dirac's formalism is more elegant. For one thing, it gives both the fermion and the boson distributions directly from the same math, while the statistical approach requires two different proofs with different assumptions. The field theory approach also requires no invoking of random statistics at all; it is entirely deterministic. But the earlier approach had already been widely accepted before quantum theory in Dirac's approach was developed.

2.8 Are Particles Real?

It is quite often the case in physics, and science in general, that, in the early days of experiments, scientists cast about for simple mental models to describe their work, without elaborate mathematical justification. Once a mental model gains predominance, it continues to have a life of its own even after a more exact theory is formulated. The idea of chemical "bonds" is one such idea, which has persisted as a useful concept although quantum field theory gives a more exact description. The notion of particles is another. In many contexts, thinking in terms of particles makes it much easier to describe experiments. However, when such a concept is pushed too far, unnecessary philosophical conundrums can arise.

Two physical examples show how the particle picture can sometimes confuse things. The first is the *Unruh paradox*. Suppose that we have a sealed box with perfect mirrors inside, with electromagnetic energy inside that bounces off the mirrors and can never escape. We now subject this box to acceleration near a massive object like a star. It can be shown (Unruh 1974; Arageorgis 1995) that, if two different observers look at this box, from two different vantage points and moving at different speeds and accelerations, they will give different answers for the number of photons inside the box. This is known as the *Unruh paradox*. How can this be, if photons are like little billiard balls that have a definite identity and location? Do some of them become invisible to different observers? At what speed do they disappear and reappear?

There is a simple resolution if we are not wedded to the idea of particles as indivisible objects. As we have discussed, the photons in a system register the amount of energy in a wave, that is, excitations of the field. The amount of energy in a box depends on the point of view of the observer, according to Einstein's theory of relativity.

Another physical system that casts doubt on the fundamental existence of particles is the case of coherent states of bosons. States with a definite number of particles (called *Fock states*, defined mathematically in Sections 12.1 and 13.1) are not the only possible physical states allowed for bosons. One can write any number of other allowed physical states as superpositions of Fock states. One such state, called a *coherent state*, which is perfectly physically possible and is produced all the time in lasers and other systems, has a definite amplitude and *indefinite particle number*. The math of coherent states is reviewed in Section 12.5.

It would be incorrect to say that a coherent state is "really" in a state with a definite number of particles, but we just don't know how many. That would be to say that the state is not a superposition of different number states but instead is actually randomly jumping from one state to another. There are ways to experimentally distinguish between a system that is in a true superposition as opposed to one randomly jumping between different states. In the case of coherent states of light, there are measurements that can be done that make a definite measurement of the *phase* of the wave, that *prohibit* there from being a definite number of particles; as shown in Section 12.5, if there were a definite number of particles, we could not make a good phase measurement, due to the principle of *number-phase incompatibility*.

There are other arguments as well (see, e.g., Davies 1984). One is that the mathematics of treating "point" particles leads to all kinds of problems. There is no natural size scale for fundamental quantum particles. Therefore, we must either dispense with the idea of treating them like billiard balls, or we must take them to be single points in space. But for both electrons and photons, treating them as point particles causes mathematical headaches. It has been shown that photons cannot have propagating wave functions that are point-like; at best they can have wave functions that decrease rapidly away from some center region (Mandel 1995). This is one reason why some physicists favor *string theory*, which takes line-like objects as fundamental instead of point-like objects. But regular quantum field theory does not require particles to be point-like at all.

Finally, there is a *utility* argument that calls into question the centrality of the particle picture. Many physical systems exist in which thinking in terms of particles is not only unnecessary but actually makes it harder to understand the system: The author of one of the most important textbooks on laser technology wrote:

> We have hardly mentioned photons yet in this book. Many descriptions of laser action use a photon picture... in which billiard-ball-like photons travel through the laser medium. Each photon, if it strikes a lower-level atom, is absorbed and causes the atom to make a "jump" upward... Although this picture is not exactly incorrect, we will avoid using it to describe laser amplification and oscillation, in order to focus from the beginning on the coherent nature of the stimulated emission process. The problem with the simple photon description... is that it leaves out and even hides the important wave aspects of the laser interaction process. (Siegman 1973)

Some might argue that these are just examples of "wave-particle duality," that is, that sometimes the wave picture is most convenient and sometimes the particle picture is. But there is not a symmetry. In every case, the wave picture can be used, with the understanding that the waves must be solutions of the quantum field equations, not classical wave equations. On the other hand, in some cases, the particle picture gives complete nonsense, as in the Unruh paradox. As mentioned at the beginning of this section, it is similar to the case of chemical bonds. In every case, the full quantum mechanical solution can be used to explain the bonding of atoms in molecules. In some cases, this behavior can be conveniently described in terms of "bonds," a concept that originated long before quantum mechanics. But in some cases, the notion of bonds can break down (see, e.g., Ochiai 2015).

References

A. Arageorgis, "Fields, particles, and curvature: Foundations and philosophical aspects of quantum field theory in curved spacetime," Ph.D. thesis, University of Pittsburgh, 1995.

D. Baker, "Against field interpretations of quantum field theory," *British Journal of Philosophy of Science* **60**, 585 (2009).

P. C. W. Davies, "Particles do not exist," in *Quantum Theory of Gravity*, S. M. Christensen, ed., (Adam Hilger, 1984).

H. Kragh, *Dirac: A Scientific Biography*, (Cambridge University Press, 1990).

A. E. Siegman, *Lasers*, (University Science Books, 1986).

D. Mackay, *Brains, Machines, and Persons*, (Collins, 1980).

L. Mandel and F. Wolf, *Optical Coherence and Quantum Optics*, (Cambridge University Press, 1995).

H. Ochiai, "Philosophical foundations of stereochemistry," *HYLE* **21**,1 (2015).

W. H. Louisell, *Quantum Statistical Properties of Radiation*, (Wiley, 1973).

P. Teller, *An Interpretive Introduction to Quantum Field Theory*, (Princeton University Press, 1995).

G. Thuillier, M. Hersé, D. Labs, et al., "The solar spectral irradiance from 200 to 2400 cm by the SOLSPEC spectrometer from the ATLAS and EURECA missions," *Solar Physics* **214**, 1 (2003).

W. G. Unruh, "Second quantization in the Kerr metric," *Physical Review D* **10**, 3194 (1974).

X. Zhang, H.-O. Li, G. Cao, et al., "Semiconductor quantum computation," *National Science Review* **6**, 32 (2018).

3 Jumpy Detectors

So far, I have presented quantum mechanics in terms of quantum field theory, which is universally accepted as the most correct fundamental theory. In that theory, particles occur as resonances of the field. From what we have seen, particles seem a bit ephemeral, since they are resonances of a field, having the same relation to the underlying fields that the ringing of a bell has to the bell itself.

But many experiments do seem to point us to the picture of particles as lumps. A Geiger counter gives clicks. A photographic emulsion exposed to weak light shows little spots. Probably the most dramatic experiment that shows particle behavior is the cloud chamber. In this experiment (or the modern high-tech equivalent), we see tracks that go in straight lines, as seen in Figure 3.1. It is quite natural to interpret these observations as detecting particles. How do we understand this type of experiment if, as we have seen in earlier chapters, particles are just certain states of waves in fields?

3.1 Atoms and Natural Length Scales

A clue that something more is going on arises if we ask why no one has ever seen a click for a radio wave particle or a human voice particle. According to quantum field theory, all electromagnetic waves, including radio waves, have the same quantization into photons, and all sound waves have a similar quantization into phonons. Why don't we get clicks for radio photons or voice phonons? The reason is that all of the waves that give clicks, spots, tracks, and so on are *high-frequency* waves. High frequency in this context means that their frequency is comparable to the electronic resonance frequencies of an atom. Atoms have a natural length scale that defines the effective size of the "lumps" that we can see when we do a particle measurement. The lumpiness that we see in clicks and tracks is a result of the lumpiness of atoms.

Every particle detector that involves saying that a particle is "here" and not "there" must define a local region in space where "here" is. In almost all particle detectors, the location of "here" is defined by the location of one or more specific atoms. The size of the atom is the natural length scale for determining locations.

What we mean by a natural length scale is a length that comes about due to intrinsic properties of the underlying quantum theory. There is no such intrinsic, natural length scale

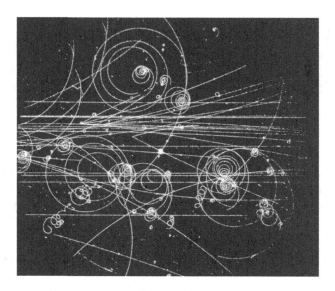

Figure 3.1   Tracks in a cloud chamber. The spiral curves are caused by the presence of magnetic field, which produces a force on charged particles. (CERN Photolab, "The decay of a lambda particle in the 32 cm hydrogen bubble chamber," (1960).)

for photons or electrons in vacuum. On the other hand, because there is a natural length scale for atoms, we are justified in treating atoms as little lumps like billiard balls.

Recall from Chapter 2 that the resonances of quantum field theory give quantized energies, which we interpret as particles. For the standard fermion fields of electrons and protons, these quantized energies correspond to quantized masses and quantized charges. As with energy, if we measure mass or charge, we will always get an integer number times a constant value, which is taken as the single-particle value.[1] Both the proton and the electron fields have the same charge[2] but with opposite signs; we write the proton charge as $+e$ and the electron charge as $-e$.

The natural size scale for atoms can be seen simply as the length scale at which two different energies are comparable: the energy of the force due to the attraction of opposite charges, and the energy of motion that comes about due to oscillation of the field. Section 9.5 gives the math for this. When we put the values of the universal constants of nature, namely the mass of an electron and proton in vacuum, the electron charge, and

---

[1] To be technical, these are the *vacuum* masses – the mass of electrons and other particles can be different if the resonances of the field are shifted by interactions among different media, a process known as *renormalization* in quantum field theory. The mass of an electron in a solid can differ from its vacuum mass by an order of magnitude or more. Also, the charge of an electron can be renormalized to fractional values of e; the discoverers of this effect, known as the fractional quantum Hall effect, were given the Nobel Prize in 1996. In each case, all the identical particles are renormalized in the same way.

[2] The fact that electrons and protons appear to have *exactly* equal and opposite charge is actually quite surprising, since they come from two completely different matter fields. This is actually a deep issue in physics. It is quite good for us, however, that this is true, since even a small imbalance of the charges would lead to a huge buildup of static electric charge everywhere.

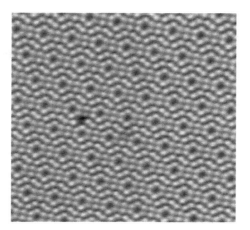

Figure 3.2 An electron tunneling microscope image of atoms on the surface of silicon. (Unisoku Co.)

Planck's constant, into those calculations, we find that the natural size of an atom is around one *angstrom*, equal to 10^{-10} m, that is, one-tenth of one-billionth of a meter. Some atoms may be a little larger, but we will not find atoms that are far different from this in size, for example, millimeters instead of billionths of meters. In recent years, this size of atoms has been verified by very high-resolution microscopes. Figure 3.2 shows a very high-resolution microscope picture of rows of atoms on the surface of a metal.

Note that this natural length scale comes fundamentally from the mass and charge quantization, which in turn come from the resonances of underlying quantum field theory. Given those numbers, one can derive a natural length scale based on the balance of energies in the wave. We do not need to invoke a billiard-ball picture of the underlying electrons and protons. All we need is the equations for the energies of waves.

Because there is no such natural length scale for electrons or photons by themselves in a vacuum, or for phonons of sound waves, there is no reason to expect that there will be lumps in systems of pure photons, phonons, or electrons.[3] In general, to have localized lumps in any system, there must be some natural length scale that gives the size of those lumps.

3.2 Electron Jumps

Let us now build a particle detector from the atoms up. Given the existence of atoms with a definite size, we consider now what will happen if a light wave shines on an atom.

---

[3] It may surprise some readers to learn that the mathematics of wave pulses, also called "wave packets," has absolutely nothing to do with the existence of photons or particles. Pulses can be generated for any wave, including classical waves, and the quantum wave equations do not change this. As discussed in Section 2.4, the particle nature of any wave comes into play in its *amplitude* – the duration of a wave pulse depends on how long the source took to emit it and can, in general, have any value.

Many people who have read popular accounts of quantum mechanics (e.g., Ferris 1988) will have heard strange things about this process, such as "instantaneous" jumps, and particles that go from one place to another without having traversed the intervening space. As we have seen, even talking about particles as objects is problematic. Saying they jump from place to place instantaneously is even more problematic, and wholly unnecessary. Just as we have seen with other effects, the effect of electrons jumping to new states can be understood entirely within the context of wave theory. In particular, the jumps of electron states when absorbing or emitting photons are direct consequences of *resonances* of waves.

The atom not only has a natural size, it also has natural resonance frequencies. As discussed in Section 2.2, confining a wave to a confined region always produces an associated set of frequency resonances, even in classical wave theory (which we called "first quantization"). This is the effect that we saw leads to musical instruments playing distinct notes. This also occurs for the electron wave around an atom.

For the natural length scale of an angstrom, the natural energy unit for an atom is around an *electron-volt*, which corresponds to an electromagnetic wave frequency around 10^{15} Hz, that is, 1,000 trillion oscillations per second (see Section 9.5 for the derivation of this natural energy scale). The resonances, or states, of an electron wave around an atom will have frequencies comparable to this.

Suppose that the electron wave for a single atom is in its lowest-frequency resonant state, and a light wave impinges on the atom. If the frequency of the light equals the difference between the lowest electron resonance and the next-highest electron resonance, the atom will start to transition to the higher resonance, because the light wave corresponds to an oscillating electric field, which exerts a force on the electron wave.

The natural timescale for the transition to a higher electronic resonance has been worked out in quantum optics[4] and is presented mathematically in Section 14.3. It shows that electronic transitions are not instantaneous. The electron wave oscillates up and down between the two states. We all have experience with this type of oscillation. If you sit on a child's swing, it will oscillate back and forth in response to you moving your feet back and forth. Your feet in this case play the same role as the oscillating electric field pushing on the electrons.

Where do "jumps" come from, then? This takes us to a crucial topic of modern physics known as *decoherence*, or *dephasing*. Perhaps surprisingly to some, the modern theory of these processes shows that electron jumps are not instantaneous, although they can occur very rapidly.

Let's start from first principles and consider the interaction of the atom with the rest of the world. An atom can give up small amounts of energy to any number of random processes, such as collisions with other atoms or radiation of light out into space. We can call all these processes *dissipative* processes. If the dissipative processes are strong enough, then the oscillations of the state of the atom will eventually reach a steady state with no oscillation.

---

[4] The theory of quantum optics, worked out in the 1960s and 1970s, was recognized in the recent Nobel Prize given to American scientist Roy Glauber. If this theory had been worked out in the 1920s, the philosophy of quantum mechanics might have turned out differently. For general textbooks on this theory, see, e.g., Louisell 1973 and Mandel and Wolfe 1995.

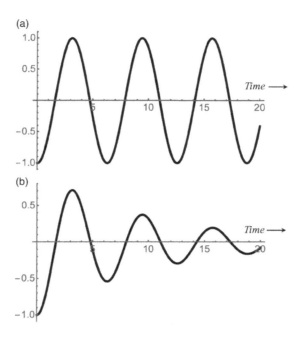

Figure 3.3 Electronic state of an atom under illumination by a light wave, as a function of time after the light is turned on, for two cases. (a) No decoherence. (b) Decoherence time approximately equal to the oscillation time. The value -1 corresponds to the atom in its lowest state, and $+1$ corresponds to the atom in a higher-energy resonant state. Numbers in between correspond to a superposition of both electronic states.

Figures 3.3 and 3.4 show the behavior of the atom in time, for four different cases of increasing dissipation; these plots were found by solving the relevant equations (worked out in Section 14.3). The way to think about these graphs is that the vertical height of the black line shows the fraction of the electron wave in two different states of an atom – the value of -1 means the electron wave is entirely in the bottom state, while the value of $+1$ means the electron wave is entirely in the upper state. A value in between these two means that the electron wave is partly in both states.

As the dissipation is increased, we see that the oscillation of the electron wave is damped out more and more. In Figure 3.4, we see that when the damping is strong, there is a switch-on behavior that has a fixed timescale, no matter how weak the light is. We can call this the *decoherence time*.

If we keep the dissipation the same, and keep reducing the intensity of the light, we will see that the atom undergoes the same switch-on behavior at all lower light intensities. The characteristic switch-on time remains the same, comparable to the decoherence time.[5]

---

[5] The crossover from oscillation to switch-on behavior occurs when the characteristic time for decoherence, that is, for random interactions with the environment, is roughly equal to the period of the Rabi oscillations, which have frequency proportional to the intensity of light (see Section 14.3). We could also show the same effect by considering incoherence in the input light instead of in the energy loss of the atom. Suppose multiple light waves, with the same wavelength but coming from different directions, hit the atom. The same switch-on

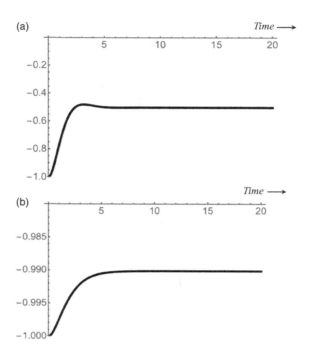

Figure 3.4 Electronic state of an atom as a function of time when the decoherence time is much shorter. (a) Decoherence time equal to about one-tenth of the oscillation period. (b) The same decoherence time, but ten times weaker illumination amplitude.

In typical atoms and solids, this characteristic time is extremely short: picoseconds to nanoseconds (i.e., trillionths to billionths of seconds). In the 1920s, this kind of time interval was unmeasurably short. But it is not instantaneous! There is also no discontinuous motion. The wave function gradually changes throughout the whole transition.

Modern experiments with "ultrafast" lasers (with light pulses as short as femtoseconds, that is, thousandths of trillionths of seconds) clearly show this type of behavior. Figure 3.5 shows an example of a *coherent control* experiment, in which one laser pulse put atoms into a higher-energy electronic state, and a second laser pulse was used to "reset" the electrons back into their ground state. If we insisted that a photon is always definitely absorbed or not, the second laser would just have added another set of photons to be absorbed, and there would just be twice as many atoms in the excited state. But in the experiment shown here, the second laser pulse created darkness, as it canceled out the coherent motion of the electrons excited by the first laser pulse.

behavior will be seen, with the characteristic time being the *correlation time* of the input light; the correlation time is a measure of the amount of time that the light stays at the same frequency without random phase shifts. The mathematics of this is worked out in Section 14.4 for an atom absorbing a single photon; again, a natural timescale is found, not an instantaneous jump.

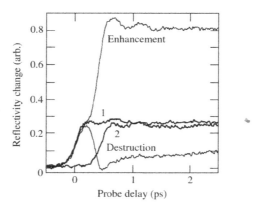

Figure 3.5

Coherent control of the optical properties of GaAs. Curve 1 is the signal in response to the first laser input alone, and Curve 2 is the response from the second laser input alone. When both inputs occur, the result can either be enhancement (top curve) or destruction (bottom curve) of the excitation, depending on the exact time delay between the two input pulses. The time units are picoseconds (ps), that is, trillionths of seconds. From Heberle 1996.

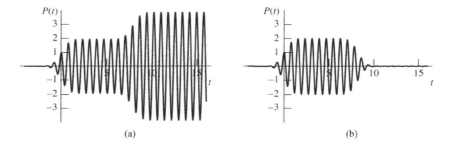

Figure 3.6

Constructive or destructive interference of the response of an oscillator to two pulses depending on the time delay between them. (a) Delay is ten times the period of the oscillation. (b) Delay is one half period longer.

The reason why this happened is that the electron wave created by the first pulse was still oscillating when the second pulse arrived. The effect of the second laser depended on how long this oscillation was allowed to continue between the first and second pulses. If the second pulse came after an integer number of oscillations, then the amplitude of the oscillation was kicked up to twice as high. If the second pulse came a half-integer number of oscillations later, then the second pulse canceled out the effect of the first pulse, due to interference (discussed in Section 1.2). This cancelation is something you can experience when you push a child on a swing. Depending on when you give a push, you can either add to the speed of the swing, or you can slow it down or stop it altogether. The oscillation of the electron wave works the same way, with the laser pulse applying the "push." Figure 3.6 shows a calculation of the wave state of the electrons in the two cases corresponding to Figure 3.5.

This type of experiment definitively shows that optical transitions are not instantaneous.[6] The intensities of the lasers for these experiments were very high, and their duration was very short, and therefore the oscillations of the electronic states in the atom occurred on timescales short compared to the decoherence time of the atoms. The historical experience of photon absorption seeming to be instantaneous was largely based on the fact that this type of experiment could not be done then; light sources had much lower intensity and much longer durations.

The vertical scale in Figures 3.3 and 3.4 goes continuously from the lower electronic resonant state of the atom to the higher one. Just because the atom has two distinct electron wave resonances, there is no rule that it can only be in one of these two resonant states. As discussed in Section 1.3, waves in general can add up, so that the total state of a system can be a superposition of two wave resonances. That is the case here. The incoming light wave sends the atom into a superposition of its lowest state and a higher resonant state. This is just the same as a tuning fork or a guitar string vibrating with two different frequencies at the same time, that is, overtones, which we discussed in Section 1.3.

Early on in the history of quantum mechanics, scientists were enamored with these jumps of the electronic states. They appeared to be instantaneous, and things that happen instantaneously seem spooky to us, even magical. Modern quantum optics shows them to be just another wave effect of fields.

3.3 The Photoelectric Effect

The discussion in Section 3.2 also explains another famous experiment, known as the *photoelectric* effect, which was used early on and is still used in some discussions of quantum mechanics, to argue for indivisible particles. The effect is the following: if you shine a light on a metal, it can cause an electric current to flow. The electric current flows only if the light wave's wavelength is shorter than a certain value (i.e., frequency higher than a certain value), and the electric current starts right at the moment when you turn on the light.

In the thinking at the turn of the twentieth century, electrons in a metal were considered to be particles like little billiard balls orbiting single atoms. Therefore, the fact that the current started flowing right away when the light shined on the atoms bothered scientists quite a bit. They reasoned that, if light is a wave and is spread out, then the amount of light energy hitting any one single atom must be very small. Knowing the amount of energy per area in the light they were shining, and the approximate size of an atom, they could calculate that the amount of time needed for a single electron to gain enough energy to leave an atom could be seconds to minutes. But the experiments showed that the electric current started flowing almost instantaneously. Therefore, scientists reasoned that the only

---

[6] It is possible, of course, to assert that what "really" happens is that the wave function evolves continuously, but the particle makes an instantaneous jump at some point in time, with probability given by the wave function. But this amounts to a bare assertion from outside considerations; nothing in the field theory or the experiments requires or even indicates this.

way this could happen would be if light always came in discrete lumps (photons), and the electrons absorbed these photons one at a time, instantly jumping off the atoms holding them.

As discussed in Section 3.2, the switch-on time for a resonant system in the limit of low light input is a constant that is independent of the light intensity; it comes from the decoherence time of the system. What the early scientists missed was that, in a metal, there are resonant states of the electrons that extend across the entire metal; many of the electrons are not localized to single atoms. Therefore, for the case of a metal, the two levels labeled -1 and $+1$ in Figures 3.3 and 3.4 should be taken as extended states over the whole metal. Just as shown in Figure 3.4, there will be a quick switch-on of the current in the metal with a time interval that is independent of the intensity of the light. The equations used to give Figures 3.3 and 3.4 give all the main effects seen in the photoelectric effect: the fraction of the electrons in the upper state (which gives the current) is proportional to the light intensity; the switch-on time is very short, and independent of the light intensity, and the transition to the upper state will only occur if the frequency of the light matches a resonance of the electronic system. For metals, there is a continuum of resonances for all frequencies higher than a certain value, known as the bandgap.

In the modern understanding, the photoelectric effect is therefore actually a very good demonstration of the wave nature of electrons, namely the extended wave states known as Bloch states (discussed in Section 2.6), and *not* the particle nature of light!

Oddly, then, the two main experiments in history that convinced people of the particle nature of light, namely the Planck radiation spectrum, discussed in Section 2.7, and the photoelectric effect, discussed here, actually prove no such thing, in the modern understanding. Both results are well understood to follow from the wave nature of quantum field theory. There may be other experiments that convince people of the billiard-ball picture of particles, but these experiments cannot be legitimately invoked for that purpose.

3.4 Avalanche Detectors, Measurement, and Randomness

We are still working toward understanding particle detectors from the bottom up. We have seen that atoms are localized lumps that have a specific size, with a natural length scale given by the fundamental constants of mass and charge. When a light wave interacts with the atom, it can cause it to jump very quickly into a higher energy state.

In the photoelectric effect, a current is generated when light hits a metal. The current comes from the whole metal and doesn't give us any information about jumps of single atoms. We can get information from single atoms, however, by a different method. Figure 3.7 illustrates this approach, known as an *avalanche* mechanism.

Let us assume that an upper electronic state of the atom is coupled to other electronic states outside the atom. In this case, when the atom is excited by a light wave, so that it is at least partly in the upper electronic resonance state, then electronic charge can leak out of the atom into these other states. This doesn't have to happen all at once: the amplitude of the electron wave on the atom can decrease continuously over time.

Figure 3.7 Illustration of an avalanche method of photon detection. Electronic charge is kicked out of the potential-energy confinement of one atom, and accelerated under an electric field until it has enough energy to kick a charge out of another atom. At that point, there are two free electron charges, which are then accelerated to hit two more atoms, and so on. Since the free charge doubles after each step, after many repeats of this process, there can be millions of free electron charges, that is, a macroscopic current.

Now suppose that a strong electric field is applied so that any electronic charge that escapes is accelerated, picking up speed. Once the electronic charge has gained a lot of speed, it has enough energy to directly excite electronic charge out of low-energy resonances in other atoms in its path, thereby freeing the electronic charge from those atoms to also be accelerated. When this happens, there will be an ever-increasing cascade of electronic charge, like an avalanche of snow on a hillside. Ultimately, even though only one atom was excited at first, an electric current can be generated, which is large enough to be measured by standard electronic devices.

This method, with different variations, is the basis of all kinds of particle detectors, such as Geiger counters, photomultiplier tubes, digital cameras, cloud chambers, and camera film (in the case of photographic film, an electron that leaves an atom causes a chemical chain reaction in a grain of an emulsion, and the chemical reaction changes the color of that grain). Avalanche detectors therefore couple the *microscopic* world (at the level of single atoms) to the *macroscopic* world. "Macroscopic" means large enough for us to see. The output of every detector (which could also be the retina of our eyes, for example) is a macroscopic motion of millions of particles, whether in the needle of a meter, the light of an LED display, many water molecules in a cloud chamber, or some other sizable thing.

As shown in Figure 3.4, the state of a single atom during illumination by a light wave reaches a steady state which is a superposition of being partly in the lowest resonant state and partly in the upper state. In the case of the photoelectric effect, this leads to a steady current. In avalanche detectors, there is an intrinsic timescale for the signal to shut off. The reason is that, at some point, the atoms in the detector will be depleted of charge. They will not be able to supply any more charge until a source from somewhere else resupplies them. This gives the familiar effect of a "click," that is, a pulse of current.

This brings us to the discussion of randomness in quantum mechanics. In any given detector, there are many atoms all with slightly different positions and local environments, and any one of them is subject to a hair trigger that will cause an avalanche. It is therefore not hard to imagine that the randomness seen in the timing of the clicks and counts of detectors could arise from random fluctuations and variations in the detectors themselves. The same effect that gave the photoelectric effect will also ensure that the average rate of clicks in this case will be proportional to the light intensity.

Empirically, scientists in the early twentieth century adopted the *Born rule* (named after German physicist Max Born), which says that the probability of getting a click from a detector at location x at any time t is proportional to the intensity of the wave hitting the detector at that time and place. This rule is not part of the fundamental quantum field theory but has been verified in numerous experiments. In fact, it has been made a rule almost by definition: if any detector does not show this behavior (and plenty do not), it is taken as a flaw of the detector, due to some secondary effect.

The Born rule is consistent with the behavior of the photoelectric effect, discussed in Section 3.2, in that, if the number of counts is very high, the rate of these counts would be the same as a current proportional to the intensity of the light falling on the detector. At very low intensities, however, the clicks and counts from a detector occur very far apart in time, and then one can only talk of the probability of getting a click or not.

Many philosophers have run very far down the road with the effect of randomness seen in particle detectors. It is sometimes stated that quantum randomness is utterly different from all other types of randomness, and is utterly *without cause*. Apart from the logical difficulties of such a thing as a temporal event with no cause, such a statement is unwarranted by either the theory or experiments of quantum mechanics. It is more accurate to say that nothing in the quantum field equations that govern the detector behavior gives us any way to predict the precise timing and location of clicks and counts. The Born rule is an extra, empirical rule, which connects between a prediction of the theory (the amplitude of a wave) and an experimental measurement (clicks at random times). As far as anyone knows, the randomness could arise from random properties of the detectors. But the *correlations* of random events at different detectors, as we will see in Chapter 4, are often *not* random.

3.5 The Uncertainty Principle

Closely related to the Born rule is the notion of *uncertainty* in quantum mechanics. Once again, the basic effect is actually a fairly simple consequence of the properties of waves. When the result is connected to the randomness of detection via the Born rule, however, it has some strange implications.

Figure 3.8 illustrates a water wave passing through a hole in a barrier. On the output side of the hole, the water wave spreads out. Therefore, although the wave crests are all going in one direction on the input side of the barrier, on the other side, the wave crests are going out in many directions. This is a simple example of *diffraction*, which is a general wave effect. The smaller the hole, the more spread out the waves will be on the other side.

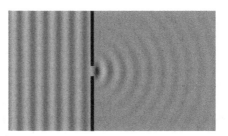

Figure 3.8 A wave passing through a hole in a barrier.

In the language of uncertainty, we can describe this by saying that the hole in the barrier gives us a definite location x of the source of the wave energy on the output side. The direction of motion of the wave coming from this source is spread out. A small hole gives a large spread in wave directions, and a large hole would give a small spread in directions. This can be stated as a tradeoff in uncertainties: the more certain we are of the location of the source, the less certain we are of it having an exact direction of motion, and vice versa.

This is normal behavior for waves, and is expressed mathematically in the formalism of Fourier transforms, discussed in Section 11.5. If we apply the Born rule, treating the wave intensities as giving us the probability of finding particles, the uncertainty principle has odd implications, however.

Suppose we detect a water-wave phonon (the proper quantum particle for sound waves) on the output side of the barrier. Since it came from the hole, we know its starting position quite accurately. But if we measure the direction of motion of all the phonons coming from the hole, we would find nearly equal probability for going in any of the wave directions. The converse is also true. On the input side of the barrier in Figure 3.8, the direction of the phonons is very accurately known – they are all going straight toward the wall. But where did they start from? We have little knowledge of that, because the wave is spread out in space.

There is therefore a tradeoff, or an incompatibility, in knowing both the starting position and the direction of motion of a water-wave phonon. This is exactly the same incompatibility found in trying to measure the position and velocity of electrons in an electron wave. If we think of electrons as billiard balls, that seems strange: we can know both the position and direction of motion of a baseball at the same time, with high accuracy – so why not electrons? The answer is not that electrons are very small. The answer is that the wavelength of the electron field can be made long enough, in some cases, for the wave properties to become important.

Not only position and velocity, but many other wave properties also have a similar tradeoff relationship. For example, measurements of different directions of spin also often have tradeoffs. Also, as shown in Section 12.5.2, a measurement of the number of particles in a wave (i.e., an accurate measurement of its intensity) and a measurement of the exact location of a wave crest of that wave (i.e., an accurate measurement of its phase) are also incompatible. Therefore, if we measure the phase of a wave accurately, we cannot insist that there is a definite number of particles in it.

References

T. Ferris, *Coming of Age in the Milky Way*, (Morrow, 1988).

A. P. Heberle, J. J. Baumberg, E. Binder, et al., "Coherent control of exciton density and spin," *IEEE Journal of Selected Topics in Quantum Electronics* **2**, 769 (1996).

W. H. Louisell, *Quantum Statistical Properties of Radiation*, (John Wiley, 1973).

L. Mandel and E. Wolf, *Optical Coherence and Quantum Optics*, (Cambridge University Press, 1995).

Nonlocality

As discussed in previous chapters, many of the things presented as mysterious in quantum mechanics are really not very mysterious at all. But there is something deeply mysterious in quantum mechanics. This is the effect of *nonlocal correlations*.

4.1 Correlation Experiments

Figure 4.1 shows a basic experiment that measures nonlocal correlations The first part is a source that can produce wave pulses that have just one photon at a time. It is not too crucial that this be exact. The source could also produce coherent wave pulses with an amplitude equal to an average of one particle per pulse, and the same experiment could be done, with only small changes to the results.

The second stage of the experiment is a beamsplitter (which can be as simple as a pane of glass) that splits the wave pulses emitted from the source into two parts, without introducing any significant dissipation or randomness. As discussed in Section 1.3, this effect of partial transmission is a standard property of waves. Finally, there are two photon detectors of the avalanche type discussed in Chapter 3. Each of these can give a click when a wave pulse hits it.

Suppose now that we do a *correlation* measurement, also called a *coincidence* measurement: we record the time that each detector clicks, and compare the times. Every time the source sends out its pulse, we record whether one or both of the detectors clicks. Most of the time, when we do this, neither of the detectors clicks. But if we measure clicks for a lot of repeated trials, we find a general rule: either one of the detectors clicks, or the other one does, but *the two detectors never click at the same time*. Figure 4.2 shows typical data for this kind of experiment. This type of measurement has actually only been possible in the past few decades, because making a source that emits only one photon on demand is not simple.

How should we understand this result? One might imagine a simple interpretation in the billiard-ball picture: a single photon is emitted, and when it hits the beamsplitter, it goes either one way or the other, and so it either hits one detector or the other, never both, even if we do the experiment exactly the same way every time.

One problem with this simple interpretation is that it does not tell us why the particle went one way and not the other. Why is there a kick to one side for some photons, and a kick to the other side for other photons? We know the properties of the beamsplitter very

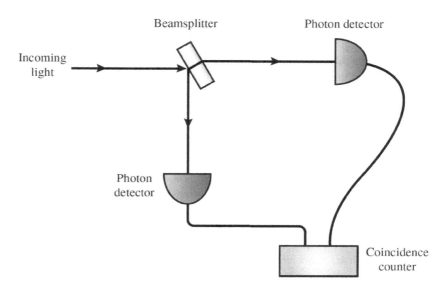

A simple beamsplitter experiment.

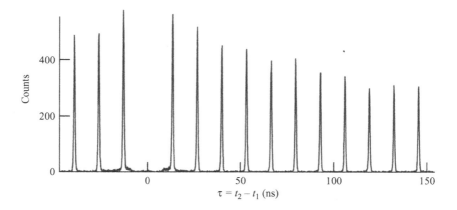

Correlation of photon detection at two different detectors when a series of single photons are sent into a beamsplitter. The lack of a peak at time $t_1 = t_2$ indicates that the two detectors never click at the same time, which would correspond to one photon converting into two photons, violating energy conservation. From Santori 2002.

well, and we know that it does not have any properties that kick photons to one side or the other randomly. From basic optics theory, we know to a very high degree of certainty that the beamsplitter simply splits the wave pulse into two, and each of these copies continues on its way.

I have argued in the previous chapters that we don't need to think of particles like little billiard balls. Let's stick with interpreting this experiment in terms of waves, then. In this case, the two wave pulses leaving the beamsplitter continue on their way to their destinations. One of them hits a detector, and that detector registers a click with some probability, which could depend on some internal randomness in the detector, which is commonly known to occur.[1] If one of the detectors clicks apparently randomly, the correlation measurement we've been discussing implies that the behavior of the *other* detector is *not* random. If the first detector has clicked, the second detector will definitely *not* click.

How does one detector know not to click if the other one has? One might imagine some signal that went out from the first detector to the second, to cause it to not click, although this would be an odd hypothesis – why should one detector communicate to another detector this way? But even supposing it were possible, we can put the detectors so far apart that no signal could have gotten from one to the other, even traveling at the speed of light, before the measurements of both detectors were recorded.[2]

One thing to notice about these correlations is that, however they may come about, they are related to *energy conservation*. If both detectors clicked, that would correspond to the energy of two photons being absorbed. But only one photon was emitted, which means that, if both detectors clicked, the total energy of the system would be increased by an additional photon's energy. The whole system seems to conspire across long distances to prevent that from happening, to keep the total amount of energy unchanged.

In the case of a particle track in a cloud chamber (see Figure 3.1), there is a similar effect. A wave coming into a cloud chamber hits many of the atoms in the chamber, and when one atom is ionized, it causes an avalanche to produce a macroscopic condensed water vapor. Once this has happened, the behavior of the other atoms in the chamber is not random. Only atoms in a straight line behind the first atom will be ionized, as the incoming particle gives up small fractions of its energy to each, while atoms at other locations in the chamber remain unaffected.

In each case, instantaneous communication seems to have happened across a long distance. In the case of the beamsplitter experiment with two detectors, communication seems to happen between the detectors faster than the speed of light. In the case of the cloud chamber, the atoms in a track seem to conspire to suppress a track from appearing anywhere else in the chamber, even though the incoming wave presumably hits all the atoms on one side of the chamber equally.

This type of behavior is known as *nonlocal correlation*. The word "nonlocal" here means that something seems to have affected something else in less time than it would take any signal to get there, even traveling as fast as possible, at the speed of light. In Einstein's theory of relativity, all causes are "local;" things only affect other things nearby, within the range that can be reached by a signal traveling at the speed of light (or slower).

Notice, however, that in the experiment described here only *correlations* are nonlocal. We can't use this scheme of two detectors to send a message to someone else faster than the speed of light, because we have to passively wait for a random detection of a photon at

---

[1] All detectors have *noise*, which is generated by thermal fluctuations, among other things. By contrast, optics theory tells us that the *beamsplitters* do *not* introduce randomness; they split the wave coherently.

[2] This is not merely a thought experiment. The speed of light is about 30 centimeters per nanosecond; the detectors can be placed a few meters apart, and modern laboratories can use electronic clocks that measure the arrival time of the photons with accuracy of a nanosecond or less (e.g., as seen in Figure 4.2). Therefore, the measurements can be done and recorded in less time than it takes light to go from one detector to the other.

one detector. This is a general result for all types of correlation experiments: nothing in the quantum correlations violates the theory of relativity as it relates to direct causation. But it violates physicists' intuitions that they adopt when they learn the theory of relativity. As is well known, Einstein called these nonlocal correlations "spooky actions at a distance"[3] and was very unhappy with them.

4.2 Why Physicists Want to Preserve Relativity

Many people who have heard of Einstein's theory of relativity wonder why physics seems to make an arbitrary boundary, saying that nothing can go faster than the speed of light. Why should the speed of light be chosen, of all things, to make this cosmic speed limit?

One reason physicists believe Einstein's theory of relativity is that it has been experimentally verified. But perhaps more importantly, the theory of relativity is a beautiful theory, which physicists call "elegant," like Dirac's theory of quantum mechanics, as discussed in Section 2.7. The theory of relativity combines many phenomena, including light, magnetic field, the laws of motion, momentum, and energy, into one set of simple equations. Dirac's theory is built on Einstein's, and eventually laid the foundation for all nuclear and particle physics. Einstein's extension of his theory to include gravity, called "general relativity," also made great strides in unifying different effects, and underlies all of our modern astronomy (and plays an important role in the accuracy of the global positioning system (GPS) used in all of our cell phones).

As we discussed in Section 1.4, the speed of light is actually a secondary consideration in the foundations of relativity, although many textbooks talk about it first. At a deep level, the main assumption of relativity is *locality*; the assumption that something that happens here, locally, cannot affect something somewhere else until there has been time for a signal to travel from here to there, and it cannot be affected by something somewhere else until a signal from there has had time to travel here. Once we establish this principle, the actual speed that it takes for causation to go from one place to another is just a matter of experimental measurement.

We can see why this is appealing through the simple thought experiment mentioned in Section 1.4. Imagine that you are in an isolated vacuum in outer space, on a rocket ship, going at a constant speed. You have a flashlight, and you shine it out the window. No matter which direction you point it, you see the flashlight act the same way.

The fact that your flashlight acts the same way no matter which way you point it indicates that the laws of physics you experience only depend on what you do *locally*. When you are going at a constant speed, you cannot tell if you are moving unless you look at some distant stars to see how you are moving relative to them (thus the name, "relativity"). This is the

---

[3] E.g., as quoted in Born 1971, Einstein said, "I cannot seriously believe in it because the theory cannot be reconciled with the idea that physics should represent a reality in time and space, free from spooky action at a distance."

same effect you notice on an airplane – if the airplane is moving at constant speed, you cannot tell if you are moving unless you look out the window at the distant ground.

If the light acted differently when you pointed it forward, in the direction of your motion relative to the distant stars, then we would say that the laws of physics for you were *non-local* – in this case, affected by those distant stars. You might not have a problem with the stars affecting your flashlight, but if we moved the stars *very* far away, infinitely far away to all intents and purposes, then it seems wrong that they should affect the behavior of your flashlight here.

4.3 One Explanation That Won't Work: The Local Hidden-Variables Hypothesis

In Section 4.1, we discussed an experiment in which a photon wave pulse is split into two at a beamsplitter, and then when those two parts of the wave hit detectors, there seems to be faster-than-light communication that ensures that only one detector will register a photon count and not both.

As mentioned in Section 4.1, one way to explain this that immediately comes to mind is to say that, at the beamsplitter, a photon definitely went one way or the other. In that case, the split of the wave into two outgoing pulses is just a useful fiction; the photon was really going one direction and not the other the whole time, long before the detectors got involved. In that case, there would be nothing actually going faster than the speed of light: both observers just acquired knowledge of what had already happened back when the photon hit the beamsplitter.

Surprisingly, this is one explanation that we can definitely rule out! This is known as a *local hidden-variables* theory. This approach was favored by Albert Einstein. Although in Section 4.1 we said that there is nothing we know of about the beamsplitter that would cause a photon (or wave pulse) to randomly go in one direction and not another, one could argue that we don't know everything, and maybe there is some underlying factor we don't know about, a "hidden variable," which causes different photons hitting the beamsplitter to act differently.

There is a long history over the past 60 years of ingenious experiments and calculations to prove that this cannot be the case. These are mostly based on a proposal by Einstein himself, along with two coworkers, called the *Einstein–Podolsky–Rosen* (EPR) experiment (Einstein 1935). Figure 4.3 shows one version of this.[4]

In this experiment, instead of the single-photon emitter discussed in Section 4.1, we use a two-photon emitter, which always emits two photons at the same time. This type of source is well understood and can be made routinely in laboratories.

---

[4] Variations of the quantum photon counting experiments discussed in this section were the subject of the 2022 Nobel Prize in Physics for Alain Aspect, John Clauser, and Anton Zeilinger; see www.nobelprize.org/prizes/physics/2022/.

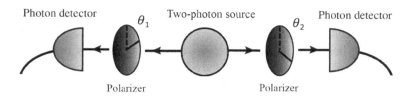

Figure 4.3 Layout of an EPR-type experiment, in which pairs of photons are sent in opposite directions by a two-photon source.

The experiment relies on two properties of the emitted photons. The first is that they are both *polarized* in the same way. Polarization is a basic property of light waves. As illustrated in Figure 4.4, the electric field in a light wave points sideways relative to the direction of motion of the wave. The electric field direction is called the polarization direction of the wave; it can point horizontally, vertically, or at any angle in between.

The second property of the photon pair in this experiment is that it is in a superposition of all possible polarization states; that is, all possible angles relative to horizontal and vertical. (See Section 16.1.1 for the mathematical representation of the EPR states, and the polarization rules used here.) There is no preferred polarization direction from the source.

The photons in a pair are sent in opposite directions to two detectors. In front of each of the detectors, we place a *polarizer*, which has the property that, if the polarization of the light wave is along the polarizer direction, the light will be transmitted through, while if the light is polarized perpendicular to the polarizer direction, it will be absorbed. If it is polarized in some direction in between these two directions, it will be partially transmitted; for example, if the light is polarized 45° relative to the polarizer direction, it will be 50% absorbed.

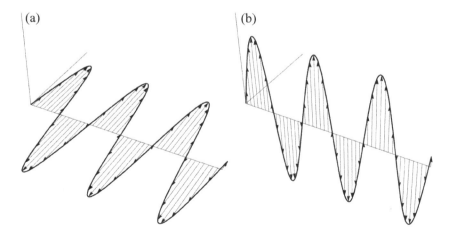

Figure 4.4 Two directions of polarization for a light wave. The arrows represent the direction and strength of the electric field at one instant in time. (a) Horizontal polarization. (b) Vertical polarization.

Consider now several possible measurements. If we look at just one side of the apparatus, we will see the photon creating a click at the detector 50% of the time, no matter what the direction of the polarizer is, because the wave is in a superposition of all possible polarizations, which means the polarizer always eliminates 50% of the light intensity.

Now suppose that we have definitely recorded a click at one of the detectors. What is the probability of getting a click at the other detector, responding to the other photon from the same pair? (This is another example of *correlation* experiment.) Since the first photon passed through the polarizer, when it was detected, it must have been polarized in the direction passed by that polarizer. But since the two photons are correlated to always have the *same* polarization, that means that we can predict the probability of the other photon passing through its polarizer with high accuracy. In particular, if the first photon went through a polarizer that passed vertical polarization, then if the second polarizer is also set to pass vertical polarization, the second photon will be detected with 100% probability. If the second polarizer is set to only pass horizontal polarization, then the second photon will never be detected, since a vertically polarized photon hitting a horizontal-passing polarizer is always eliminated.

This is the measured result of experiments – if the two polarizers are "crossed," that is, with their directions at an angle of 90° to each other, then there is never a click at both detectors for the same pair.

Now consider some more thought experiments in which we guess what is going on. First, suppose that the source "really" emits pairs of photons that are definitely some polarization or the other from the very start, and this polarization angle fluctuates randomly. In this case, it is easy to see that sometimes there will be clicks at both detectors. Sometimes the photons will both be in a polarization state of 45° relative to the horizontal, which is also 45° relative to the vertical. Fifty percent of the photons will pass through the polarizer on one side, and 50% of the photons will pass through the polarizer on the other side, which means there will be a 25% probability of having both detectors click. But the experiments give 0% probability of this happening! Therefore, our hypothesis that the source really emits pairs of photons with definite polarization, not in a superposition, must be wrong.

Imagine instead that the source does emit a state that is a superposition of all possible polarizations, and some unknown hidden variable at each polarizer causes the photon hitting to be absorbed or to pass through, on a random basis. If this effect is *local*, that is, if it depends only on what that polarizer's properties are, and not on state of the other polarizer, which we could put miles away, then each photon will pass through its polarizer half the time, so that the probability of both detectors clicking when they are crossed will also always be exactly 25% – in disagreement with the experiments. Therefore, this hypothesis must also be wrong.

One could imagine that somehow the two polarizers signal each other to arrange what they do together. But we can *change* the positions of the polarizers at any moment in time. Therefore, we could change one of them at the last second, just before a photon hits it, and put the polarizers so far apart that no signal can travel from one to the other before the detectors act. Since we still see the same correlations, this scenario must involve nonlocal correlations of the polarizers, which the hidden-variable interpretation was supposed to avoid.

The EPR experiment is often coupled to *Bell's inequality*, discussed in Section 16.1.2, which proves mathematically that no set of classical objects with only local interactions can have the experimental results seen in the EPR experiments.[5]

The hidden-variables hypothesis suggested at the beginning of this section is therefore disproved by disagreement with the experiments. Various authors have attempted to find "loopholes," which would allow hidden variables to exist somewhere, but all of these feel "rigged" in some way; that is, they posit effects of nature that are invisible and seem only to exist to deceive us.

4.4 The Copenhagen Interpretation

Because of the failure of the type of hidden-variable approach already discussed, the main consensus among physicists in the early twentieth century settled on the *Copenhagen* interpretation (named after the laboratory in Denmark at which Niels Bohr and other famous quantum scientists worked), which is probably still the majority view today among physicists, though not as universally accepted as it once was.

This view relies crucially on the assumption that the field, and the waves in it, are not "real," in contrast to all that the past chapters of this book have argued. In this book, I have argued that the field and its waves are more real than the particles – the field is like a bell and the particles are the ringing of that bell. Instead, in the Copenhagen interpretation, a quantum wave is an ephemeral entity that just represents our knowledge of the system, with no real existence. For shorthand, we can call this a *knowledge wave*.[6]

The appeal of this viewpoint is that a "knowledge wave" can vanish in a puff of smoke, so to speak, since it is not real. If we do a measurement that tells us the location of a particle, or some other property, our knowledge changes, and we can simply discard what partial information we had before.

In this approach, we interpret what happens in the beamsplitter experiment of Section 4.1 as follows: the knowledge wave, described mathematically by the standard wave equations of quantum mechanics, hits the beamsplitter and splits into two parts going in the two different directions. However, when one of the detectors registers a particle, the knowledge wave is instantaneously everywhere altered to a new knowledge wave that agrees what we now know. This is known as *collapse* of the wave function. Figure 4.5 illustrates the general notion of wave-function collapse. A wave is initially spread out, allowing for the possibility of detection of a particle over a wide range of possibilities. After a detection event has occurred at one place, the wave function is taken to instantaneously change to a new form, bunched up where the detection happened.

---

[5] Note, however, that Bell's theorem does not apply to *all* classical objects. Chapter 17 gives an example of a classical wave system that violates a Bell inequality.

[6] The term "knowledge wave" is not common in the literature, but the Copenhagen interpretation denies the ontological reality of the waves, treating them as nothing but solutions of equations that describe our knowledge. Sometimes the wave function is called "epistemological," which, of course, means that it is a knowledge wave. See, e.g., Aharonov 1993; for an example of use of the term "knowledge wave," see Sapogin 1980.

(a)   (b)

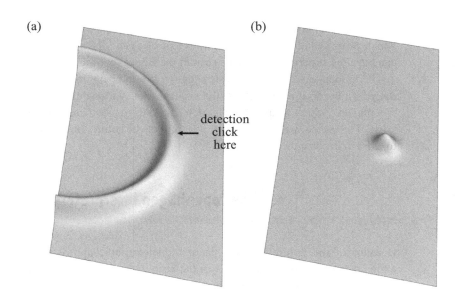

detection
← click
here

Figure 4.5 (a) A quantum mechanical knowledge wave spread out in space. (b) After a particle is detected, the wave "collapses" to a localized area where the particle was detected.

This approach has the advantage that it gives calculations that agree with the experiments. It enforces the Born rule by stating that the probability of collapse into a given new state is proportional to the fraction of the wave intensity in that state before the collapse happened.

There are numerous philosophical and practical objections to this view, however. Many of these center around the fact that *knowledge* has been inserted into the theory directly. In Chapter 7, we will discuss some of the ways this has been used in religious arguments, but without going there yet, many questions arise immediately. What do we mean by knowledge? Does it have to be human knowledge? Does an animal (e.g., a cat) count? Does a sophisticated machine? Just anything large?

Suppose we stick with human knowledge. If a wave function doesn't collapse until someone knows about it, there are very strange implications for how we view reality. The famous "Schrödinger's cat" scenario, illustrated in Figure 4.6, is an example of this. Suppose that a quantum particle is put into a superposition of two states by a beamsplitter. After this, it is either detected at one detector or the other; if one state is detected, a poison vial is opened by a machine, which kills a cat (known as *Schrödinger's cat*), and if the other state is detected, the cat lives. In the Copenhagen view, if no one looks inside the box with the cat, so that there is no human knowledge of the result of the experiment, then there will be a superposition of a living and a dead cat, because the cat is (presumably) also made of quantum matter. The cat will remain until a superposition of these two states until someone looks inside.

It can get stranger. Whose knowledge counts? In the scenario of Figure 4.6, suppose a person gains knowledge about the state of the cat by looking at the cat, but that *person is*

Figure 4.6 The famous Schrödinger's cat thought experiment. A single-particle detection experiment like that of Figure 4.1 is wired so that, if the particle passes through the beamsplitter, nothing happens, while if it is reflected, it actuates a machine that kills the cat in the box. In the Copenhagen approach to quantum mechanics, the cat goes into a superposition of being both dead and alive until someone looks inside the box, at which point the cat "collapses" into one of the two outcomes. Artist: Alexey Kavokin.

inside the closed box, and no one else looks into *that* box. Since that person is also made of quantum matter, does that person remain in a superposition of *seeing* the cat alive and seeing the cat dead, until a second person observes the situation? Why not, if people are made of the same quantum particles as cats?

Even stranger, consider the cosmic radiation from outer space that hits all of us all the time. Some of this radiation can cause cancer by hitting a DNA molecule, which mutates. According to quantum mechanics, a superposition of all possible radiation hits each person. In the Copenhagen interpretation, a person will go into a superposition of having cancer and not having it (actually, many superpositions of getting cancer at different times), and this superposition might not collapse until decades later when the person first learns of the diagnosis of cancer! In this scenario, as well as in many others that could be imagined, the time of the observation is determined by the collapse, because there is a continuum of possible times when a diagnosis can occur. Thus, knowledge appears to be caused by collapse, and not the other way around.

These and other problems with the Copenhagen interpretation have led some scientists to adopt the attitude of "Shut up and calculate!" That is, they simply don't ask what is real about the waves and particles. One can solve the field equations perfectly well and then use the Born rule to identify the final amplitudes of the waves with the probabilities of measurements at detectors. What the particles or waves were doing before the detector measured them is taken as irrelevant.

This is sometimes called quantum pragmatism and is a variation of a more sophisticated philosophy known as *positivism*, associated with Ernst Mach in the nineteenth century. Positivists officially remain agnostic about the reality of anything that is not directly measured. Mach himself applied this approach to atoms, even before quantum mechanics existed. He viewed them as "useful fictions," because assuming they existed led to very successful mathematical predictions for heat flow and gas dynamics, but because at the time no one could actually see an atom, Mach did not commit to whether they were real or not.

Pictures like Figure 3.2 eventually gave us direct images of atoms. But long before we had such pictures, scientists universally believed in the existence of atoms, because the theory based on them was so useful that there was no reason to doubt their existence. In general, physicists resist positivism; they resist being told, so to speak, "Thou shalt not look there!" Many important physics discoveries were first found by someone simply asking, "What is really going on?" As we have seen, Einstein imagined numerous "thought experiments" in developing his theory of relativity, which were never actually done. Had he taken the positivist attitude, we might never have gotten his theory of relativity.

This positivist approach may eventually also lose its appeal due to experimental input. Like the Copenhagen interpretation, it treats "detectors" and "measurements" as processes off the books of the quantum field equations. Detectors are made of quantum particles, however, and modern physics is blurring the boundary between the quantum system being studied and the detector observing it. We can write down the exact quantum wave functions of systems that are fairly macroscopic.

4.5 Are Fields Real?

As we have already discussed in Chapter 1, the viewpoint that electromagnetic waves and sound waves are real, but quantum matter waves are not, does not make sense because these fields are treated exactly the same way in quantum field theory. Some authors argue that the fact that matter waves are complex, that is, have two components instead of one or three, makes them fundamentally different from other waves, but as discussed in Section 9.1, this is just a difference in how many numbers we need to keep track of. Others have argued that because the electric field is directly measurable by a force, while quantum matter waves do not give a measurable force, therefore matter waves are unreal. But the electromagnetic field gives the probability of finding a photon in exactly the same way as a matter wave gives the probability of finding an electron, via the Born rule; the electromagnetic field just has an additional property of giving a force.

The Copenhagen model treats both the electromagnetic field and the quantum matter field in the same way, as generating "knowledge waves" that have no reality except for giving the probability of finding a particle. If we say that the electromagnetic field is not real, and is just a theoretical construct, then by rights we should say that sound waves are not real, and are also just a construct for computing the probability of detecting phonons. As proven mathematically in Sections 12.2 and 12.4, phonons and photons are treated exactly the same way in quantum theory.

Rejecting the reality of sound waves grates against us, however. Sound waves seem awfully real, more real than phonons. Furthermore, water waves on the ocean are just another type of sound wave: all vibrational waves in gases, liquids, or solids are treated in quantum mechanics the same way and are quantized into phonons in the same way. So, if we are to take the particle as the real thing and the wave as imaginary, then when we look at a water wave on the ocean, we should conclude that we are seeing an interesting theoretical construct, but, of course, only the phonons are real!

The example of water waves on the ocean also raises a question for the Copenhagen interpretation. It is physically possible according to the quantum field equations to have a water wave in a superposition of states with different phases, for example, with the wave simultaneously at a crest and a trough. The Copenhagen interpretation says that the reason we never see this is because it collapses into one or the other possibility when we look at it. But a wave on the ocean could be half a mile high, with a wavelength of dozens of miles, completely dwarfing any person. Why should that small person's knowledge collapse such a huge wave? This example negates any attempts to interpret the measurement process in quantum mechanics as intrinsically due to a large person interacting with a microscopic system. According to the Copenhagen approach, it can easily be the reverse – a small observer interacting with a large quantum system.

Working the other direction, we can say that, if water waves are real, then sound waves are real, and so then electromagnetic waves are real (having exactly the same type of field equation); and if electromagnetic waves are real, then matter waves are real, having the same type of field equation, but with a minus sign in a certain place, as shown in Section 13.1. Table 4.1 lists various different types of waves. All of them are vibrations of some field, and all of them can be associated with particles that give the quantized amplitudes of the waves, in quantum theory.

As mentioned at the beginning of this section, some authors argue that matter waves are not real, but electromagnetic waves are. A minority school takes the opposite approach, and argues that only the *matter* field is real, and that the electromagnetic field isn't (e.g., Mead 2000). In classical electromagnetism, charged objects are the "sources" of all

Table 4.1 Types of waves and their associated particles.			
Wave	Associated particle	Field	Type
Light radio microwaves	Photon	Electromagnetic field	Relativistic boson
Electric charge	Electron	Electronic matter field	Relativistic fermion
Nuclear matter	Quark	Nuclear matter field	Relativistic fermion
Sound	Phonon	Air	Nonrelativistic boson
Water wave	Phonon	Water	Nonrelativistic boson

Figure 4.7 A standard process by which a photon in vacuum (represented by the incoming squiggly line) can turn into an electron–positron pair, due to the coupling of the electromagnetic and matter fields, and then back into a photon. The electron is represented by a line with an arrow going forward, while the positron is represented by an arrow going backward.

electromagnetic field. We could then view the matter field as real, and the electric field as just giving us a description of the complicated forces between charged particles. Since the field theory treats both types of field on an equal footing, we are free to treat either one, both, or neither as real.

Although it is true that classical electromagnetism treats charged particles as sources for electric field, it is also true that, in modern quantum electrodynamics, photons are sources of charged particles! As shown mathematically in Section 13.2, Dirac's theory implies the existence of antimatter, which in turn implies that electrons do not exist permanently; they can be annihilated by combining with antimatter, and they can be created by photons exciting electron–positron pairs out of the vacuum state. Figure 4.7 shows a common *Feynman diagram* from quantum field theory which represents a photon turning into an electron and a positron, which then exist for some time before annihilating each other and releasing another photon. Quantum field theory says that this process happens all the time, everywhere in the universe.

Interpreting a diagram like that shown in Figure 4.7 as a particle popping out of vacuum, as though it were a little billiard ball, is not really the right picture, though. It is better to say that such a figure indicates the corrections to the total energy of the vacuum state due to the interactions in which the electromagnetic field acts as a source for matter waves.

Many physicists never learn the math of Feynman diagrams, but they are just a shorthand for various math calculations that arise in quantum mechanics. For those who have some background in the math of quantum mechanics, the basics can be learned in just a few chapters, given in Part III of this book. Section 15.3 discusses in depth how to interpret Feynman diagrams.

References

Y. Aharonov, J. Anandan, and L. Vaidman, "Meaning of the wave function," *Physical Review A* **47**, 4616 (1993).

M. Born, *The Born–Einstein Letters: Correspondence between Albert Einstein and Max and Hedwig Born from 1916–1955, with Commentaries by Max Born*, (Macmillan, 1971), p. 158.

A. Einstein, B. Podolsky, and N. Rosen, "Can quantum-mechanical description of physical reality be considered complete?" *Physical Review* **47**, 777 (1935).

C. Mead, *Collective Electrodynamics*, (MIT Press, 2000).

C. Santori, D. Fattal, J. Vučković, G. S. Solomon, and Y. Yamamoto, "Indistinguishable photons from a single-photon device," *Nature* **419**, 1 (2002).

L. G. Sapogin, "A unitary quantum field theory," *Annales de la Fondation Louis de Broglie* **5**, (1980).

5 Alternative Interpretations of Quantum Mechanics

The situation a hundred years after the development of quantum mechanics is that local hidden-variable theories are considered dead, the Copenhagen interpretation is strongly disliked by many physicists, and positivism is unsatisfying (mostly). As a result, rival interpretations of quantum mechanics have started to be taken much more seriously.

In this chapter, we survey some of these alternatives. Although some may sound strange, each has some appeal and reasoning. Though none has won a consensus, each has some dedicated adherents.

5.1 The Many-Worlds Hypothesis

The many-worlds hypothesis of quantum mechanics, suggested by Everett and Wheeler in the 1960s, has gained popularity in recent years, especially with science fiction writers but also with a small number of practicing physicists. Going back to the beamsplitter experiment discussed in Section 4.1, this hypothesis says that when the superposition of the photon going either to the left or the right never ends – the detectors go into a superposition of having clicked and not clicked, and eventually the entire universe interacting with those detectors goes into a superposition of the two possible outcomes. This leads to a view much like the science fiction trope of "parallel universes." Forever after, there are multiple universes with all possible outcomes of all quantum events.

The philosophical problems with this hypothesis are nearly endless. For example, since the splitting of the universe quickly leads to an infinity of possible universes, with all possible variations, it seems to make all of our choices meaningless, since for every choice we make, in some other universe, we will presumably have made the opposite choice.[1]

But the mathematical underpinnings of the many-worlds view are not nonsense. In fact, one can make the strong statement that, if one adheres purely to the quantum field theory with no alterations, then the many-worlds hypothesis must be true, because the field theory gives no mechanism to prevent superpositions of systems with large numbers of atoms, no matter how large.

---

[1] Larry Niven (1979) explores many of the philosophical conundrums of this view, including a fictional short story in which belief in the many-worlds view leads to mass suicide. David Wallace (2014) has argued that this need not happen, because people can adopt a game-theory approach to decision making, but this subtlety might be lost on the general public.

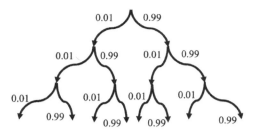

A branching diagram of superpositions after measurements. The numbers give the quantum "weight factor" for each branch.

However, the field theory is not all that we have. We also have the Born rule, which, as discussed in Section 3.4, is a rule about the statistics of measurements, added to the quantum field theory, and confirmed by all experiments. The many-worlds approach has no way of deriving the Born rule.

The problem of probability. To see why the many-worlds approach does not give the Born rule, consider the diagram in Figure 5.1, which shows the multiple branching paths of experience for some person. Whenever a superposition of states is generated, multiple paths are generated, one for each possible observed state. The quantum field theory assigns a "weight factor" to each of these possible paths, determined by the amplitude of the wave involved in each superposition.

In standard quantum mechanics, this "weight factor" gives the probability of each possible path via the Born rule. We imagine running the same set of measurements thousands of times, and each time the person follows one path. After counting things up, we find that a person is most likely to have an experience that follows a path with high weight factor, according to the Born rule.[2]

In the many-worlds view, there is no randomness, and no stochastic averaging. Every path is deterministically followed. The weight factors in this case do not give probabilities; they give the weight of that path as a fraction of the whole universe. In the standard approach, a large weight factor means there is "more of" one possibility than other because it is more likely. By contrast, in the many-worlds view, there is no probability at all; there is "more of" one path than another because there is literally more weight of that alternate universe than there is of the other, in the grand manifold of all possibilities.

To put it another way, suppose that some measurement yields a quantum state which is a superposition of two possibilities with very different weights, which we can write in a simplified version as

$$0.01 \left(\begin{array}{c} \text{universe with me} \\ \text{seeing outcome 1} \end{array} \right) + 0.99 \left(\begin{array}{c} \text{universe with me} \\ \text{seeing outcome 2} \end{array} \right).$$

In the many-worlds view, a person seeing outcome 1 does not feel less of a person; that person feels just as much a person as someone seeing outcome 2. Only an outside observer

---

[2] Although clearly distinguished paths are assumed here for simplicity, there is nothing in this argument that requires a countable number of paths; the same argument applies if there is a continuum of many paths with tiny differences between them.

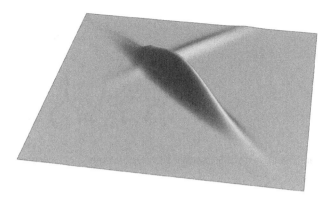

Figure 5.2   Two waves in superposition.

would say that the person seeing outcome 1 has less "weight" as a fraction of the whole universe. This means that, for people *inside* universe 1 or universe 2, the weight factors are effectively stripped off and inaccessible to their knowledge.

A person following a path in the diagram shown in Figure 5.1 would therefore calculate the probability of an event happening by asking how many times it occurred in the past *on that path*, compared to other events that happened *on the same path*. Lacking any experience or knowledge of the overall weight factors for various paths, that person could only count the total number of options possible in any given measurement.[3] For the great majority of paths, this would give an incorrect prediction for the probabilities of Born's rule and, in particular, would give much higher probability to very low-amplitude events. For example, suppose that the person sent a wave pulse through a beamsplitter 100 times, and the weight of one direction in the outgoing superposition was 99% and the weight in the other direction was 1%. In almost all the paths through the branching history, there would be nearly equal numbers of both outcomes, while there is only one path with the correct Born statistics of 99 photons detected in one direction and only 1 in the other direction.

To make the many-worlds approach agree with the experiments, one must modify it by adding the Born rule ad hoc. In other words, one can simply posit that a person's consciousness is attracted to paths with high amplitude. This can be argued as a "natural" choice,[4] but

---

[3] Some have criticized this "frequentist" approach (e.g., Wallace 2014), because any statistical approach never gives absolute certainty that one has gotten the right rule; one might have just observed a set of events that was a statistical outlier. But in practice, counting past experiences to make inductive conclusions is how scientists and engineers work, and, of course, how the Born rule itself was settled upon, historically.

[4] David Wallace (2014) has argued that the Born rule can be recovered as a betting strategy in game theory, namely that a person should bet on outcomes that have more weight – I should bet on ones with more of me, so to speak – in the future. This approach does not explain why any one person, looking at actual past experience, should expect to find that the Born rule has always been obeyed to high precision. Ultimately, Wallace argues (per private conversation) that the Born rule for our lived experience is a very natural choice to make but is indeed an additional assumption beyond the assumption of unitary evolution.

Frank Tipler (2014) has argued that the Born rule can be derived from a "density of universes" based on the Bohmian quantum potential, defined mathematically in Section 16.3. However, as discussed in Section 16.3, this formalism cannot be used for massless or relativistic particles. The notion of a "density of universes" is also somewhat unclear, and not part of normal quantum field theory, which simply gives an amplitude of waves. For further discussion, see Hsu 2012.

there are other "natural" choices one could make, such as saying that a person's conscious-ness always follows the path of highest amplitude (which would involve no randomness at all) or that the probability of a path is proportional to the absolute value instead of the square of the amplitude as required by the Born rule. At the end of the day, it is only the lived experience that, when we look into our past, we see the Born rule followed to a high degree of precision, which leads us to affirm it, in the many-worlds view.

Most physicists are so familiar with the Born probability rule that they assume that it must somehow be preserved in the many-worlds approach. But there are no random collapses in the many-worlds approach, and that is its main appeal to many physicists. In this context, it doesn't make sense to say that a higher wave is "more probable" than a lower wave. In the many-worlds scenario, if there are two outgoing worlds in a superposition, one is not more probable than the other. Both exist, and one has more *spectral weight*, that is, higher wave amplitude, than the other, like the two waves shown in Figure 5.2.

If one takes the approach that the only right way to count probabilities is to count the total number of paths in a diagram like that shown in Figure 5.1, then the many-worlds view is quantitatively falsified by the experiments. But it is not clear that this type of path-counting is the only way to get probabilities, and it is not necessary. Instead, one can simply put the shoe on the other foot and ask: On what basis can the Born rule be derived, in the many-worlds approach? The answer is that it cannot be derived from the mathematical formalism of quantum mechanics alone; it is an addition, both in the many-worlds approach and in the Copenhagen approach. Once one realizes this, the many-worlds view loses much of its appeal. If one is to have random jumps on the basis of an empirical rule, why not just have jumps to one possibility out of many, instead of to all possibilities?

The spectral weight problem. Another problem with the many-worlds view can be called the "spectral weight problem." This arises from the endless splitting of the universe into alternate versions.

The many-worlds hypothesis does not say that new universes are created with every quantum option; that would violate conservation of mass and energy, and is not what the equations of quantum mechanics imply. Instead, the many-worlds hypothesis says that, at each quantum choice, the existing universe is divided into two or more parts, which then forever go on their way without any further interaction. In wave terminology, each new fractional part has smaller and smaller spectral weight.

To see why this is a problem, consider a simple particle-counting experiment like that shown in Figure 4.1. In my own optics lab, we have routinely counted single photons up to millions of counts. (A typical rate of single photon counting with a photomultiplier is 50,000 counts per second, and experiments can easily run for 60 seconds or more.) Suppose that we ran the experiment of Figure 4.1 for a million counts. We have then just created $2^{1,000,000}$ alternate scenarios. That is, at the end of the experiment, the fractional weight of each of the alternate worlds in the many-worlds approach is equal to the original world's weight divided by 2, then divided by 2 again, and so on, for a million times. In engineering terms, the weight of the "signal" of any one alternative path has been reduced by *six million decibels*. Ask any engineer how easy it is to find a signal that has been diminished by six million decibels. Typically, a signal reduced by even 100 decibels is irrecoverable. The reason is that, in all real systems, there is background "noise," that is, randomly fluctuating, extraneous signals, and when the strength of a signal goes far below the average strength of the noise, it is permanently lost, to all intents and purposes.

And that is just one counting experiment run in my lab, one time! There are thousands of experimentalists who have run particle-counting experiments around the world for over a hundred years, running such experiments hundreds or thousands of times. And since the many-worlds view does not make a sharp distinction between measurements in laboratories and other events that affect the larger world, millions and billions of other divisions of the universe happen all the time, whether or not we count them. For example, cosmic rays continually fall from outer space that have some probability of hitting a DNA molecule in an animal or a person, leading to cancer. In the many-worlds view, the entire universe is divided down again and again into fractions to allow each possibility.

For the many-worlds approach to work, all of the information of each sub-universe must be perfectly maintained without loss, even as the fractional weight of each universe becomes vanishingly small. There must be no "noise" that messes up the perfect clock-like accuracy of the behavior of each sub-universe. Section 16.2.1 presents some of the mathematics of the spectral weight problem in greater detail.

Absence of evidence. Finally, it cannot be passed over quickly that there is absolutely no evidence for any of the parallel worlds proposed in this view, even when looking with the most sensitive of scientific instruments. To what degree should this matter? As Alvin Plantinga[5] and others have argued, absence of evidence does not count as evidence of absence unless we have some good reason to expect that there should be evidence. In the case of the many-worlds hypothesis, if we believe the equations of quantum mechanics are exact, then we will not expect to see evidence of these other worlds. But our general experience with the real world is that none of our equations is ever perfectly exact. Our *experiential* expectation is then that there should at least be some tiny echo of these other worlds, especially if we are in a world with small spectral weight compared to another with much greater spectral weight.

For example, if our world has a single one-in-a-million event that does not happen in an otherwise identical world, then that other world will outweigh ours by a factor of a million. Why should we expect that the other world, and many others much weightier than ours, existing in the same space as ours, to be absolutely undetectable? The experience of physicists and engineers dealing with waves in the real world is that some tiny factor always exists that messes up what the pure mathematical equations predict, giving "cross-talk" between different signals. That is, when trying to detect a particular wave signal, one always gets some small amount of other signals mixed in. In a normal engineering scenario, therefore, the existence of other parallel worlds would give at least some tiny noise in our world, like the muffled voice of a person in another room.

As an extreme example, imagine that we do an experiment like the Schrödinger's cat experiment, but instead of killing a cat if we get a certain detection outcome, we launch a nuclear weapon to blow up the moon. According to the many-worlds view, the enormous

---

[5] Plantinga (2001) uses the example of looking in a tent. Not seeing a St. Bernard dog in the tent is evidence for its absence, but not seeing any "noseeums," tiny insects that people can hardly see, is not evidence for their absence.

gravitational disturbance of the moon splitting into two would not be felt at all by anyone in a parallel world, even though they existed at the same location in space.[6]

The only way for the infinitude of other worlds coexisting in our location, including those with broken moons, is to be absolutely undetectable, and the only way to not have universes of tiny spectral weight vanishing into background noise is if the behavior of quantum mechanics is perfectly *unitary*. We will discuss this property further in Chapter 6, but for the moment, we can think of it as having no deviations whatsoever from running as a perfectly accurate machine. For the many-worlds approach to work, the behavior of the universe must be *exactly* accurate, down to signal reductions of trillions of decibels and more. While this is formally the case in the equations, to hold the many-worlds view, one must have an absolute commitment to the belief that these equations are perfect descriptions of reality.

5.2 Bohmian Pilot Waves

Because of the problems with the Copenhagen and many-worlds views, some physicists have proposed interpretations of quantum mechanics that use the same equations and give the same experimental predictions as the Copenhagen approach, but have a different story of what actually happens. In this section and in Section 5.3, we look at a few of these. Since these are not widely held among physicists, the reader can safely skip these sections and jump ahead to Section 5.4, which discusses proposals to alter the equations of quantum mechanics. Each of the approaches surveyed here and in Section 5.3 has some appeal, however, and therefore they deserve a detailed critique.

One approach that has grown in popularity in recent years is the "pilot wave" theory. In this approach, originally proposed by American physicist David Bohm, both waves in the quantum fields and particles exist as separate entities. (For a review, see Holland 1995.) The particles move through space like billiard balls, but they are attracted to regions with high wave amplitude and repelled from areas with low amplitude. The wave in this case is called a "pilot wave" because it directs the motion of the particles.

As shown in Section 16.3, the Schrödinger equation for quantum particles with mass can be manipulated to give an equation for a "probability current." This probability current can be viewed as a sort of fluid that flows where the wave has the highest amplitude, and which carries the particles along with it. There is an appeal to this, because the particles that flow in this current stream never disappear or jump discontinuously from one place to another. But there are several drawbacks and oddities to this approach.

---

[6] Penrose and Diósi have pointed out that defining a superposition of different curvatures of space in the context of general relativity is ambiguous, and from this, they have argued for spontaneous collapse of objects of sufficient mass; for reviews, see, e.g., Van Wezel 2008 and Diósi 2022.

Either there is spontaneous collapse (see Sections 5.4 and 6.5) or superpositions of large masses like the moon are possible without any detectability; otherwise, a "Cavendish Schrödinger's cat" experiment could be done in which a large mass was sent in one of two different directions depending on a quantum choice, and the motion of the mass in the "alternate universe" could be detected by its gravitational pull.

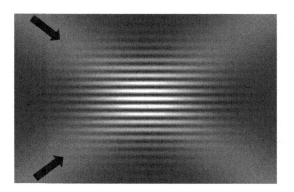

Figure 5.3 Intensity of the sum of two crossed beams with the same wavelength entering from the left side, in the directions indicated by the black arrows. The gray scale gives the total wave intensity at every point in space downstream. The dark horizontal lines give regions of zero probability for finding a particle.

First, as shown in Section 16.3, the mathematics of the probability current relies crucially on the use of the classical formula for kinetic energy of a particle with mass. This means that there is no equivalent approach for particles with no mass, such as photons. Some adherents of this approach treat the photons as off the books, so to speak, and treat only massive particles as real, but the approach also doesn't work for particles with relativistic mass at high speeds. Others take the Bohmian probability current equations to apply to photons anyway, as a basic assumption, even though the wave equation for photons can't be used to derive them.

The Bohmian current equations also have some bizarre implications for particle behavior. For example, Figure 5.3 shows two coherent beams crossing through each other, giving an interference pattern in the middle. Since there is nothing in the middle to affect either of the beams, the standard view of quantum mechanics is that the beams simply pass through each other, as any two waves would such as water waves or sound waves. In the Bohmian rules for particle flow, however, no particle can ever cross a region with zero wave amplitude. Therefore, particles coming in from the lower left, following the lower incoming beam, can never exit out the upper right, because the interference pattern in this case has broad lines of zero amplitude, seen as the horizontal dark lines in Figure 5.3. Where do the particles go, then? In the Bohmian view, they snake around, making sharp turns in free space, to stay inside the bright regions of the interference pattern, until they exit out the bottom right. Particles entering from the upper left also do the same thing, never crossing a completely dark region, until they exit out the upper right. Yet there was nothing in the crossing region that ever put a force on the particles; they passed through empty space.

Even more bizarre, in the case of a wave confined between two boundaries, such as the guitar string shown in Figure 1.9, a particle can never cross through one of the nodes, according to the Bohmian current flow equations. Therefore, in every region between two nodes, the particles are standing still! This gives a completely different account of the particles from quantum field theory. For example, in a laser, the photons bounce back and forth between the two mirrors at the ends. In the Bohmian view, the photons do not do this, but instead are sitting still. In other words, the Bohmian hypothesis implies a complete separation between the particles derived from quantum field theory and the particles that follow the pilot wave. In effect, it requires us to have two sets of particles, which have completely different behavior, for a single field.

The Bohmian approach also does nothing to make it easier to understand the nonlocality of quantum mechanics or the collapse of the wave function. The Bohmian formulation is explicitly nonlocal, with correlations between particles very far apart. The different results that are seen in different detector clicks are attributed, in the Bohmian approach, to different initial locations of the particles, which then flow deterministically to their final destinations. But nothing in the formalism tells us why a particle is initially at one location or another – the initial location of the particle within the wave ends up being a random variable, so that the randomness of the detector clicks is just pushed back earlier in time to the randomness of the initial particle locations.

5.3 Variants of Positivism

As mentioned in the previous chapter, many physicists simply have the attitude, "Shut up and calculate." This is a simple form of *positivism*, which says that we simply should not ask questions about things we can't directly measure. There have been several proposals for interpreting quantum mechanics that, at the end of the day, are more sophisticated forms of positivism, telling us not to ask certain questions. We will not survey all of these here; Appendix A lists some of these.

Note that sometimes it is right to say that questions don't make sense. For example, asking "Can God make a rock so heavy he can't lift it?" is the same as asking "Can God be omnipotent and not omnipotent?" which is simply asking if a self-contradiction can be true. In the same way, "What is the color of an object in the absence of light?" and "What is the sound of a piano when it is silent?" are simply badly defined questions, since color is defined as a property produced by light, and sound is incompatible with silence.

The rejection of certain questions in positivist approaches to quantum mechanics is not like that. We can conceive the situation we are asking about, unlike a logical self-contradiction, but positivism says not to worry about it.

One version of this, called "consistent histories," proposed by physicist Robert Griffiths (2003), says that after a measurement has been recorded, we can construct a self-consistent history of what was happening at all earlier times. We don't ask what would have happened for the same initial conditions if we had measured a different final outcome.

The basic approach is to imagine a series of measurements that would give a definite result at every time from the beginning state of the system to the final detected state. Even though those measurements were never actually done, we can assume, once we know the final result of a measurement, that, at all earlier times, the system really had the properties indicated by that sequence of measurements.

At one level, this is just a variation of the Copenhagen interpretation that aims to give a sensible story about the past. But it also can lead to some bizarre implications for how we interpret some experiments. Consider again the experiment with crossed beams shown in Figure 5.3. In the consistent histories approach, if we detect an outgoing particle at the upper right, we can create a history of the particle in which it definitely started at the lower left and then moved continuously in a straight line up to the detector where it was recorded at the upper right. But as seen in Figure 5.3, there are regions in which the quantum mechanical wave function says that there is zero probability of the particle ever being detected. How does the particle pass through these regions in a continuous fashion?

One could argue that the interference pattern was not really there when the particle passed through, since it is just a "knowledge wave" used to predict probabilities. But an ingenious experiment was done (Afshar 2007) in which the experimenters made an interference pattern like the one shown in Figure 5.3, and then placed wires at certain points in the dark regions, to block any photons there. Since those regions had very tiny light intensity due to the interference pattern, there was almost no change in the final experimental results at the detectors on the right side when the wires were there. But if a particle had been "really" traveling in a straight line from the lower left to the upper right, then it would have scattered off the wires, giving a different experimental result. The experimenters were able to see what this pattern would be by blocking the beam coming from the upper left, thus forcing all the particles to come from the lower left.

In the consistent histories approach, one can construct a history for a particle in the case when the wire grid is there, but this history will be very different from the one constructed if the wire grid is not there, even though the two cases had the same initial conditions and gave the same experimental outcome. Somehow, in this approach, the particles know to behave one way during their transit and not the other way, depending on whether the wire grid is there, even though no energy is given to or taken from the wire grid.

Griffiths has argued[7] that the presence of the wire grid amounts to a phase measurement. Because of number-phase uncertainty (see Section 12.5.2), if a wave packet from one source had definite phase, it could not have a definite number of just one particle. Therefore, the constructed history before the waves passed through the wire grid is that there were two wave packets from the two sources with a definite phase relationship between them; it only makes sense to speak of particles *after* the phase was measured at the wire grid. But in the experiment, the two sources were designed to have given a definite phase relationship of the wave packets (e.g., by using a beamsplitter to split a single, coherent wave packet). Therefore, the phase measurement by the wire grid provides no new information; indeed, the experimenter knows exactly where to place the wires at the nodes based on the prior knowledge of the phase relationship of the waves from the sources. Therefore, it is natural to take the view that the relative phase was well defined at *all* prior points in the path of the wave.

If we take the view that quantum mechanics is fundamentally a wave theory, and particles are not fundamental, as argued in Chapters 1 and 2, then this experiment has a much simpler explanation. The wave propagates from the two sources through the interference

[7] R. Griffiths, private communication.

pattern and then on to the detectors on the right, and only at the detectors do we need to apply the Born rule to turn the wave function into a probability of particle counts. This requires that we accept nonlocality, however, because we could generate the two coherent wave pulses by splitting a pulse from a single-photon source. In that case, only one of the detectors on the right will click, and never both, no matter how far away they are from each other.

Another variant of positivism is John Von Neumann's idea of "quantum logic." This essentially says that we should not be worried about things that are contradictions by normal logic. While that may seem irrational, it has some appeal by analogy with computers. We can program normal logic into computer circuits, but we can also program other rules. Why not just say that quantum mechanics has some other rules? The problem, of course, is that human logic is not *merely* a system programmed into computers; it is how we think. To assert a self-contradiction is to be incoherent, to say nothing at all.

As discussed in Section 4.4, although positivism can make life easier, it is generally unsatisfying to many scientists, because science has often progressed when people did ask basic questions about the consistency of theories, and often things that weren't measurable became measurable. Other versions of positivism are listed in the Appendix.

5.4 Spontaneous Collapse

So far, all the interpretations of quantum mechanics we have surveyed do not make any alterations to the basic equations of quantum mechanics, namely the Schrödinger equation and the Born rule. This is because these equations have been very successful for predicting many experimental results. There are also proposals to make alterations or additions to existing quantum theory. These proposals are generally called *spontaneous collapse* models. In this approach, some actual alteration of the equations of quantum mechanics is assumed, which gives the Born rule not as a result of human knowledge but as the result of some intrinsic instability in the physical system. This behavior may be closely related to a classical kinetic effect known as *chaos*, which describes systems that are hypersensitive to small changes in their initial state (discussed mathematically in a simple way in Section 10.2).

This approach can also be called a *nonlocal hidden-variables* view, as opposed to the local hidden-variables view discussed in Section 4.3. In other words, instead of seeing the unknown physical element that gives the randomness as localized in a particle, a nonlocal approach says that some system-wide property can give the observed behavior.

The main problem with this approach is that it requires new mathematics, and no one has yet persuaded the physics community of a comprehensive, quantitative mathematical version of quantum mechanics using this approach. Several spontaneous collapse models have been proposed, but there is not yet any widely accepted mathematical model. There is, however, a mathematical technique known as *quantum trajectories* that invokes spontaneous collapse to make quantitative predictions, which makes it very tempting to find a way to make spontaneous collapse a real part of the physics. One way this might come

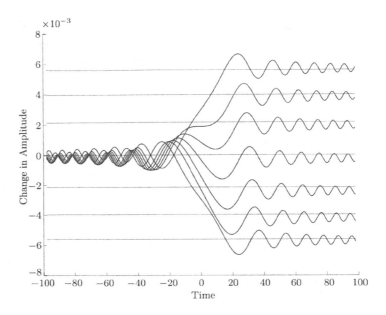

Figure 5.4 The behavior of the wave function of a superconducting metal ring over time, when it is in the presence of another, identical superconducting ring, allowing magnetic field from each ring to pass through the other. The different curves are for very slightly different initial conditions. From Mead 2000.

into the equations is through the effect of gravity, which at present has not been integrated into quantum mechanics. We will return to discuss models of spontaneous collapse in more depth in Sections 6.4 and 6.5.

One version of spontaneous collapse is known as the *transactional interpretation* (e.g., Cramer 1986, 2016; Kastner 2012). In this approach, it is assumed that there are waves traveling backwards in time. This is not as strange a proposal as it might first sound; it is well known that many types of wave equations allow backwards-in-time waves as solutions; these are known as *advanced waves*. These solutions are typically removed in classical physics from consideration as unphysical. But sometimes solutions of equations that seem unphysical turn out to have real, physical implications, such as Dirac's electron states with negative kinetic energy, discussed in Section 13.2.

In the transactional interpretation, one imagines a scenario in which a source (such as an atom) emits an electromagnetic wave going forward in time, which then hits several possible absorbers. Then each of them emits backwards-in-time waves that go back to hit the original emitter. Out of this, an instability arises that selects just one of the many possible cases of a photon going from the emitter to an absorber.

What is intriguing is that a mechanism for this type of instability can be shown mathematically in some simple systems, such as two superconducting rings emitting coupled to each other by electromagnetic field (Mead 2000). As seen in Figure 5.4, this system has "jumpy" behavior that looks a lot like the jumps of particle detectors, even though it follows standard wave theory exactly. However, the examples for which calculations like this have been done have so far only been very simple systems; the math that gives the instabilities has not been worked out for the general three-dimensional case.[8]

Physicists' intuition. In general, many physicists have the intuition that any macroscopic object, not just a human brain, should lead to the same type of measurement outcome. The problem with this intuitive view, as we will see in Chapter 6, is that it cannot be derived from quantum mechanics as we know it. Something, which might be human knowledge or something else, is presently "off the books" of the equations of quantum mechanics. Therefore, although many physicists think intuitively in terms of spontaneous collapse, there is no widely accepted mathematical model for it.

We will return to discuss proposals for spontaneous collapse in more detail in Sections 6.4 and 6.5; a quantitative proposal is given in Chapter 20. In general, it is premature to say that spontaneous collapse theories cannot work. While it involves the daunting challenge of mathematical alterations to quantum mechanics, the conundrums of the other views we have looked at, from Copenhagen to many-worlds to Bohmian pilot waves, make it appealing to look for alternatives. In many ways, the situation is similar to that faced by Einstein at the turn of the twentieth century, when the perceived paradoxes of electromagnetic theory cause him to look for alterations of the equations that would still agree with known observations and experiments.

References

S. S. Afshar, E. Flores, K. F. McDonald, and E. Knoesel, "Paradox in wave-particle duality," *Foundations of Physics* **37**, 295 (2007).

Y. Aharonov, J. Anandan, and L. Vaidman, "Meaning of the wave function," *Physical Review A* **47**, 4616 (1993).

J. G. Cramer, "The transactional interpretation of quantum mechanics," *Reviews of Modern Physics* **58**, 647 (1986).

J. Cramer, *The Quantum Handshake: Entanglement, Nonlocality and Transactions*, (Springer, 2016).

J. Cramer and R. E. Kastner, "Quantifying absorption in the transactional interpretation," arXiv:1711.04501.

L. Diósi, "On the conjectured gravity-related collapse rate $E_\Delta/\hbar$ of massive quantum superpositions," *AVS Quantum Science* **4**, 015605 (2022).

P. C. W. Davies, "Extension of Wheeler–Feynman quantum theory to the relativistic domain. I. Scattering processes," *Journal of Physics A* **4**, 836 (1971).

---

[8] John Cramer and Ruth Kastner (e.g., Cramer 2017; Kastner 2021), both citing Davies (1971), have argued that the math of the S-matrix for many-particle quantum mechanics, which gives imaginary terms for the energy of particles, implies intrinsic nonunitarity in quantum mechanics, which can be used to generate spontaneous symmetry breaking. As discussed in Section 16.4, the imaginary energy generated by the S-matrix method does not actually imply nonunitary time evolution. The introduction of imaginary self-energy in the unitary S-matrix theory is actually a subtle point; in many textbooks, a term $i\varepsilon$ is introduced without explanation in the denominators of the propagators, but as discussed in Section 16.4.2 of this book, this term can be derived rigorously within the unitary theory. Essentially, it arises from truncating a perturbative expansion of the S-matrix to second order, which gives a term that corresponds to Fermi's golden rule for outflow into an external environment, which is treated as "off the books."

R. Griffiths, *Consistent Quantum Theory*, (Cambridge University Press, 2003).

P. R. Holland, *The Quantum Theory of Motion*, (Cambridge University Press, 1995).

S. D. H. Hsu, "On the origin of probability in quantum mechanics," *Modern Physics Letters A* **27**, 1230014 (2012).

R. Kastner, *The Transactional Interpretation of Quantum Mechanics: The Reality of Possibility*, (Cambridge University Press, 2012).

R. E. Kastner, "The relativistic transactional interpretation and the quantum direct-action Theory," arXiv:2101.00712.

C. Mead, *Collective Electrodynamics*, (MIT Press, 2000).

L. Niven, *All the Myriad Ways*, (Del Ray, 1979).

A. Plantinga, *Warranted Christian Belief*, (Oxford University Press, 2001), Chapter 14.

F. Tipler, "Quantum nonlocality does not exist," *Proceedings of the National Academy of Sciences* (USA) **111**, 11281 (2014).

D. Wallace, *The Emergent Multiverse*, (Oxford University Press, 2014).

J. Van Wezel, T. Oosterkamp, and J. Zaanen, "Towards an experimental test of gravity-induced quantum state reduction," *Philosophical Magazine* **88**, 1005 (2008).

6

Decoherence and Collapse

As we have seen throughout this book, instead of treating quantum waves as spooky "knowledge waves," it is far more natural to treat all waves as physically real, including matter waves – as real as the sound and light waves you encounter every day. This means that we have to deal with the reality of nonlocal correlations.

There seems to be no way out of having nonlocal correlations, no matter what version of quantum mechanics we use, if we require a theory that agrees with the experimental data. Why not just embrace that? The type of nonlocality observed in quantum mechanics does not change our basic rules of causality; we cannot, for example, send signals to other people faster than the speed of light. Only correlations of events appear to occur faster than the speed of light, and we cannot predict those, only measure them.

If we accept nonlocal correlations as real, then the only remaining problem in quantum mechanics is how to account for the randomness in quantum measurements. As discussed at the end of Section 5.4, most physicists have the intuition that measurements, or collapse, of quantum wave functions occur due to interactions with any large object, not just a human brain. In this chapter, we will discuss why that is not so easy to put into the equations. In fact, there is simply no way to do it using only the existing equations of quantum mechanics.

6.1 Unitary Theories Will Not Work

Practicing physicists work every day with wave equations for photons, phonons, and electrons, and these waves seem very real and tangible to them, giving well-defined, quantitative predictions for experiments.

A tacit assumption of many physicists is that, if we knew enough about how to write down quantum mechanical equations for large systems such as machines and people, standard quantum mechanics would give us all the results we see in experimental measurements. However, that can't be – there must be something additional, off the books of the quantum mechanical equations we know. The reason is that the standard equations of quantum mechanics involve only *unitary time evolution*, and the Born rule from experiments (introduced in Section 3.4) intrinsically gives nonunitary behavior.

The term "unitary evolution" here means that the equations obey the principle of superposition, discussed in Section 1.3. In particular, if a system starts out at some point in time in a superposition of two wave states, then at all later times, the state of the system will be

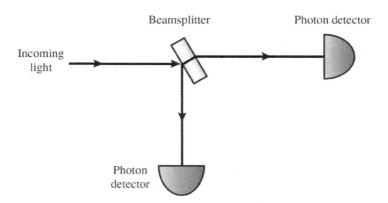

Figure 6.1 The beamsplitter experiment with two detectors.

a superposition of what each of the wave states would have done by itself, if the other one wasn't there.[1]

Let us return to the beamsplitter experiment of Section 4.1 to discuss what this means. Figure 6.1 shows a simplified version of this. A wave pulse with amplitude corresponding to a single photon hits the beamsplitter and divides into two. Each of these two wave pulses now travels to a detector; in other words, there is now a superposition of the photon going each direction. According to the equations of quantum mechanics, the photon on either path deterministically causes its detector to click and record a photon count. The total system will then be in a superposition of both detectors having clicked. There is nothing in the equations of quantum mechanics that makes one of the detectors click and the other not. Because these equations have unitary time evolution, the system at all future times will be in a superposition of the consequences of each detector clicking.

As discussed in Section 5.1, this is the main appeal of the many-worlds hypothesis. In that approach, there is no collapse of the detectors; forever after, the universe has a superposition of whatever events follow from each detector clicking, including dead cats or people believing and acting on different results. But as discussed in Section 5.1, the many-worlds view gives no way to obtain the Born rule for probabilities, other than simply asserting it, ad hoc. And if we are going to assert the Born rule, we may as well assert it for just one universe instead of an infinite number of them.

It is because we have the Born rule, but the equations of waves in quantum fields don't give it to us, that the Copenhagen view became popular. Something "off the books" gives the random collapses; in the Copenhagen view, the thing that is off the books of the equations is human knowledge.

Human knowledge could just be a stand-in for something macroscopic, however. After all, for us to know about something, it has to have a reasonably large effect; for example, a cascade of many electrons in the retina of an eye, or the cascade of electrons that gives rise

---

[1] *Unitary* time evolution is different from linear intensity dependence. As shown in Section 14.5, many systems, such as optical materials illuminated by powerful laser beams, have *nonlinear* intensity dependence, but their quantum mechanical wave functions still have unitary time evolution.

to an electrical current that causes a photomultiplier or Geiger counter to record a count, as discussed in Section 3.4. It could be that our knowledge is just a by-product of the effect of something being large enough to see.

But if large, macroscopic objects obey the equations of quantum mechanics, how can we get something nonunitary, off the books, with these large objects? What is the cutoff between microscopic and macroscopic, anyway?

Science does not stand still, and progress has been made on this question, although there is not yet a consensus about anything nonunitary in quantum mechanics. While we may not know what that is, we do have a way to clearly discuss the difference between microscopic systems, on the one hand, and macroscopic systems that act like collapses or measurements that give human knowledge, on the other hand. This distinction arises from the modern theory of *decoherence*. We have already touched on this topic briefly in earlier chapters, but we will now focus on this topic.

6.2 Decoherence

In the past 50 years, enormous progress has been made in understanding the dynamics of quantum mechanical processes, in particular the process of *decoherence*, also called *dephasing*. The difference between coherent and incoherent waves is illustrated in Figure 6.2. When many waves are in phase with each other, they add up to one large wave. This is known as *coherence*. When they are shifted randomly in their phase, they mostly cancel out. There are many processes that can give shifts to the phase of waves, which are random to all intents and purposes. Decoherence processes are any processes that give random phase shifts leading to lack of coherence. These might include random bumps from atoms in the environment outside the system of interest, or interactions with other parts inside the system of interest.

Why does decoherence matter? The crucial result of decoherence processes is that they lead to *irreversibility*. This is a well-known effect in our daily experience – friction causing something to slow down and stop but never to start moving and accelerate; something cooling off by radiating heat to its environment; an egg falling and breaking but never reassembling itself. It is closely connected to the second law of thermodynamics, which says that entropy (a measure of disorder) always increases, and closely related to our sense of an *arrow of time*, that time runs forward and not backward.

The second law of thermodynamics and the arrow of time were originally justified in terms of probabilities; the argument was that a system with many atoms or particles is much more likely to move into a more disordered state, even though it is physically possible to move into a more ordered state. We now know that the second law of thermodynamics is much more fundamentally connected to the quantum process of decoherence. A system of many wave interactions can be shown to move deterministically toward decoherence, which leads to greater entropy of the whole system, without using any notions of wave-function collapse or randomness, and without taking any position at all on the reality of particles; the irreversibility occurs entirely on the basis of unitary wave-function evolution

(a)

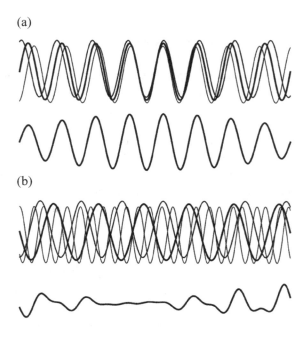

(b)

(a) Top: Several waves that have a high degree of coherence. Bottom curve: The sum of these waves. (b) Top: Several waves with random phase shifts. Bottom curve: The sum of these waves.

in large quantum mechanical systems. Section 18.2 reviews the math of this. The only exceptions we know of are systems that are restricted to only have a few degrees of freedom, in which case the system stays in the same state or goes through simple cyclical motion,[2] and systems of many bosons that undergo *enphasing* due to stimulated transitions (discussed in Section 21.1), such as lasers, superconductors, and superfluids. But even superfluids eventually dissipate energy irreversibly to their surroundings (see, e.g., McDonald 2020).

To see why decoherence leads to irreversibility, consider a coherent wave moving across a lake, which hits a rocky shore. As a thought experiment, we assume that the water has no heat loss and no friction of any kind; in other words, we suppose that all the motion of the wave proceeds deterministically. In a very short time after hitting a rocky shore, the wave will have scattered into many little waves moving in all directions, as seen in Figure 6.3. At this point, although no loss of energy or mass has occurred, the wave will effectively be lost, or "dissipated," because it has been spread around in many different, small motions.[3]

In principle, someone could arrange to have thousands of small waves, all started in the lake in just the right way so that after some time, they added up to give the exact reverse

[2] This is known as *integrable* motion; such systems can lead to very surprising behavior (see, e.g., Kinoshita 2006).

[3] Some readers may be familiar with another example of wave dissipation which is much closer to the assumption made here of no energy loss. A coherent laser light bounced off a rough surface such as a wall produces a "speckle" pattern that is, to all intents and purposes, random. The light waves producing this pattern come from reflected waves with very little energy loss into the wall.

Figure 6.3   Multiple random reflections of a water wave from rocks. (Creative Commons C0 Public domain.)

of the incoming wave, moving back outward. But for even a relatively small lake with just a few random rocks, this would be nearly impossible to calculate and predict, even for a person armed with a supercomputer. The two types of motion – coherent motion ending up in dissipation into many scattered waves, and motion adding many small waves to assemble a coherent wave – are completely different, leading to an asymmetry in regard to time. We identify time running forward as associated with processes like this wave spreading out and dissipating. The same thing happens when, for example, we speak to make a sound wave, which is then absorbed in the walls of a room. No energy has been lost, but the sound has turned into heat, which is made of many incoherent sound waves.

Although this type of dissipation process is natural in the world around us, scientists and philosophers historically found it hard to reconcile with the equations of physics, whether classical or quantum mechanical, because at the microscopic level, all of these equations are time-reversible. It can be proven (see Section 18.1) that any finite system with unitary time evolution must eventually return back to the same state that it started in, given enough time. This is called the *Poincaré recurrence theorem*, after the French mathematician Henri Poincaré.

Poincaré's theorem might lead us to expect to see lots of reversible cycles. But even in only moderately complicated systems, the time for a Poincaré cycle to complete can be much longer than the age of the universe. And it may be that our universe is infinite or, at least, large enough that most waves moving outward are never reflected back to us. In an infinite system, Poincaré cycles do not occur. In an infinitely large lake with no rocks, for example, a circular wave would simply keep spreading out forever, and never come back.

Our sense of time running forward, and the second law of thermodynamics, is therefore directly related to wave radiation into the rest of the universe. Almost everything radiates heat into outer space, where it can travel into the far reaches of space in the distant future. Only if that radiation could all return to its starting point would a Poincaré cycle ever be completed.

Decoherence by coupling a wave to a larger system also explains the irreversibility of measurements performed by detectors. In Section 3.2, we discussed how an electromagnetic wave coupling to an electron can cause it to oscillate between two states, spending time in a superposition of both, but eventually dissipation due to coupling to the outside world will lead it to settle into a steady state.

But that isn't the whole story of quantum measurements. The equations of quantum mechanics allow for the final outcome to be a superposition of different quantum states. As we saw in Chapter 3, though, detectors seem to make jumps into one or another possible final state, and don't stay in superpositions. How can we get these different outcomes?

6.3 Environmentally Induced Selection

One of the most useful developments in the theory of decoherence has been to allow us to define what we mean by a measurement, without needing to refer to human knowledge. In general, measurements occur when there are interactions that lead to strong decoherence.

To discuss this in more detail, let us define some terminology. There are two ways that interaction with a complicated, large environment can lead to irreversible behavior. One is called a T_1 process, in which energy is lost to the environment. In this case, the equations of quantum mechanics give irreversible behavior that is easy to understand – things lose energy until they settle down into the state of lowest energy, like a ball rolling down to the bottom of a hill.[4]

In many cases, however, there are two or more available states with the same energy. In this case, there is no process of energy loss that forces the system to go to one state or the other. Irreversibility can still enter in, though, because the two states can become decoupled from each other; that is, they can lose any well-defined phase relationship between them. This is known as a T_2 process.

One can visualize these two different types of processes by imagining two waves vibrating simultaneously on a guitar string, with two different frequencies (e.g., the main tone and an "overtone"). In a T_1 process, one of the waves (the overtone) will eventually disappear, as shown in Figure 6.4(a), leaving behind just the lower-frequency wave. In a T_2 process, after some time has passed, both waves will still be there, but there will have been random fluctuations in their oscillations, as illustrated in Figure 6.4(b), so that there is no predictable phase relationship of the two waves.[5]

---

[4] Quantum physicists talk about irreversible processes like this as an increase of "entanglement entropy" (e.g., Calabrese and Cardy 2004) because the total entropy of the whole system, including the two wave states and the environment, technically does not increase, but if one looks at just a subset of the whole system, one can define a quantity that behaves just like classical entropy that increases.

[5] These effects are often discussed in quantum mechanics using a grid known as a *density matrix*, which, for two wave states, has the following form:

$$\begin{pmatrix} \rho_{11} & \rho_{12} \\ \rho_{21} & \rho_{22} \end{pmatrix},$$

where $\rho_{11} = |\psi_1|^2$ gives the fraction of energy in state 1, $\rho_{22} = |\psi_2|^2$ gives the fraction of energy in state 2, and $\rho_{12} = \psi_1^* \psi_2$ (which, for our purposes, has just the same information as ρ_{21}) is the *correlation function* of the two waves. If there are random fluctuations in the waves, this product will average to zero. If the two waves keep a definite, coherent phase relationship, this product will average to a nonzero value. Section 14.3.2 gives a mathematical model for T_1 and T_2 processes.

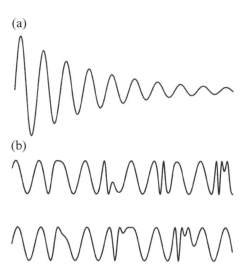

Figure 6.4
(a) A pure T_1 process, in which a wave decays away, giving its energy to the local environment. (b) A T_2 process, in which random phase shifts (seen here as quick accelerations of the wave oscillation) cause two waves to have no definite phase relationship.

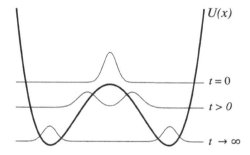

Figure 6.5
A simple model of the evolution of a wave function in the presence of environmentally induced selection. The system is assumed to have a fixed potential-energy landscape given by the heavy black line, with two local minima, plus random interactions with an external environment. At time $t = 0$, a wave is localized at the center of the system, with the intensity plotted by the top thin gray line. At later times, this wave evolves both lower in energy and also into two wave packets in separate regions.

Now let us apply these concepts to a situation that resembles a measurement. Figure 6.5 illustrates a simple model in which a wave that starts at the center moves out into two separate regions. When energy is lost to an external environment, the wave settles into these two localized regions and does not return back to the center.

In one sense, this is still a single wave, made up of a superposition of two localized states. However, the existence of T_2 processes means that the two localized waves in the different regions will have no definite phase relationship with each other. Effectively, they become completely decoupled from each other. The environment plus the specialized geometry of the system have selected out two independent outcomes.

This is a model for what happens in measurements. Although the system obeys perfectly unitary equations of quantum mechanics, there is an irreversible change into two decoupled possibilities. Wojciech Zurek has called this process *einselection*, for *en*vironmentally *in*duced selection, and has written extensively on the topic (e.g., Zurek 2005).

In real measurements, there is an additional step, in which the system collapses into just one of the two final states, according to the Born rule. The unitary equations of quantum mechanics can't produce this outcome. But although einselection doesn't take us the full way to understanding why only one or the other happens, it helps enormously in defining what we mean by a detector, or a macroscopic process that acts like a measurement. We can define a measurement as any process that leads to decoherence of two or more different outcomes. This might be a human-designed detector, such as a photomultiplier, in which an excited state of an electron leads to an avalanche of electric current, while the nonexcited state of the electron does not, but it can also be something that happens routinely in the world. For example, a molecule may have a stereo symmetry, that is, a left-handed version and a right-handed version, which are identical except that one is the mirror reflection of the other.[6] Both of these states of the molecule have the same energy, and therefore, in principle, molecules like this could exist in superpositions of both states, but, in practice, they never do, because interactions with the environment select out one or the other of these two states, via the einselection process (Day 2009). The same thing presumably happens when a gamma ray either does or does not hit a DNA molecule in a person's body, causing cancer.

6.4 Quantum Trajectories and Spontaneous Collapse

Let's take a step back and assess where we stand, when we take into account the modern understanding of decoherence. The following are well established in known quantum theory:

- The *irreversibility* of measurements can be well understood as part of the general irreversibility of large systems evolving according to the equations of quantum mechanics.
- The *selection* of certain states as possible outcomes of measurements can be well understood as a result of environmentally induced decoherence processes that are well described by the equations of quantum mechanics.

On the other hand, there are some things that are *not* predicted by the equations of quantum mechanics as we have them:

[6] Living systems have many molecules that have this type of symmetry, but all the molecules of this type that exist are the right-handed version. The origin of this asymmetry has been a mystery, akin to the symmetry-breaking problem discussed in Section 7.4, because nothing would favor one over the other chemically, from first principles, at the origin of life.

- The equations of quantum mechanics imply that decoherence is accompanied by a vast superposition of all possible outcomes, while our experience is that many systems jump to definitely one or another of the possible outcomes, according to the Born probability rule. The Born rule cannot be derived from the equations of quantum mechanics as we have them, whether in the Copenhagen view, the many-worlds view, or any other view that strictly adheres to the equations of quantum mechanics as we have them. Decoherence theory does give a very suggestive result, however, that the relative weight of the incoherent outcomes matches with the probabilities of the Born rule.

- The random quantum jumps predicted by the Born rule are accompanied in many cases by *nonlocal correlations*, in which two or more detectors far away from each other coordinate to give consistent outcomes; for example, if there is only one photon, only one detector will undergo a jump in response to it. The equations of quantum mechanics do not predict which detector will respond, but given the Born rule for quantum jumps, the equations of quantum mechanics do accurately predict the measured nonlocal correlations.

Quantum trajectories. In Section 5.4, we looked at the hypothesis of spontaneous collapse as a way of understanding the Born rule. It is often the case in physics that things that start out as "useful fictions" for theoretical calculations evolve over time into real things in the minds of most physicists. This has been the case in the past with such things as action at a distance, fields in a vacuum, atoms, and so on. In Chapters 1 and 2, we discussed how quantum fields have become "real" in the minds of most physicists.

Something of the same nature has been taking place in the past decades in regard to spontaneous collapse. A powerful theoretical method called *quantum trajectories* has been developed that invokes spontaneous collapse as a "useful fiction." In this approach, the standard equations of quantum mechanics are solved to give the state of a system as it changes over time, and then at random times during the calculation, jumps are made to force the system into certain states, consistent with the Born probability rule. These jumps don't depend on any knowledge of any observer but occur under the same conditions of environmentally induced selection discussed in Section 6.3. The mathematical details are given in Section 19.4.

This calculation method was developed primarily to make the computations easier for complicated systems. The scientists developing this method made no commitment philosophically to the reality of the quantum jumps but just found that assuming their existence made the calculations much easier and very accurate in describing real systems.

What if we make the leap that there really are spontaneous collapses, or jumps, in the physical world, as the quantum trajectories method assumes? There are many appeals. We would not have to worry about the mental gymnastics of "knowledge waves," infinite alternate universes, Bohmian particles that don't move while they orbit atoms, or other oddities. We know that it would give the right mathematical predictions, because the quantum trajectories method already uses this as a useful fiction to give correct experimental predictions.

The only drawback is that any quantitative theory of spontaneous collapse would require an alteration to the equations of quantum mechanics as we have them now, and that is not

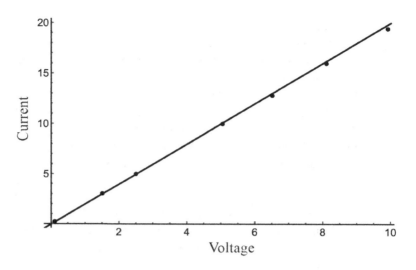

Figure 6.6 A typical plot of data (solid circles) versus a linear theory, namely Ohm's law (straight line).

so easy to do. Yet it is not far-fetched to suppose that the equations of quantum mechanics are not complete. The present theory of quantum mechanics assumes that the behavior of the universe is strictly unitary, so that superpositions of states remain perfectly in the same superpositions, after trillions of interactions. This unitary time evolution is a type of *linear* behavior. The experience of scientists and engineers, however, is that nothing we encounter in nature is really perfectly linear – all of the linear equations we use are actually approximations for things that are actually nonlinear to at least some small degree.

Linear and nonlinear behavior. In general, linear behavior means that two things are proportional to each other, so that one can draw a line to describe the relationship between the two quantities. For example, Ohm's law in electronics says that the electrical current that flows in a resistor is proportional to the applied force, namely the voltage of battery or power supply. Figure 6.6 shows this relationship drawn as a straight line. But it is well known that Ohm's law is just an approximation, which breaks down under certain circumstances, for example, for very high current. Slight deviations of the actual data from the linear mathematical rule, indicated by the way the dots in Figure 6.6 deviate from the straight line, are a common experience for scientists and engineers.

A great number of equations of physics have linear behavior but are known to only give that behavior over a certain range of conditions. For example, the optical properties of some materials change when they are exposed to high light intensity. Not only that, but even in vacuum, quantum theory gives nonlinear corrections to the optical wave equations at very high intensities (see, e.g., Mourou 2011). Many other linear equations have had to be adapted to allow new terms as experiments studied more extreme conditions.

The standard equations of quantum mechanics say that the frequency of oscillation of a wave is strictly linearly proportional to the energy in the wave (for the math, see Section 9.1). No exceptions are made, down to the tiniest length scale or largest energy scale.

It would not shock most physicists to find that there is some small nonlinear term that has been left out. However, such a term would have to be subject to some constraints. We can list several conditions for a reasonable theory of spontaneous collapse, which would make spontaneous collapse not just a useful fiction:

- It should conserve energy on average, if not in every instance. (It is already known to be the case that energy conservation can be violated in standard quantum mechanics for short periods of time, as discussed in Section 11.3.)
- It should not change the predictions of quantum mechanical equations for the behavior of wave functions when decoherence processes are weak.
- It should agree with the Born probability rule, on average.
- It should explicitly allow for nonlocal correlations.
- In addition, most physicists would add another condition, which may be called an aesthetic criterion: it should not seem especially "rigged" with unusual elements.

6.5 Quantifying Spontaneous Collapse

Various researchers have proposed models of spontaneous collapse, which are often called *GRW* theories, after the work of authors Ghirardi, Rimini, and Weber (Ghirardi 1986, later refined in Ghirardi 1990), or *continuous spontaneous localization* (CSL). Diósi and Penrose have proposed that gravity may play an underlying role in causing the effect (e.g., Diósi 1989; Penrose 1996).

One approach posits that, on a random basis, the wave functions of particles change from whatever they have been to a new, *localized* wave function, confined to some small region in space. The place at which the localization happens is random, with a probability given by the Born rule; that is, this type of localization is more likely to occur at places with higher strength of the wave function. If many particles are bound together, the average position of the whole set (the "center of mass" of the group) localizes in the same way, with even greater likelihood, proportional to the number of particles involved.

There are some problems with this approach. First, the random localization is not related to decoherence, which, as we have seen in the earlier sections of this chapter, play a major role in modern measurement theory. Second, and perhaps more fatally, these theories do not conserve the total energy of a system. Every time there is a jump to a localized region, there is an effective motion of the particle, with an associated kinetic energy. This implies a constant increase of the total energy over time, which acts like an energy source that constantly heats things.[7] Yet conservation of total energy is a foundational assumption in modern physics. Furthermore, recent experiments have looked for this energy source and

---

[7] An astute reader might wonder whether the spreading out of a wave packet over time, which happens in normal quantum mechanics and is known as *dispersion*, violates energy conservation in the same way. This is not the case, because the mix of wavelengths in the wave packet are not actually getting longer; they are just getting out of phase with each other.

have not found it, which means that it must be smaller than the amount predicted by some versions of the theory, if it exists at all (Donadi 2021).

Perhaps the greatest drawback, from the perspective presented in this book, is that the spontaneous collapses give spatially compact objects that look suspiciously like little billiard balls. As we have seen throughout this book, quantum field theory doesn't require such a view.

A field-theoretical approach. Chapter 20 describes my own work on spontaneous collapse, which is based explicitly on treating the wave functions of quantum field theory as real (Snoke 2021, 2022; Snoke and Maienshein 2023). In a nutshell, this proposal has the following features:

- Every fermionic resonance tends to randomly pop into one of its two allowed states ("0" and "1") instead of staying in a superposition. (Fermionic resonances were introduced in Section 2.4.)
- The probability of popping into one of these two states is proportional to *two* factors, namely (1) the rate of decoherence, *and* (2) the strength of the wave function in each of the two states. The first factor ensures agreement with the effects of decoherence discussed earlier in this chapter, and the second factor ensures that the Born rule is satisfied.

Energy is conserved in this model because it always selects between states with the same total energy. This approach requires that we introduce nonunitary evolution into quantum mechanics – as discussed in Section 6.1, a unitary theory cannot give random collapses of the wave function. Of course, the Copenhagen interpretation of quantum mechanics also assumes nonunitary behavior, but the nonunitary behavior in that case is "off the books," so to speak; it only comes in when we apply the Born rule *after* evolving the wave function using fully unitary equations.

This model is inspired by the behavior of classical nonlinear systems. This topic has been extensively studied since the 1970s and has generated a great deal of quantitative theory, including the theory of *chaos*, that is, deterministic systems that are nevertheless effectively unpredictable (see Section 10.2), and biological systems with switching behavior (e.g., Heltberg 2016). A general feature of many of these systems is that adding two features, namely *nonlinearity* and *noise*, can lead to "popping" behavior in which a system jumps suddenly from one type of behavior to another. "Noise" here refers to random fluctuations from an inhomogeneous, external environment. Figure 6.7(a) shows an example of the well-known effect of "mode hopping," in which a laser jumps randomly from one frequency to another and back again, via a combination of nonlinearity and noise. Noise in systems like this can act both to stabilize some states and to cause some states to jump away from what seems to be stable behavior.

Figure 6.7(b) shows this same type of behavior for a single ion, held in isolation using electromagnetic confinement. The state of the ion can be detected by its effect on the light passing by it. Note that the data of Figure 6.7(a) is described by a purely *classical* model,

(a)

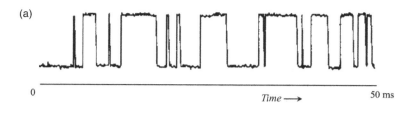

0   *Time* ⟶   50 ms

(b)

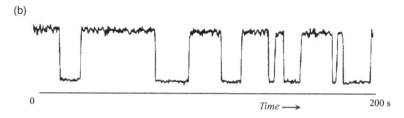

0   *Time* ⟶   200 s

Figure 6.7 (a) Mode hopping between two laser wave states, due to the purely classical effects of nonlinear coupling and random noise. From Mork 1990. (b) Quantum jumps of a single ion held in a magnetic trap in vacuum, detected by its effect on light emission while the system is continuously illuminated by a laser. From Sauter 1988.

with classical noise and nonlinearity, while the data of Figure 6.7(a) is from a *quantum* system. Is it so far-fetched to imagine that the behavior of the quantum system may be, at its root, caused by a similar mechanism as the classical example? Random noise is known to exist in many quantum systems; as we discussed in Section 6.3, quantum systems experience noise when they encounter large, macroscopic systems. And, as argued at the end of Section 6.4, physicists generally expect that there will always be some tiny nonlinearities in any real system.

As shown mathematically in Chapter 20, this combination of noise and nonlinearity can give the type of nonunitary "popping" behavior seen in quantum detectors. The nonlinearity introduced for this model is not of the type seen in well-known interactions of different fields, such as nonlinear optics, discussed in Section 14.5. The nonlinearity introduced for this model introduces nonunitary dynamics, in which noise in the system (i.e., time-varying fluctuations due to inhomogeneity in the environment) is amplified to give a definite collapse into one particular outcome or another.

This model allows us to understand the beamsplitter experiments discussed in Sections 4.1 and 6.1 in the following way. After a photon wave is split at a beamsplitter, there will be a superposition of an electron in an excited state in two different detectors. Each of these two overall states has the same total amount of energy. If the type of nonlinear process assumed in this model occurs, then decoherence will cause an electron in one of the detectors to pop into a definite state instead of remaining in a superposition. Either one or the other of the detectors will register a count. The nonlocal correlations of quantum mechanics will then ensure that, if one of the electrons pops into a specific state, everything elsewhere that is associated with that state must fall into place consistently, also.

6.6 Living with Nonlocality

The model of spontaneous collapse described in Section 6.5 has the advantage of explaining the randomness of jumps, but it doesn't remove the nonlocality of quantum mechanics. Can we live with that?

Nonlocal correlations are strange because they defy our normal expectations for causality. But keep in mind that our sense of causality itself depends on the effect of decoherence. As discussed in Section 6.2, our sense of time, in which we put causes before effects, arises from the irreversibility that comes from decoherence, which scrambles waves and doesn't reassemble them. Therefore, one can say that the quantum wave theory is a more fundamental concept than causality.

None of the various interpretations of quantum mechanics eliminate the problem of causality and nonlocality. In particular, the Copenhagen interpretation, the increasingly popular many-worlds hypothesis, and the spontaneous collapse model discussed in the previous section all have problems for our intuition about causality.

Consider the two diagrams shown in Figure 6.8, which are called *space-time diagrams*. Lines on this type of graph indicate the paths of objects through space and time, which are known as *world lines*. Figure 6.8(a) shows the experiment in Section 4.3, in which two photons are sent in opposite directions, from the perspective of someone who is at rest with respect to the two detectors. The world lines of the two detectors are the gray, fuzzy lines. As seen by a person who is sitting at rest relative to the two detectors, each of these world lines moves straight upward on this diagram, which corresponds to moving forward in time but not in moving in space. The two possible paths of the light going out from the beamsplitter are shown as the solid black lines. The slope of these curves, which shows the distance traveled in a given amount of time, is determined by the speed of light. In Figure 6.8(a), the detector on the left is a little bit closer to the source, and registers a photon at time t_1, before the second detector records a photon at time t_2.

Now consider the same set of events as seen from a different perspective. Figure 6.8(b) shows the same set of events as seen by an observer moving past the detectors at high speed. The equations of relativity (known as the *Lorentz transformations*) tell us exactly how to transform the perspective of Figure 6.8(a) into this one. An odd implication of the theory of relativity is that, if two events occur far enough apart in space (called *spacelike* separation), the laws of Einstein's relativity imply that observers moving at different velocities will have different experiences of which of the two events happened first.

In this perspective, the detector on the right encounters the path of a light pulse first, at time t_2'. From the point of view of the person in the frame of reference of Figure 6.8(b), the sequence of events is that the detector on the right detects a photon, which then controls what outcomes can be measured by the detector on the left.

Suppose that there are two people, each locally observing one of the detectors. In the Copenhagen approach, we would say that in the perspective of Figure 6.8(a), the knowledge of the person on the left forced a collapse of the person's experience on the right. But in the perspective of Figure 6.8(b), the knowledge of the person on the right forced a collapse of the experience on the left!

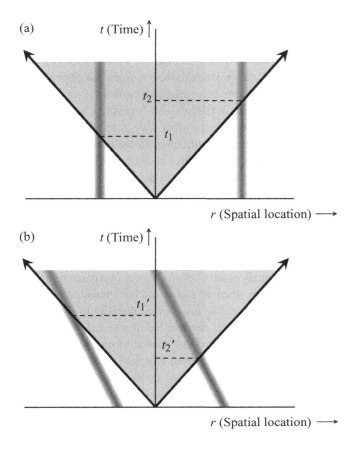

(a)

(b)

"World diagram" of a sequence of events, starting with a source that creates a nonlocal correlation, with two subsequent detection events. The diagonal black lines in each diagram give the "light cone," which contains all of the events that can be causally connected to the source. The histories of two detectors are plotted as the fuzzy gray paths. Both diagrams represent the same set of events from different perspectives. In perspective (a), the detector on the left triggers first, which then determines what the detector on the right can detect. In perspective (b), for an observer moving rapidly past the two detectors, the detector on the right triggers first, which then controls what the detector on the left can record. All observers agree on the events that happened but cannot agree on an absolute definition of time by which they could order the events.

The same problem arises for models of spontaneous collapse. Suppose we adopt the perspective of Figure 6.8(a) and say that the detector on the left feels a local fluctuation that causes it to commit to one outcome instead of another. The nonlocal correlations of quantum mechanics imply that this collapse immediately leads to the collapse of the detector on the right, as well. But if we changed perspective to that of Figure 6.8(b), we would say the reverse – that a local fluctuation at the detector on the right instigated both detectors to collapse. There would be no disagreement about the events that happened, nor any disagreement about the probabilities – the probability of getting a count on the right is equal to the probability of *not* getting a count on the left – but there would be disagreement about the causality, namely which event really caused the other.

The many-worlds hypothesis has the same causality problem. We can ask: "When does the splitting of the two worlds occur, from the perspective of the observers?" It cannot occur when the photon leaves the beamsplitter in a superposition state, because that would be equivalent to each of the two worlds having a hidden variable that makes the photon definitely commit to one or the other state before being detected. We know from Section 4.3 that this gives incorrect predictions for the experiments. Instead, in the many-worlds interpretation, the divergence of the experience of one world or the other occurs when there is strong decoherence, that is, when a detector with macroscopic size is involved. But this takes us to the same problem as the Copenhagen and spontaneous collapse approaches. In the perspective of Figure 6.8(a), we would say that the universe was split into two due to the action of the detector on the left, and the detector on the right just followed along, while in the perspective of Figure 6.8(b), we would say that the detector on the right caused the split. Because the splitting of the universe depends on the actions of the detectors, which can be as far apart as we like, an event at one detector can control the splitting that occurs at the other, even when no signal can have traveled at the speed of light or less from one detector to the other. The issue of nonlocality in the many-worlds approach is discussed with more mathematical detail in Section 16.2.2.

The bottom line is that every version of quantum mechanics that agrees with the experiments has nonlocal correlations that give us headaches when we think about their causal relationship. Causality is so much part of our experience that it is hard to even think or talk about something like nonlocal correlations. Yet, as discussed in Section 6.2, causality depends on the direction of time, which we define by having that causes precede effects. Therefore, the direction of the arrow of time is a consequence of irreversibility, and irreversibility is a result of decoherence. We thus see that causality is just a special case of the more general effect of decoherence in quantum mechanics.

The quantum field as a drumhead. Let's go back to the diagrams of Figure 6.8 and think anew. The diagonal, black lines indicate the *light cone*, which contains all the points in space that can be reached from the source by a signal traveling at the speed of light or slower. According to quantum field theory, in the perspective presented throughout this book, the quantum field exists everywhere in space and time, including the inside of this light cone. The quantum wave function of this field gives the amplitude of the field oscillations at every point in space and time, from which we deduce the probabilities of detecting single particles. The wave function of the quantum field also contains more information than this: it also provides all manner of *correlation functions* of the wave at different points in space.

If we change our relativistic point of view, as was done in going from Figure 6.8(a) to Figure 6.8(b), all these field values are mapped to some new values by the well-defined equations of relativity.[8]

---

[8] Penrose and Diósi (e.g., Penrose 1996, 2002, 2014; Diósi 1987, 2022) have pointed out that this mapping is well defined only if there is a well-defined curvature of space–time, when general relativity is taken into account. The ambiguity of defining the curvature of space–time when there is a macroscopic superposition of matter is the basis of their argument that macroscopic superpositions that lead to different curvatures cannot be stable, and therefore spontaneous collapse must occur.

Instead of thinking of objects moving through empty space, it is better to think of the quantum field inside the light cone in Figure 6.8 as a continuous sheet like a drumhead. At one end, the source is like a drumstick that bangs on that drumhead, and at the other end are two regions with friction, corresponding to the detectors. The whole drumhead finds a way to oscillate, given these constraints.

No matter which relativistic perspective we take, the quantum field as a whole contains the correlation information introduced at the source. This correlation is not without cause: it was caused by the source, which is causally connected to everything within the light cone. The field as a whole responds to the interaction of the light with both detectors. Instead of worrying about the causality of single particles, we can adopt the perspective that the entire field is responding to its constraints.

References

P. Calabrese and J. Cardy, "Entanglement entropy and quantum field theory," *Journal of Statistical Mechanics Theory and Experiment*, **6**, P06002 (2004).

C. Day, "Month-long calculation resolves 82-year-old quantum paradox," *Physics Today* **62**, 16 (2009).

L. Diósi and B. Lukács, "In favor of a Newtonian quantum gravity," *Annalen der Physik* **499**, 488 (1987).

L. Diósi, "Models for universal reduction of macroscopic quantum fluctuations," *Physical Review A* **40**, 1165 (1989).

L. Diósi, "On the conjectured gravity-related collapse rate $E_\Delta/\hbar$ of massive quantum superpositions," *AVS Quantum Science* **4**, 015605 (2022).

S. Donadi, K. Piscicchia, C. Curceanu et al., "Underground test of gravity-related wave function collapse," *Nature Physics* **17**, 74 (2021).

G. C. Ghirardi, A. Rimini, and T. Weber, "Unified dynamics for microscopic and macroscopic systems," *Physical Review D* **34**, 470 (1986).

G. C. Ghirardi, P. Pearle, and A. Rimini, "Markov processes in Hilbert space and continuous spontaneous localization of systems of identical particles," *Physical Review A* **42**, 78 (1990).

M. L. Heltberg, R. Kellogg, S. Krishna, S. Tay, and M. H. Jensen, "Noise induces hopping between NF-κB entrainment modes," *Cell Systems* **3**, 532 (2016).

T. Kinoshita, T. Wenger, and D. S. Weiss, "A quantum Newton's cradle," *Nature* **440**, 900 (2006).

R. G. McDonald, P. S. Barnett, F. Atayee, and A. S. Bradley, "Dynamics of hot Bose–Einstein condensates: Stochastic Ehrenfest relations for number and energy damping," *SciPost Physics* **8**, 029 (2020).

J. Mork, M. Semkow, and B. Tromberg, "Measurement and theory of mode hopping in external cavity lasers," *Electronics Letters* **26**, 609 (1990).

G. Mourou, "Nonlinear optics: From quartz to vacuum," *OSA Technical Digest: Nonlinear Optics: Materials, Fundamentals, and Applications*, NWB3 (Optical Society of America, 2011).

R. Penrose, "On gravity's role in quantum state reduction," *General Relativity and Gravitation* **28**, 581 (1996).

R. Penrose, "John Bell, state reduction, and quanglement," in *Quantum (Un)speakables: From Bell to Quantum Information*, R. Bertlmann and A. Zeilinger, eds., (Springer, 2002).

R. Penrose, "On the gravitization of quantum mechanics. 1. Quantum state reduction," *Foundations of Physics* **44**, 557 (2014).

Th. Sauter, R. Blatt, W. Neuhauser, and P. E. Toschek, "Quantum jumps in a single ion," *Physica Scripta* **T22**, 128 (1988).

D. W. Snoke, "A model of spontaneous collapse with energy conservation," *Foundations of Physics* **51**, 100 (2021).

D. W. Snoke, " Mathematical formalism for nonlocal spontaneous collapse in quantum field theory," *Foundations of Physics*, (2022).

D. W. Snoke and D. Maienshein, "Experimental Predictions for Norm-Conserving Spontaneous Collapse," *Entropy* **25**, 1489 (2023).

W. Zurek, "Probabilities from entanglement, Born's rule $p_k = |\psi_k|^2$ from envariance," *Physical Review A* **71**, 052105 (2005).

7

Quantum Mechanics and Our View of Reality

Many expositions of quantum mechanics present a view in which our sense of reality is undermined; we are told that quantum waves are not really there, or that things remain only in a state of potentiality until we look at them, and so forth. From all we have seen so far in this book, what views are really required by the data? What does quantum mechanics say about the nature of reality?

This takes us to the point of intersection between physics, philosophy, and religion. Physicists have a long history of "kibitzing" on the topic of religion. This can occur in two different ways. On the one hand, some physicists have argued against the need for religion by saying that the physical world is all there is, and that there is nothing more. As a variation of this, some argue that, while a spiritual realm might exist, there is never a need to invoke its existence to explain any of our experience, in which case the principle of Ockham's razor would imply we should dismiss religion as irrelevant. As physicist Stephen Hawking famously asked, "What place, then, for a Creator?" (Hawking 1988).

On the other hand, some physicists have argued that quantum mechanics supports certain types of religious views. As we have seen, in some formulations of quantum mechanics, the human mind plays a crucial role in a way that cannot be explained by any physical law. This has led to many authors invoking quantum mechanics as evidence for a spiritual reality. We need to be careful, however, not to make quantum mechanics say more than it really does.

7.1 The Tao of Copenhagen

In the 1970s and 1980s, several books about quantum mechanics appeared that caught the popular imagination.[1] These presented a view that integrated quantum mechanics with eastern religion, especially Zen Buddhism. They went a long way toward establishing an intellectual foundation not only for Zen Buddhism, but more generally for the movement known variously as new-age spiritualism, neo-paganism, spiritual mysticism, or just "spirituality." Many people now believe that the physical theory of quantum mechanics firmly establishes a mind–spirit–matter connection. This view was popularized in the 2000s by the movie *What the Bleep Do We Know?*, connected to the Ramtha School of Enlightenment, which promotes transcendental mediation. That movie not only insisted that quantum

---

[1] In particular, Capra 1975, Zukov 1979, and Talbot 1983.

mechanics proved their views but also expressed anger that anyone in the modern world would disagree.

The argument of these books essentially amounts to the following. First, it is assumed that the Copenhagen interpretation of quantum mechanics, discussed in Section 4.4, is absolutely proved by the experiments. The authors can't exactly be faulted for this, since it was the prevailing paradigm in physics throughout the second half of the twentieth century.

Second, based on the notion of fields as "knowledge waves" in the Copenhagen interpretation, a special role is seen for human consciousness. Human knowledge, in this case, is not just a passive observation of an external reality but actually a creator of reality out of a realm of the merely possible. As Fritjof Capra wrote:

> Quantum theory thus reveals a basic oneness of the universe. It shows that we cannot decompose the world into independently existing smallest units. As we penetrate into matter, nature does not show us any isolated "building blocks," but rather appears as a complicated web of relations between the various parts of the whole. These relations always include the observer in an essential way. The human observer constitutes the final link in the chain of observational processes, and the properties of any atomic object can be understood only in terms of the object's interaction with the observer. (Capra 1975)

There is a materialistic version of this role of the observer, which is that observers are so big that they always have effects on the microscopic systems that quantum mechanics deals with. This perspective breaks down, however, when we realize that many quantum mechanical systems can be quite large. For example, as pointed out by the physicist George Uhlenbeck many years ago (Uhlenbeck 1973), superfluids can be macroscopic enough to fill a bucket or a vat, and yet have well-defined quantum wave-function properties such as phase. (The properties of superfluids are discussed in Section 8.5.) Furthermore, as discussed in Section 4.5, very large water waves or sound waves are in fact macroscopic states of phonons, and could exist in very large superpositions of different phases, for example, a superposition of the water at both a crest and a trough. The Copenhagen interpretation therefore does not simply say that observers are always very large and therefore disturb small systems. It says that the act of knowing, itself, causes collapse of quantum waves, even waves much larger than the observer.

Third, from this connection between the knowledge of the observer and the physical world, some authors have gone on to conclude that free acts of human choice can change the course of nature. This is not just the choice to exert some kind of physical force but also the pure act of knowing or thinking, itself. This opens the door to all kinds of beliefs in extra-sensory perception (ESP), bending reality to our will by simply wishing it, and so on. Such things are clearly not part of the Copenhagen interpretation, but because the Copenhagen approach seems to imply a mind–matter connection, some people see ESP and such as the natural next steps.

Christian and other theistic philosophers have also used the Copenhagen interpretation to draw religious conclusions. While they would not ascribe to humans the ability to bend nature to our will just by thinking, they have sometimes taken the requirement for an observer in the Copenhagen perspective to imply the existence of a cosmic Observer or Mind outside the physical world, which can be equated with God.[2]

Postmodernism and quantum mechanics. With less explicitly religious overtones, quantum mechanics has also been used as a foundation for postmodernism in philosophy. This view says that we cannot know any truth objectively; instead each person creates his or her own meaning based on value choices and commitments. As sociologist Dan Handelman wrote:

> No longer may we assume with ease that nature (and culture) exist "out there," to be mapped and discovered without evaluating our own roles and operations at one and the same time. The particle physicist, Werner Heisenberg ... put it this way: "When we speak of a picture of nature provided by contemporary exact science, we do not actually mean any longer a picture of nature, but rather a picture of our relation to nature." (Handelman 1998)

This perspective does not explicitly say that people control reality with their thoughts, but it invokes the lack of reality of the wave function in the Copenhagen interpretation of quantum mechanics to argue that the physical world is only in a state of mere possibility, without firm reality, until someone gains knowledge of it. This is sometimes argued as due to the inevitable effects of the observer on the observed world. Of course, a classical physicist would also have said that people affect the world by their actions, including in their experiments to observe things, but in the Copenhagen interpretation, the world is viewed as much less concrete, changed dramatically by our mere knowledge.

In the 1990s, physicist Alan Sokal created a firestorm when he published a "hoax" article in a prominent postmodernist journal (Sokal 1996).[3] In this article, he argued along postmodernist lines against belief in reality itself, invoking quantum gravity. He then publicly retracted the article and argued that it proved that postmodernism is bankrupt. As he argued later, if one is going to be a progressive opposing nuclear weapons, it matters very much whether nuclear bombs are real and can kill people!

Probably the most effective response to Sokal was an article in *Physics Today* by Mara Beller (1998), pointing out accurately that the founders of the Copenhagen view could well be described as promoting the same postmodern views that Sokal mocked. Much of their writing indeed seems mystical, such as the following quotes by physicist Niels Bohr:

> Everything we call real is made of things that cannot be regarded as real. (Bohr 1938)

> The opposite of a correct statement is a false statement. But the opposite of a profound truth may well be another profound truth. (Bohr 1968)

> It is wrong to think that the task of physics is to find out how nature is. Physics concerns what we say about nature. (Bohr 2000)

The founders of the Copenhagen view do seem to have engaged in some very fuzzy thinking. But as one wit responded, they can be forgiven their philosophical vagueness,

---

[2] For example, William Lane Craig (1996) has written, "Why must the observer be human? Could God act as a Cosmic Observer, who collapses the wave-function in any measurement, or who would collapse any wave-functions in the universe with respect to any possible measurement?".

[3] For an extended discussion of the debate, see Tauber 2009 and my review of Tauber's book in Snoke 2011.

because they also created the physical theory that led to modern technology, computer chips, and so on, while postmodern philosophers have only produced some obscure poetry and plays! In fact, the founders of quantum mechanics were themselves influenced by early currents of thought that eventually became postmodernism, especially by the philosopher Hegel, who promoted the idea of "synthesis" of apparently contradictory concepts, and by Nietzsche, who promoted the idea of the "superman" creating his own truth.

Postmodernism does not require a religious perspective, but it is often used to support religious or spiritual claims like those described at the beginning of this section, because it supports the idea of choosing what we want to believe, instead of asking what is really true about the world, whether we like it or not. As Gary Zukov said:

> According to quantum mechanics there is no such thing as objectivity. We cannot eliminate ourselves from the picture. (Zukov 1979)

> We are in the greatest evolutionary transformation in the history of our species. We are expanding beyond the five senses. We are becoming aware of ourselves as immortal souls, as powerful creators and co-creators. We are becoming aware that we experience what we create, and there is no escape from that. (Zukov 2019)

These notions, sometimes not seen as "religious," but merely "spiritual," have become deeply accepted by much of the population. Although many who hold these views do not explicitly study or invoke quantum mechanics, there is a general sense that it somehow gives support to this view of the world.

How sure can we be of the Copenhagen view? These religious and philosophical views already discussed rely heavily on the assumption that the Copenhagen interpretation of quantum mechanics is correct. But as we have seen in previous chapters, it is far from proven that the Copenhagen interpretation itself is correct; in fact, it is not even clear whether it is currently the majority view among physicists; alternatives such as many-worlds, Bohmian pilot waves, and spontaneous collapse have increasing numbers of adherents. We therefore would do well to not build whole structures of religion and philosophy based on the Copenhagen view.

If the spontaneous collapse hypothesis is true, then we have a much more mundane picture of the natural world. There would still be some uncomfortable implications of quantum mechanics, namely that some things are unpredictable and some things have nonlocal correlations, which makes it difficult to write down a clearly established causal chain for every set of events. But an explicit role for human (or other) knowledge in the behavior of the physical world would be the same as any interaction with a macroscopic object in shaking out quantum superpositions into one of several possible outcomes.

Unpredictability of events is not philosophically difficult; it is already known that purely classical systems can lead to effectively unpredictable behavior. As shown in Section 10.2, it is possible to construct a perfectly deterministic system, called a *chaotic* system, in which a later outcome cannot be predicted unless one has perfect knowledge of the original state, which is never the case in scientific experiments. This is simply a result of hypersensitivity of a system to its initial conditions.

As we saw in Chapter 6, nonlocality is a much more problematic concept than randomness. Solving the nonlocal correlation problem was one of the main motivations for the Copenhagen view, originally – if "knowledge waves" are not real, scientists at the time felt that it doesn't matter if they have faster-than-light behavior. But as we saw in Section 6.6, the Copenhagen interpretation doesn't solve the nonlocality problem of correlations between random events happening faster than the speed of light. Since causality problems still exist with the Copenhagen framework, this removes much of the motivation for holding to it.

A minimalist Copenhagen view. Suppose, however, that the Copenhagen interpretation is correct, and that there is something about knowledge that is truly off the books of the equations we use for the physical world. How many of the spiritual and philosophical claims discussed at the beginning of this section would necessarily be implied?

A "minimalist" Copenhagen view could look much like the spontaneous collapse view. In this approach, every time a conscious being gains knowledge, the nonphysical whatever-it-is, which we can call "spirit" as shorthand, could just act to supply a random kick toward one state or other, without any conscious activity of the observer. The wave functions of quantum mechanics could still be real, with macroscopic superpositions, but this extra little trigger would lead to collapse into certain states. This would allow macroscopic superpositions of objects which then disappear upon observation, but not Schrödinger's "cats," because cats (presumably) are conscious observers.

As with the spontaneous collapse view, randomness in this minimalist Copenhagen view would merely reflect our ignorance about whatever fluctuations are coming in from the nonphysical, spirit world, which could have its own set of deterministic rules. Even knowledge-based collapse, as strange as it is, does not imply that people can change reality by an act of will.

The many-worlds hypothesis and reality. As discussed in Chapter 5, some physicists consider the many-worlds hypothesis a viable alternative to the Copenhagen view. The many-worlds approach strictly rules out the Copenhagen view of quantum waves as unreal; it treats the waves as entirely real, and every wave superposition as lasting forever, no matter how small one fraction might be.

The viability of this view gives us another reason to not put too much weight on the Copenhagen interpretation, with its emphasis on human knowledge, but it has its own strange implications for how we think about reality. Instead of having reality exist only in state of possibility until we look at it, the many-worlds view says that every possible reality happens.

As we saw in Section 5.1, there are many reasons to doubt the viability of the many-worlds view, not least of which is that we have no evidence whatsoever of any alternate universes.

The bottom line is that, as strange as quantum mechanics is, it need not be taken as so *very* strange. We can distinguish between things that are *required* of any workable theory of quantum mechanics, in particular randomness and nonlocality, and things that are *hypothesized* in the context of explanations for these, such as Schrödinger's cats, the non-reality of things until someone looks at them, parallel universes, and so on. While some people may find one or more of these to be credible for various reasons, we cannot say that they are proven or even strongly evidenced by the equations and experiments of quantum mechanics.

7.2 Free Will and Quantum Mechanics

When we think about reality, of course a large part is our view of ourselves. Some authors have invoked quantum mechanics to explain the free will of human consciousness. The argument goes essentially as follows: (1) If there is a cause for our choices, then we are not morally responsible for them. (2) But we are morally responsible, and feel that we make free choices. (3) Therefore, there must be some way for our choices to have no cause whatsoever. (4) Quantum mechanics supplies this because random things happen without any cause.[4] Let's look at each of these premises closely, in reverse order.

Regarding premise (4), as we have seen in this book, it is far from clear that quantum mechanical processes are "purely" random. It may be the case that the randomness of quantum mechanics is like that of the classical chaos discussed in Section 10.2; that is, deterministic behavior that is extremely sensitive to its initial conditions, so that it is effectively unpredictable to anyone without perfect omniscience. The proposal of spontaneous collapse presented in Chapter 6 may involve this same kind of deterministic randomness.

One argument that the randomness of quantum mechanics cannot be like this is that experiments done with identical particles can give different results. But it is very hard to prove that two experiments were done truly identically. Even if the particles being tested could be proved to have exactly identical initial conditions, any detectors involved in the experiment would have millions of interacting particles that would be hard to put into exactly the same initial conditions for each experiment. Since a complete experiment involves decoherence due to interaction with a macroscopic detector of some sort, the whole system must be truly identical from one measurement to the next in order to prove that the exact same conditions gave two different outcomes.

Some would assert as an axiom, however, that the randomness of quantum mechanics has no cause. It is not entirely clear what this could mean. One way to imagine it is by analogy with classical chaos. In a system with classical, deterministic chaos, we can define some range X of the initial conditions that leads to a range Y of the final outcome. If we want to predict Y, we will need to know that the initial conditions are within the range X. When a system is very sensitive to its initial conditions, the range X could be very small – the more sensitive the system, the more accurate our knowledge of the initial conditions must be. One could argue that, in a "perfectly" random system, the range X is zero for every outcome Y – that is, there is no level of accuracy of knowledge of the initial state that will allow the prediction of Y.

This brings us to premise (3). If such "perfect" randomness existed – a perfect random number generator, so to speak – would it help us to feel more morally free? Why? Should I feel better about my decisions being the result of a random number generator than if they are the result of some other type of machine?

---

[4] For example, Roger Penrose (1996) has suggested that the spontaneous phase symmetry breaking that occurs in Bose–Einstein condensation (BEC) may have an analog in the human brain, which allows us to go beyond what deterministic machines can do. See Section 8.5 for a discussion of Bose–Einstein condensation, and Section 7.4, which discusses spontaneous symmetry breaking.

There are only four possibilities when it comes to our choices: (a) our choices are determined by an easy-to-predict mechanism, (b) our choices are determined by a hard-to-predict mechanism, (c) our choices are determined by an impossible-to-predict mechanism, or (d) our choices are godlike, an eternal uncaused cause. In the case of option (d), there is no logical problem with an uncaused cause, as theologians and philosophers have known for centuries (see, e.g., Sproul 1984), but such an uncaused cause must either be God or must be outside the domain of control of God. Every being with uncaused free will would be at the level of God, in a sort of polytheistic universe.

If it is not the case that we are gods and have forgotten it, then one of the first three possibilities (a)–(c) must be true, but all three of these collapse into the same category. In each case, something outside myself is the cause of my state. The only real question then is whether the ultimate outside cause is impersonal and machine-like (whether random or predictable), or an intentional being, which we can call God.

All of the same considerations apply even if part of me exists in a parallel spirit world, unobservable in the natural world. Some philosophers have argued that having an immaterial, spiritual nature gives us free will (e.g., Moreland 2009). But if there is such a parallel domain, we still have to ask all the same question: Is that spirit world ultimately governed by predictable laws, by unpredictable random events, or by a God with purposeful intentions?

Being controlled by a random number generator doesn't make me feel especially more free than being caused by a predictable set of laws, or by an intentional God. Let us therefore take a step back to premises (1) and (2) at the beginning of this section. Premise (2) seems obviously true, that we do have a moral nature. Some determinists have argued that morality is an illusion, but as C. S. Lewis and others have argued (e.g., Lewis 2009), no one seems able to hold that view consistently. We all get mad when someone does something to oppress us or marginalize us. When we do, we can't feel that the person was just the product of amoral forces. We may not use the language of morality, but if not, we use the words of disease and sickness just like the words of morality: "That person is *sick!*"

That leaves us with premise (1). Is it possible to define a moral nature in such a way that prior causes don't invalidate it? Many thinkers have wrestled with this question, notably the American theologian Jonathan Edwards (1997) and British brain scientist Donald Mackay (1980).[5]

We can start by asking why causation matters so much to us in the first place. Our thinking about our moral nature can be illustrated by the diagram in Figure 7.1. When we think of a free choice, we think about having a desire for one option over another, and acting on it. If, after that, something outside me swoops in and subverts my intention, bringing about a different outcome from the one I wanted, I feel my freedom has been thwarted, and I am not responsible for the outcome.

---

[5] See also Feinberg 2018 and Chalmers 1996. In theological terminology, the perspective of Edwards and Mackay is called *Calvinism*, as opposed to *Arminianism*, the perspective that human free will is beyond the causal power of God. As discussed here, Arminianism reduces to either the notion of people as godlike or that they are ultimately controlled by impersonal natural phenomena such as quantum mechanical fluctuations (see, e.g., Sproul 1986; Hunt 2004; Packer 2008).

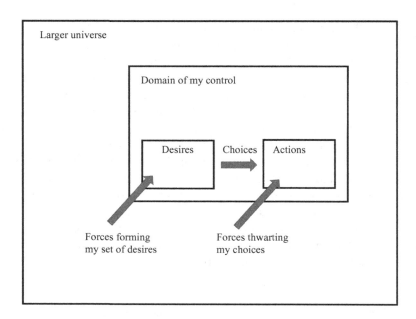

Figure 7.1   Layers of experience in our decision making.

We rarely ask ourselves, though, why we wanted what we wanted in the first place. We take our desires as a given; indeed, if I did not want what I want, as my highest goals, I would be a different person.

We can therefore ask: Given who I am, do I have a domain of control over which I can affect outcomes based on my desires and choices, as illustrated in Figure 7.1? The answer, empirically, is clearly yes. People do make choices. This is not incompatible with saying that at a deeper level, who I am, including what I desire, is not something I created entirely myself. There could have been causes that made me who I am, which do not remove the real existence of my domain of control now that I exist. In this case, I can still say that I am free if I have a domain of control.

This question, of course, is the subject of much debate not just about humans but also about the artificial intelligence (AI) community. We can be confident that the machines we make do indeed have actions determined by the laws of nature (even if they involve random number generators). Does this rule out automatically that they can never have free will? The above discussion indicates that we cannot categorically rule out that they could.[6] But it may be that people never accomplish the task of making such AI systems; although computer systems can accomplish many tasks when they are well-defined, progress toward anything that really copies human thought remains abysmal. The reason may not be because there is an intrinsic problem in being made of physical stuff. The problem may simply be that engineers and biologists have consistently underestimated the complexity of the human

---

[6] Hofstadter (1979) argues that the complexity of human thought processes, akin to classical chaos, could give free will even in an overall deterministic system, and that the same could occur in an AI system.

brain and body. We typically think of machines as complex if they involve 10 or more different levels of hierarchy – in computer language, for example, this might correspond to a subroutine that invokes another subroutine, which invokes another, and so on, 10 levels deep. What if humans have 100 levels of hierarchy, or 1,000 or more? Modern biology is constantly uncovering amazing and subtle aspects of living systems and brains.

In summary, it is far from proven that quantum mechanics is not deterministic; it could just involve extremely unpredictable processes, like classical chaos. And it is far from clear that, if it did involve some type of "pure" randomness, it would enhance our sense of self. Our fear of being "machine-like" is mostly based on our experience with relatively simple machines. Perhaps extremely well-designed, complex, deterministic systems should be described as artworks, not machines. Those exquisite artworks could still have freedom within their domain of control.

7.3 Can Quantum Fluctuations Create Something from Nothing?

Another deep question about reality is, "Where did it all come from?" Many authors (e.g., Ross (1995); Davies (2008); Collins (2013)) have written on the history of physics and astronomy that led to the establishment of the Big Bang in the history of the universe, and the philosophical issues that came with it. There was initially strong opposition among physicists to the Big Bang theory, and only overwhelming empirical evidence convinced the scientific community to adopt it.

The primary opposition to the Big Bang theory can be called aesthetic in nature. For centuries, those studying the natural world have looked for, and often have found, aspects of the laws of nature that they have considered "beautiful." It may be surprising to some non-experts that physicists can consider their mathematical equations and theories to be beautiful, but often they do. In fact, quite often one theory is preferred over another mainly because it is more "elegant," even if both describe the same data, or even if the more elegant theory doesn't explain the data quite as well.

It is hard to define exactly what makes a theory aesthetically appealing to physicists, but two ingredients are *symmetry* and *simplicity*. The second aspect, simplicity, has some overlap with the principle of Ockham's razor, which states that theories are to be preferred if they do not have lots of complicated exceptions and elements invented just to fit the experimental data. This reflects in part the sociological fact that people who don't really understand something tend to make up stuff as they go along. This bias toward simplicity also reflects in part the expectation in western civilization historically that God would create things well, without a lot of patch-up jobs. It also reflects some degree of utility, that simpler theories are easier to remember and to use than complicated ones.

The other aspect of a beautiful theory, symmetry, has also been a driving aesthetic force in the science world. This bias goes back to the philosophy of Plato in ancient Greece that supposed the existence of a world of pure ideas, or "Platonic ideals," that have their own existence whether or not a physical world exists to embody them, and perhaps even generate the physical world.

The use of the symmetry aesthetic has a mixed record in the history of science. In the days before Kepler, many astronomers assumed that the orbits of the planets must be perfect circles, because circles are properly symmetric. This view was supported by the fact that their orbits are very nearly circular, but in the end, the observational data showed that they could not be perfect circles. This led to years of counterproductive attempts to fit the data with circles inside of other circles, the notorious "epicycles" (see, e.g., Gingerich 1975). Newton's laws were celebrated, in part, because they replaced this type of symmetry with other types of symmetry – instead of things moving in circles, his theory presented a universal law in which every force has a balancing, equal and opposite force, and the strength of each force is proportional to the surface area of a perfect sphere with a radius given by the distance between two objects.

Newton's laws were so successful that later scientists and philosophers, such as Laplace, argued that they were self-evident absolutes. Of course, Newton's laws had to be modified by Einstein, but Einstein's laws have a type of symmetry of their own. As discussed in Sections 2.7 and 13.2, Paul Dirac was a strong believer in aesthetic symmetry arguments, and out of this bias he came up with the successful prediction of the existence of antimatter, as well as the symmetric relationship of fermions and bosons in quantum field theory that has been verified in many experiments (discussed mathematically in Section 13.1). More recently, some physicists have searched for "magnetic monopoles," which would be a type of magnetic charge that played a symmetric role to that of electric charge in the equations of the electric field, but so far, the experiments looking for this have failed.

The opposition to the Big Bang hypothesis came in large part from this symmetry aesthetic; the famous astronomer Arthur Eddington, for example, considered the idea of a beginning "repugnant."[7] It seemed to most scientists in the early twentieth century that the most appealing scenario for the history of the universe was an eternal universe, with the same average properties at all times. The observational data couldn't be made to agree with that simple view, however, and over the next several decades it became clear that there was a beginning of the universe. This still bothers many physicists, again, mostly for aesthetic symmetry reasons, and for this reason there have been various proposals to embed our finite-lived universe inside a much larger, eternal "macroverse" (see, e.g., Linde 1987; Collins 2013). In this scenario, universes like ours pop up and then decay inside a larger macroverse like bubbles in a glass of seltzer water, while the average behavior of the macroverse stays the same. It is hard to define "time" in such a scenario, but in general these models still have some parameter which extends infinitely and eternally, that has the type of symmetry envisioned for our universe by people like Eddington before the Big Bang model came along.

It is beyond the scope of this book to survey theories of cosmology. But in general, all of the proposed models of cosmology have a fundamental problem: for aesthetic reasons, physicists want to have perfect symmetry of some sort, while our universe is manifestly *not* symmetric – the stars in the night sky are not symmetrically distributed, nor are the trees

---

[7] LeMaître said, "Sir Arthur Eddington states that, philosophically, the notion of a beginning of the present order of Nature is repugnant to him" (LeMaître 1931).

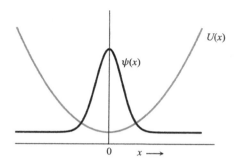

Figure 7.2 Dark line: A matter wave function in a bowl-like region of the potential energy, or "trap," shown as the gray line.

in a forest, or almost everything else. How can we get something so nonsymmetric from something perfectly symmetric?

Quantum fluctuations. To solve this symmetry problem, some cosmologists have invoked the idea of a *quantum fluctuation* to make something happen in a perfectly symmetric universe. Some authors have called this "something coming from nothing" (e.g., Krauss 2012), although as discussed in Section 1.4, the quantum field that has the purported fluctuations is not "nothing;" it is very real.

Many physicists talk loosely of quantum fluctuations as though they produce spontaneous time-varying behavior. Let's discuss exactly what is meant by this term.

In general, a wave function can have a spread of nonzero amplitudes across some range of values of a parameter. For example, Figure 7.2 shows the wave function of a matter wave in a confining region, sometimes called a "trap." Suppose that we could investigate this wave function using a needle-like detector that gives a yes-or-no particle detection event at any value of x that we want. According to the Born rule discussed in Section 3.4, we will get a particle detection event at position x with a probability proportional to the square of the amplitude of the wave function at that location x.

Suppose that we repeat this measurement many times for many different locations of x. We can then ask what quantum mechanics and the Born rule predict for the results. First, what will be the most likely position x to detect a particle? This is clearly $x = 0$ for the example shown. We could then go further to ask what the typical deviation from $x = 0$ is. The math for this is worked out formally in Section 11.5, and gives, as we would expect from looking at the picture in Figure 7.2, some range of the deviation around $x = 0$.

This is often called in quantum mechanics textbooks the range of "fluctuations" of the value of x. But that is not really correct. In the absence of any external probe, the wave function shown in Figure 7.2 does not change in time (a nonchanging state is called an *eigenstate*, in technical language). If we wanted to, we could describe it as an unchanging superposition of many waves localized at all possible locations x. This superposition does not fluctuate over time. The randomness in the value of x comes only when this system interacts with something outside, namely the needle-like detector we used. In the Copenhagen interpretation, that gives a collapse of the wave function based on the knowledge of an observer; in the spontaneous collapse model presented in Chapter 6, the interaction with the outside world gives random kicks to the detector and the wave function that lead to avalanche events. Without outside influences, the equations of quantum mechanics say that the wave function will stay in exactly the form shown, with no variation in time.

The ground state of a vacuum is an eigenstate in the same way. In the context of the Big Bang, this means the vacuum does not change in time – it does not "fluctuate." Sometimes physicists talk of "particles popping into existence and going back into the vacuum," as though the vacuum was constantly changing in time, but this is not what quantum field theory says. Rather, it says that there is a superposition of many possible states that make up one, unchanging state.[8]

It simply makes no sense, then, to invoke a quantum mechanical "fluctuation" as the cause of the universe. Theists, of course, can identify the Big Bang with the free act of God, outside the laws of physics. Indeed, many physicists noted early on that the Big Bang, as a starting point in time, looks suspiciously like the Creation of the Bible, and it is perhaps no coincidence that the first astronomer to take this possibility seriously was Georges LeMaître, who had been trained as a Catholic priest. If one rejects the idea of a creative act by God, then one can simply hypothesize some ad hoc time-symmetry-breaking cause (which would be aesthetically "ugly" to many physicists), but one cannot derive this from quantum mechanics itself on the basis of quantum fluctuations.

7.4 Spontaneous Symmetry Breaking

The aesthetic approach to physics has had successes over the years, which have led many physicists to the hope of Einstein, Dirac, and others that this principle could eventually explain everything in physics theory without free parameters. For example, as discussed in Section 7.3, Dirac's aesthetic sense directly led to the prediction of antimatter. But even this was a mixed success. The equations written down by Dirac imply an exactly equal amount of matter and antimatter. Yet, if that was the case, we could not exist, because the matter and antimatter would annihilate each other. In our observed universe, there is an imbalance, with more matter than antimatter. That requires the laws of physics to have a seemingly arbitrary free parameter, namely the amount of deviation from perfect symmetry of matter and antimatter.

If we assume that the laws of physics are perfectly symmetric, how can we get non-symmetric parameters? Starting in the 1960s, some physicists have argued that a well-known effect in quantum mechanics called *spontaneous symmetry breaking* could do the trick.

The theory of spontaneous symmetry breaking was first developed to explain the behavior of magnets, and eventually was generalized to explain all kinds of effects in modern technology, including lasers and superconductors.[9] The basic effect is as follows. We

---

[8] In Section 4.5, we looked at a Feynman diagram, which represents particle interactions in quantum field theory. Section 15.3 gives the math of how to interpret these diagrams correctly, in particular in the vacuum state.

[9] For a review of the theory of spontaneous symmetry breaking in magnets, superconductors, and lasers, see Snoke 2020, Chapters 10 and 11.

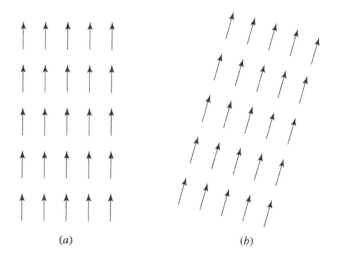

(a)   (b)

Figure 7.3   The lowest-energy state of a set of magnetic atoms, for two different directions of the spins.

imagine that each atom is spinning about an axis. At high temperature, the directions of the spins of atoms in a magnetic material are randomized to point in many different directions, so that the average of the magnetic field generated by the atoms is zero. At low temperature, however, the spins interact with each other, and can lower their energy by aligning their spin axes with each other. The lowest energy state will be when all the spins are aligned, as shown in Figure 7.3(a). But there is nothing in the equations that says which direction all the spins will point. The state shown in Figure 7.3(b) will have the same total energy, and will be just as favored as the state in (a).

Figure 7.4(a) shows the energy of the system as a function of the amount of magnetization, calculated for two different temperatures. Above the critical temperature T_c, the lowest energy state has $m = 0$, that is, no average magnetization. Below T_c, the system can move to lower energy by having an average value of m which is nonzero, corresponding to most of the spins aligned with each other. But there is no preferred direction for them to point – m could be either positive or negative. Figure 7.3(b) shows the lower curve of (a) in two dimensions, to account for the fact that the spins can swing freely in any direction. As seen in this figure, the lowest energy corresponds to a circular trench. There is no preferred state in this trench; instead there is a continuum of possible directions for the spin, all with the same energy.

This effect has been used as the paradigm for getting asymmetry out of symmetry. As seen in Figure 7.4(a), both the curves above and below the critical temperature are symmetric around $m = 0$. However, the point of $m = 0$ of the lower curve is *unstable*. One can think of the state of the system like a small marble sitting on top of a hill, which is the point $m = 0$. Any little jostle will send the marble down the hill to land at one place or another with nonzero m, unpredictably.

Some people argue that our universe is like this magnetic system. Overall, it is perfectly symmetric, which appeals to our aesthetic sense, but under the right conditions,

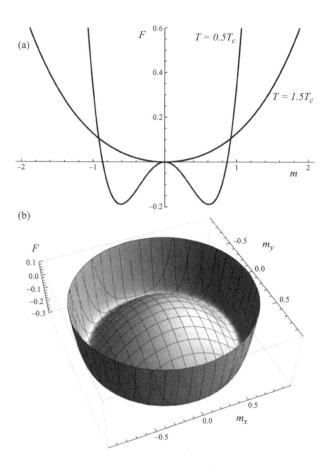

Figure 7.4 (a) The free energy function of a magnet as a function of the net magnetization m, at two different temperatures, above and below the critical temperature T_c for spontaneous magnetization. (b) A prediction of the energy similar to the lower curve of (a), plotted in two dimensions. From Snoke 2020.

this symmetry can be spontaneously broken, leading to arbitrary values of various free parameters.

There is a fundamental problem with this approach, however, similar to the problem with invoking quantum fluctuations to get something from nothing, discussed in Section 7.3. If the system is symmetric, it can't generate asymmetry. In the context of the physics of magnets, in which the mathematics of spontaneous symmetry breaking was first developed, it is easy to suppose there is some small, stray perturbation from outside the system giving a jostle in one direction or another, which can then be amplified by the unstable character of the system. But if our universe is all there is, and it is symmetric, what could play the role of this stray perturbation from "outside"? As we saw in Section 7.3, quantum fluctuations can't do it.

In general, even in a real-world laboratory, it is possible for systems to remain at unstable, symmetric points for arbitrarily long periods of time. For example, in my lab we had a laser that exhibited spontaneous symmetry breaking – it went from a constant-wave state to pulsing with very short pulses every few nanoseconds (known as *mode-locking*). This effect was discovered years ago by laser technologists who accidentally found that when they bumped the laser, it would jump into the pulsing state. In my commercially made laser system, there was a button to push to have an electronic mechanism hit the laser with a little stick, to give it a slight "jostle" to cause it to jump into the pulsing state. If we didn't give it that jostle, it would stay in the non-pulsing state indefinitely.

No matter how tiny, there must be some nonsymmetric entity from outside that interacts with the symmetric system, if it is to become asymmetric. The scenario of spontaneous collapse of quantum states presented in Section 6.4 does not change this picture. In such a scenario, there still must be some small asymmetry in the initial state to get a large asymmetry later.

The pull is strong to want all the laws of physics have Platonic symmetry and simplicity. But we manifestly live in a nonsymmetric world. Perhaps we need to see that there can be more than one type of beauty. An artist who makes a perfectly symmetric glasswork indeed makes something beautiful, but sometimes art can be beautiful precisely because it involves many free design choices by the artist which were not constrained by rules of symmetry or other rigid rules. To put it another way, while classical music was based on exact rules and has a certain beauty, jazz can be beautiful too.[10]

Does the many-worlds interpretation give spontaneous symmetry breaking? Figure 6.5 in Chapter 6 presented a system that has much in common with the symmetry-breaking scenario shown in Figure 7.4. In that type of system, the unitary evolution of quantum mechanics can give two, uncoupled outcomes. In the Copenhagen or in the spontaneous collapse models, something nonunitary collapses the state to just one of these possibilities. That clearly requires something asymmetric to select out one or the other. But in the many-worlds approach, both outcomes continue on forever in a superposition. Could something like that occur in the early universe, so that we just happen to be in one of many asymmetric universes in a macroverse with overall symmetry? For every universe like ours, could there be an "anti-universe" that balances it?

This involves subtle calculations of the evolution of quantum systems. In general, the answer is no. A mathematical discussion is given in Section 16.2.3, but we can get the basic picture by considering whether we could just treat the early universe as a superposition of two halves: one universe with half the matter on the right side, and another universe with

---

[10] The analogy to classical music versus jazz is not accidental. The desire to embed everything within a system of purely orderly behavior goes back to the 1700s. Europeans became enamored with the success of clocks and machines, and tended to see these as paradigms for everything good, including music (namely, their strongly rule-based classical music), social structures, and the natural world. A similar movement arose in theology, with the view of the "clockmaker God" for whom it would be unseemly to intervene in nature via miracles, because these would be improvisational, like jazz. This view remains common to the present (Rossiter 2015). The emphasis on predictable order arose in part due to European reaction against the superstitions of their prior culture and of pagan cultures. Allowing anything not orderly and rule-based, such as a miracle by a volitional God, might mean going back the old ways of the Medieval age, in their minds. Allowing "jazz" to be beautiful, that is, unique, nonruled-based actions, opens up the possibility of a nonsymmetric universe as well as nonreproducible miracles by God.

the other half of the matter on the left side. We cannot, because the homogeneous state of the whole universe includes quantum states with long wavelength, which extend through both sides. If we eliminate the possibility of particles in such long-wavelength states, we change the physical state of the system, including its temperature.

The separation into two disconnected outcomes discussed in Section 6.3, crucially relies on "rigging" the system to isolate one outcome from the other. In particular, it requires a potential-energy profile that is not spatially homogeneous, which has the special feature of pulling apart the wave into two regions. A homogeneous system will not break up in the same way.

Some physicists are content to allow explicit symmetry-breaking terms into the equations of the early universe for some parameters, such as the mass of particles, but few are comfortable with the idea of putting explicit terms that break spatial symmetry.[11] But the only way to get spatial asymmetry in the universe from an earlier, purely homogeneous universe is to have something in the equations or the initial conditions that explicitly has spatial inhomogeneity.

7.5 Did We Create Ourselves?

One other possibility for breaking the symmetry of the early universe has been floated by some of the authors of the "new age" philosophy discussed in Section 7.1. In this proposal, the universe evolves in many-worlds fashion until conscious humans evolve to exist, at which point their observation of the universe causes the universe to collapse, Copenhagen-style, into one of the possible universes in which they exist. A variation of this is the "final anthropic principle," which posits that humans evolve until they become God, at which point they create the universe retroactively (see, e.g., Leslie 1983; Barrow 1987).

At face value, this scenario is circular logic: we exist because the universe produced us, and the universe exists because we produced it. In the self-creation by self-observation scenario, there is a fundamental paradox about what causes what.

One could make this scenario into a viable one, however, by envisioning it as a cyclical universe, in which the godlike observer of the universe creates a subsequent universe, which then evolves forward in time until a new observer is generated.[12] In this case, the cycle itself would be a "brute fact," or uncaused cause, of the type mentioned in Section 7.2. As in the case of an uncaused God, one can posit that there is some asymmetric, nontrivial aspect of the physical world that exists eternally. It is not clear, however, that an eternally reappearing god has greater aesthetic appeal than an eternal, single God.[13]

The big picture. As we have seen, many of the mystical views of reality promoted using quantum mechanics rely crucially on the central role of *observation* in the Copenhagen

---

[11] This is sometimes called the "Copernican principle": the assumption that there is nothing special or "rigged" about our universe.

[12] A variation of this scenario is presented in the science fiction short story by Isaac Asimov, "The Last Question" (Asimov 1994).

[13] For further critiques of the oscillating universe model, see Earman 1987.

view. This is not just because observers tend to have affects on what they observe – as mentioned in Section 7.1, every classical scientist would have agreed with that, although they tended to expect that an observer could have a very small effect. Rather, in most versions of the Copenhagen interpretation, the waves of quantum mechanics have no "ontological reality" until someone looks at them.

As discussed at length in the first few chapters of this book, quantum waves are fundamentally like other waves we know, such as water waves and sound waves, and therefore we should be wary of treating them like mere figments of our imagination. The role of the observer in quantum mechanics may be just due to the fact that large, macroscopic systems tend to have strong decoherence, which may in turn lead to spontaneous collapse. Although that is not proven, it is as viable as other interpretations of quantum mechanics, and therefore it is premature to invoke the Copenhagen view as certain. Furthermore, as discussed in Section 7.1, a minimalist version of the Copenhagen interpretation could effectively operate the same way as a spontaneous collapse model.

References

I. Asimov, "The last question," in *The Complete Stories of Isaac Asimov*, (HarperCollins, 1994).

J. D. Barrow and F. Tipler, *The Anthropic Cosmological Principle*, (Oxford University Press, 1987).

M. Beller, "The Sokal hoax: At whom are we laughing?" *Physics Today* **51**, 29 (September 1998).

N. Bohr, "Celebrazione del secondo centenario della nascita di luigi galvani," *Rendiconto Generale*, **15**, 68 (Tipografia Luigi Parma, 1938).

N. Bohr, as quoted by H. Bohr, in *Niels Bohr: His Life and Work as Seen by His Friends and Colleagues*, S. Rozental, ed., (North Holland, 1968), p. 325.

N. Bohr, as quoted by Abraham Pais, *The Genius of Science: A Portrait Gallery* (Oxford University Press, 2000), p. 24.

F. Capra, *The Tao of Physics: An Exploration of the Parallels between Modern Physics and Eastern Mysticism* (Shambhala, 1975).

D. J. Chalmers, *The Conscious Mind: In Search of a Theory of Conscious Experience*, (Oxford University Press, 1996).

R. Collins, "Modern cosmology and anthropic fine-tuning: Three approaches," in *Georges Lemaitre: Life, Science and Legacy*, R. Holder and S. Mitton, eds., (Springer, 2013).

W. L. Craig, "Cosmos and Creator," *Origins and Design* **17**, 2 (1996).

P. C. W. Davies, *The Goldilocks Enigma: Why Is the Universe Just Right for Life?*, (Mariner, 2008).

J. Earman, "The SAP also rises: A critical examination of the anthropic principle," *American Philosophical Quarterly* **24**, 307 (1987).

J. Edwards, *The Freedom of the Will*, reprint, (Sole Deo Gloria Publications, 1997).

T. E. Feinberg and J. M. Mallatt, *Consciousness Demystified*, (MIT Press, 2018).

O. Gingerich, "Crisis versus aesthetic in the Copernican revolution," *Vistas in Astronomy* **17**, 85 (1975).

D. Handelman, *Models and Mirrors: Toward an Anthropology of the Public Events*, (Berghahn, 1998).

S. Hawking, *A Brief History of Time*, (Bantam, 1988).

D. Hofstadter, *Gödel, Escher, Bach: An Eternal Golden Braid*, (Basic Books, 1979).

D. Hunt and J. White, *Debating Calvinism: Five Points, Two Views*, (Multnomah, 2004).

L. Krauss and R. Dawkins, *A Universe from Nothing: Why There Is Something Rather than Nothing*, (Atria Books, 2012).

G. LeMaître, "The beginning of the world from the point of view of quantum theory," *Nature* **127**, 706 (1931).

J. Leslie, "Observership in cosmology: The anthropic principle," *Mind* **92**, 573 (1983).

C. S. Lewis, *Mere Christianity*, (HarperOne reprint, 2009).

A. Linde, "Particle physics and inflationary cosmology," *Physics Today* **40**, 61 (1987).

D. Mackay, *Brains, Machines, and Persons*, (Collins, 1980).

J. P. Moreland, *The Recalcitrant Imago Dei: Human Persons and the Failure of Naturalism*, (SCM Press, 2009).

J. I. Packer, *Evangelism and the Sovereignty of God*, (Intervarsity Press, 2008).

R. Penrose, *Shadows of the Mind* (Oxford University Press, 1996).

H. Ross, *Creator and Cosmos*, (Navpress, 1995).

W. Rossiter, *Shadow of Oz: Theistic Evolution and the Absent God*, (Pickwick, 2015).

D. Snoke, "Review of A. Tauber, *science and the quest for meaning*," *Journal of the History of Medicine and Allied Sciences* **66**, 418 (2011).

D. W. Snoke, *Solid State Physics: Essential Concepts*, 2nd ed., (Cambridge University Press, 2020).

A. D. Sokal, "Toward a transformative hermeneutics of quantum gravity," *Social Text* **46/47**, 216 (Spring/Summer 1996).

R. C. Sproul, J. Gerstner, and A. Lindsley, *Classical Apologetics*, (Zondervan Academic, 1984), Chapter 7.

R. C. Sproul, *Chosen by God*, (Tyndale House, 1986).

M. Talbot, *Mysticism and the New Physics*, (Bantam, 1983).

A. Tauber, *Science and the Quest for Meaning*, (Baylor University Press, 2009).

G. E. Uhlenbeck, "Problems of statistical physics," in *The Physicist's Conception of Nature*, J. Mehra, ed., (Springer, 1973).

G. Zukov, *The Dancing Wu Li Masters: An Overview of the New Physics*, (William Morrow, 1979).

G. Zukov, in an interview with Marti Glenn, produced by East Beach Productions for La Casa de Maria in Santa Barbara, 2019.

8 Quantum Mechanics and Technology

Before we finish our survey of quantum mechanics, it is worth asking whether there is any practical use for all the strangeness we have seen. At one level, the answer is that all of our modern technology depends on quantum mechanics, so much so that we don't even notice it. All of the calculations that are done for semiconductor circuits on the chips that underlie computers, cell phones, and so on are based on the wave nature of electrons, the quantum properties of photons and phonon that go into the heat flow equations, and the fermion nature of electrons that gives the Pauli exclusion effect. In another sense, though, these parts of our modern technology only use a few aspects of quantum mechanics. Therefore, they are often called *semiclassical* methods. These are surveyed in Sections 8.1 and 8.2.

In the past few decades, scientists around the world have started to work toward new applications that would fully utilize all of the strange intrinsic properties of quantum systems. These are surveyed in Sections 8.3–8.4. One long-term, but realistic, goal is to produce a quantum computer that would far outstrip any existing computer.

Some versions of quantum computers involve superconductors, which are fascinating examples of applied quantum mechanics in their own right. As mentioned several times in this book, superconductors and superfluids defy the normal expectation that quantum mechanics only matters on microscopic scales. Superconductors have macroscopic quantum wave functions that can be as big as a person, or even much bigger, in principle. Superconductors and lasers are examples of spontaneous symmetry breaking, in which a small fluctuation is amplified to become macroscopic coherence, as discussed in Section 7.4. In Section 8.5, we will discuss some of the odd implications of this.

8.1 Quantum Mechanics in Your Pocket: Computer Chips and Nanotechnology

It is fair to say that the fields of modern chemistry, materials science, and semiconductor physics, which underlie all of our modern technology, are "applied quantum mechanics." Students in these fields learn to use quantum mechanics in many ways.

As already discussed in Sections 1.3 and 2.1, the wave nature of electrons causes them to have well-defined resonances, or "states," in atoms, molecules, and solids, in what is known as "first quantization" (as opposed to second quantization, which produces the particles in these states, as discussed in Section 2.2). Let's look into a little more detail about these states.

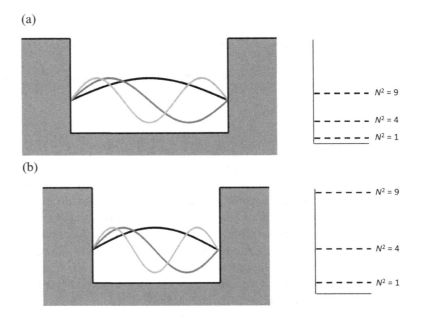

(a)

(b)

Figure 8.1   (a) A wide trap for electrons, with several of the waves that fit in this trap. The set of dashed lines on the side indicates the "energy levels" corresponding to these waves. (b) The same but with a narrower trap.

Figure 8.1 shows two "traps" for electrons, with different widths. Several of the waves that fit in these traps are also shown. These waves are determined by the condition that the wave function must have zero amplitude at the edges of the trap. Each of these waves is called a "state" of the electrons.

One of the rules for quantum mechanics, discussed mathematically in Section 9.1, is that waves with shorter wavelength have higher energy. The energy of electrons increases as $1/\lambda^2$, which means, for example, that, if an electron's wavelength λ gets shorter by a factor of 2, its energy increases by a factor of 4. This causes the different electron wave states to have different energies, as shown on the right side of Figures 8.1(a) and (b). We therefore can also call these states "energy levels."

Although we have used square traps here, it is generally true that only certain wavelengths of electrons can fit in confined regions. As discussed in Section 2.1, this is the reason for the quantized energy levels of atoms, also called "orbitals," of chemistry.

We also see in Figure 8.1 the general effect known as *quantum confinement*, namely that putting an electron in a smaller space tends to increase its energy, because that makes its wavelength shorter. Not only does the average energy of the states tend to go up, but the difference in energy between the allowed energy levels also increases.

This leads to a way in which small electron systems can become *two-dimensional*, or even one-dimensional. Figure 8.2 shows a typical layered structure made of different materials such as that often used in a computer chip. The "quantum well" is a thin layer of material that allows electron flow, sandwiched between two insulating layers. Modern technology allows the thickness of these layers to be controlled to an accuracy of nanometers

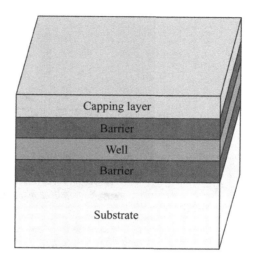

Figure 8.2 A typical layered structure, with a "quantum well" between two insulating layers. Electrons are confined to purely two-dimensional flow in the well, if the layer is thin enough.

(billionths of a meter). When the thickness of the quantum well is just a few nanometers, the energy jump from one electron wave state to the another will be large compared to the thermal energy of the electrons, that is, the typical amount of energy they can get from heat in the surrounding material. (Section 9.2 gives the simple math for this.) The electrons in this case will settle into the lowest energy state. They then can only have motion in the two-dimensional plane perpendicular to the vertical direction. It is also possible to confine the electrons in one of the sideways directions in which case they are free to move only in the one, remaining direction; this is called a "quantum wire," with purely one-dimensional motion. (See Section 9.6 for a mathematical discussion of universal behavior in quantum wires.)

Quantum wells and quantum wires are two examples of how the properties of particles in solids can be altered to look as though the laws of physics had changed – in this case as though the universe were two-dimensional or one-dimensional instead of three-dimensional. Other examples of altered fundamental properties that can be made using specially designed solids include making electrons with mass much heavier or much lighter than in vacuum and even negative mass (see Snoke 2020, Chapter 2) or zero mass (e.g., Ahn 2018), photons with mass, which repel each other like electrons (Snoke 2020, Section 11.13), and electrons with charge different from the universal value of electron charge in vacuum (Jain 2003). These cases of particles with altered properties point out once again that the notion of indivisible particles is misleading; in each of these cases, the proper particles are defined as the resonances of the field, taking into account all of the interactions in that field.

Transistors. Quantum confinement is increasingly important in real-world technology, as engineers work to squeeze more and more circuits onto a chip, making the size of each circuit element smaller. Figure 8.3 shows the design of a *metal–oxide–semiconductor*

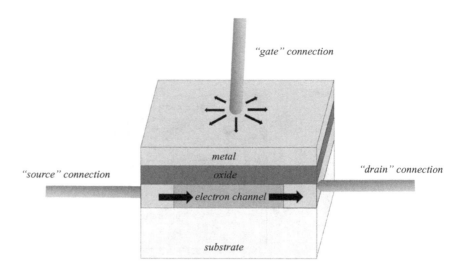

"gate" connection

metal

"source" connection   oxide   "drain" connection

electron channel

substrate

Figure 8.3   The design of a MOSFET. Electron flow is indicated by the black arrows. The main current of electrons flows from the "source" of the electrons to the "drain." The electrons coming into the "gate" of the device do not flow through the insulating oxide barrier; instead, they repel electrons in the conducting channel and cut off the electron flow there.

field-effect transistor (MOSFET), which is the main device in most computer processors. (For further discussion of transistors, see Snoke 2015, Chapter 5.)

The basic action of a transistor is to be a switch; that is, to have two definite states, either on or off. In the MOSFET shown in Figure 8.3, the current in the channel from the electron source to the drain is switched on and off by the presence of electrons in the gate layer. The thin channel region can have dimensions of the order of nanometers in modern devices. The oxide layer is an insulating barrier to prevent electrons in the gate from going into the conducting channel; their role is just to repel the electrons in the conducting channel, shutting it off. If the oxide layer is very thin, however, electrons can undergo quantum mechanical *tunneling* through the barrier (discussed in Section 1.3). This is often unwanted but can also be used on purpose. For example, "flash" memory uses control of the tunneling rate to load or unload electrons in a trap for long-term memory storage.

The current flowing through a transistor can be used to control the gate of another transistor. This allows a large sequence of circuit elements to turn each other on and off. The two states of the transistor, either with current flow in the channel or with no current flow, can be treated as the "bits" of digital information, corresponding to the numerical values 0 and 1. Typically, transistor systems are set up with nonlinear behavior to cause the system to pop into one of these two states, with very little chance of remaining in between. This helps to prevent errors – all of the numbers represented by the system will be definitely either a 0 or a 1, which can then be used in a binary mathematical system. We will discuss this further in Section 8.4.

Is smaller "more quantum"? As we have seen in this section, nanometer-scale circuits have quantum-confined energy levels for electrons that are large compared to their typical thermal energies. For this reason, we can say that quantum mechanics plays a direct role in the design of the circuits. Does that mean that the mysterious aspects of quantum mechanics we have discussed in earlier chapters, such as nonlocal correlations and collapse of wave functions, become more important? Some people may think that purely by virtue of being small, strange quantum mechanical effects must happen. This is not necessarily the case at all. First of all, the nonlocal correlations discussed in Chapters 4 and 6 are connected with low numbers of particles, not with small size. If the number of electrons in a current is large, then the system can be described in terms of average behavior like a classical liquid, no matter how small the channel for that current is.

Also, recall from the discussion of Section 4.5 that systems with quantum coherence can be very large – as large as waves on the ocean. Superconductors and superfluids, which we will discuss in Section 8.5, are additional examples, which consist of matter waves described by quantum wave functions on a large scale. There is no fundamental limit on how big such coherent quantum waves can be – they could be meters to kilometers in size, in principle. These large waves can be put into superpositions of different states, including "entangled" superpositions, as discussed in Section 8.4.

The only thing that is "more quantum" about nanotechnology is that the circuits are nearly as small as atoms, which means the resonant energies of the electrons in these circuits start to become comparable to the jumps in the energy resonances of single atoms. (The math for this is given in Section 9.2.) Since these energies are much higher than typical thermal energies at room temperature, the characteristics of individual electron-wave states come into play.

8.2 Tunneling, Radioactivity, and Quantum Biology

Another aspect of quantum wave mechanics comes in to some very delicate and sensitive processes that affect our lives. This is the effect of *quantum tunneling*, introduced in Section 1.3 and mentioned in Section 8.1. Figure 8.4 shows the basic effect: any wave (whether a sound wave, a light wave, or a quantum matter wave) can be transmitted through a thin barrier even if the wave normally cannot propagate in the barrier material.

Inside the barrier, the wave function does not oscillate in the way described in Chapters 1 and 2. Instead, it undergoes *exponential decay*, in which the wave amplitude drops smoothly toward zero.[1]

A key characteristic of tunneling is that the amplitude of the wave leaking through the barrier is hypersensitive to the thickness and other parameters of the barrier. The exponential function means that tiny changes in l give very large changes in the wave, as plotted in Figure 8.5. For example, for an intrinsic tunneling length l of 1 nanometer (nm), if the thickness of the barrier is 3 nm, the fraction of the wave escaping will be 0.2%, while if

---

[1] This behavior is given by the mathematical function $\psi(x) \propto e^{-x/l}$, where l is the tunneling length, which is determined by the properties of the medium and the energy of the wave.

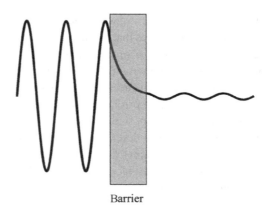

Barrier

Figure 8.4 A wave function with exponential decay ("tunneling") through a barrier, from left to right.

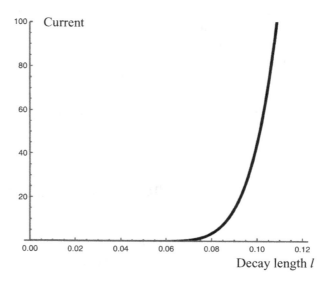

Figure 8.5 The transmitted current for a wave with exponential dependence $e^{-x/l}$, as the decay length l is increased to allow greater tunneling.

the thickness is 10 nm, the fraction of the wave escaping will be about one-billionth of the original wave!

So far, all of this has used only wave properties. When we think about particle statistics in terms of the Born rule, the wave function leaking out through the barrier gives us the probability of detecting a particle on the other side of the barrier. In the case of radioactive elements, a particle inside the nucleus of an atom must tunnel through a barrier to escape the nucleus. Because of the hypersensitivity of tunneling rates, radioactive atoms can have timescales anywhere from a few minutes to millions of years to emit particles. Small details of the differences from one type of atom to the next give huge changes in the tunneling rate.

Another example is the electron tunneling needed for the respiration process for breathing oxygen. The rate of respiration is controlled by a finely tuned barrier that allows a well-defined rate of electron transfer (see, e.g., De Vries 2015; Hosseinzadeh 2016). If the barrier was a little harder to cross, respiration would not happen at all, while if it was a little thinner, electrons would escape so rapidly that they would burn up the biological circuits.

This is an example of a "fine-tuning coincidence."[2] Another famous example is the process by which carbon was synthesized inside of stars, the original source of all the carbon we need for life. Fred Hoyle, the famous American astrophysicist, found that the natural rate of production of carbon by nuclear processes ought to be exponentially suppressed; only the amazing coincidence of two resonances in the carbon and oxygen nuclei being nearly the same allows these to be produced in the nuclear reactions in stars in roughly equal amounts (Hoyle 1982).

The "semiclassical" picture. The processes discussed in this section and in Section 8.1 use quantum mechanics, but only a few aspects of it. Namely, the wave equations of quantum mechanics are used to find the energy states available to the moving electrons, and the Pauli exclusion principle determines the allowed number of electrons in each wave state. It is easy to visualize the system as little billiard balls going into little boxes (the states) that each can only hold one ball. From these two considerations, the total flow rate of the electrons in circuits can be calculated.

This basic picture is known as the *semiclassical* approximation. It leaves out any concern about the phase of the electron wave functions, and any concern about correlations of particles in the many-particle wave function. When we only care about the total flow of electrons, we only need the total amplitude of the electron waves, which gives the total number of particles. This is a very reasonable approximation when the electrons have strong decoherence, which is normally the case for electrons in circuits at room temperature.

The question of whether quantum phase coherence plays a role in biology is a controversial one. One set of experiments appeared to show long-range coherence of electrons in the photosynthesis process (Lee 2007), but later experiments (see Kassal 2013, and references therein) indicated that the coherence in the phonon states, that is, vibrations of the molecules, may be more important, and in a realistic environment, electronic coherence will be quickly lost.

8.3 Quantum Cryptography

A method that uses the full properties of quantum mechanics, which is already in use, is *quantum cryptography*. This type of system is used along with regular computers to improve the security of private communications. It is based on the type of detection apparatus described in Section 4.3 in regard to the Einstein–Podalsky–Rosen experiment.

---

[2] For discussions of fine-tuning in the laws of nature, see Davies 2008 and Collins 2013.

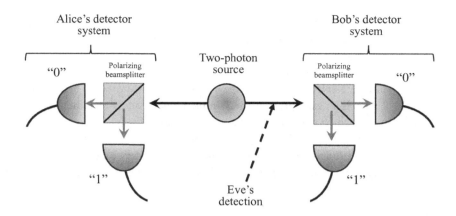

Figure 8.6 The two-photon apparatus used in a quantum cryptography.

A general need in secure communications is the ability to give out to various allowed users a password key in such a way that no one else has that key. Quantum cryptography provides a way to prove that only two people share the information of a digital password, and that no one else has eavesdropped when the passwords were sent out.

This method uses a variation of the two-photon source with two detectors discussed in Section 4.3. Figure 8.6 shows the basic configuration. This is the same apparatus as in Figure 4.3, but we now replace the simple polarizers with "polarizing beamsplitters," which have the property that, instead of simply destroying photons with one of the two possible polarizations, they send the photons of one polarization in one direction, and photons of the other polarization in a different direction. This allows a definite count of when a photon was received, no matter what its polarization.[3]

The two detection systems are commonly described as belonging to two people, "Alice" and "Bob." The quantum cryptography method can then go as follows:

- Alice and Bob set their polarizers at various different predetermined angles randomly, for example, at 0° or 45°.
- At each polarizer setting, Alice and Bob record the arrivals of photons at their detectors, until they accumulate a large number of recorded photons.
- Alice and Bob then share publicly their polarizer settings. For every case where they both had the same setting, they automatically know that whatever bit one recorded, the other recorded the same. A detection of a photon at one polarization can be identified as a "1" bit, while the detection of the other polarization is counted as a "0" bit. In this way, each person eventually has a long sequence of 0s and 1s. This string of numbers is random, since there is no way to predict the random action of the detectors, but the same for both Alice and Bob, because of the nonlocal correlation effect.

---

[3] If the detectors are not perfectly efficient, as is the case for real detectors, the users can share publicly the times when they recorded counts, without saying what the polarization of that photon was, and simply discard any data when one or the other of them did not detect any photon.

Now consider what will happen if a third person, whom we will call "Eve," tries to eavesdrop on the sequence of photons sent out. Eve does this by detecting the polarization of a photon en route, recording it, and then sending out another photon with the detected polarization toward the receiver.

In our correlated photon apparatus, there is a way to know if Eve has done this. In effect, Eve has changed the experiment to the case in which the source sends out pairs of photons that already have a definite polarization. As discussed in Section 4.3, this case gives a definite prediction for the statistics of the photons, which is different from the case of superpositions of photon states. For example, if Eve sends out photons definitely with polarizations of $0°$ or $90°$, which she has to do if she measures with a polarizer set at one of those angles, then when both Alice and Bob have their polarizers set at $45°$, they will detect photons hitting with the same polarization 50% of the time, which would disagree with the quantum mechanical prediction that they should see the same result 100% of the time. If Alice and Bob switch the angles of their polarizers randomly, there is no way for Eve to always set her polarizer at the same setting.

This means that Alice and Bob can use the following protocol to determine whether someone was listening in:

- Both Alice and Bob share publicly (i.e., through a communication channel that is insecure, which Eve may also monitor) the settings of their polarizers and the photon polarizations that they recorded for a small fraction of the total number they detected. Which small fraction of the data they use is picked randomly.
- When looking at that shared data, if they find that, whenever their polarizers were at the same angle, they did not get 100% agreement of their data, then they can conclude that there was an eavesdropper.

Of course, this relies on the assumption that Eve did not refrain from listening in just during the time when this small fraction was recorded but only listening in at a different time. Since Alice and Bob randomly pick what part of the data to share publicly, after all the recording has been done, this would be hard for Eve to do.

In practice, for Alice and Bob to share photons in this way, they would each need a dedicated optical communication channel, for example, a fiber-optic cable going from the source to each detector system. This raises the question of whether the quantum cryptography method really adds much value. If the people communicating have dedicated optical cables, they could monitor whether someone tapped into their communication simply by continuously sending light down the cables and measuring if there is any disruption of the light intensity, which would presumably occur when someone connected to the cable. But quantum cryptography adds another level of security.

Quantum "teleportation." A few years ago, many news outlets carried the story of the experimental demonstration of "quantum teleportation." While this is a fascinating effect, unfortunately, it has nothing to do with the teleportation of matter as in *Star Trek*. It is fundamentally just another way of showing the nonlocality of correlations that can occur in quantum mechanics.

The basic scheme uses the same type of two-photon source and detection system as shown in Figure 8.6. We can discuss this setup in terms of bits, in which we take one polarization to represent a "1," and the other polarization to represent a "0," and we allow superpositions of these two states. In this terminology, we can say that Alice and Bob each receive from the source an exact copy of the same superposition, which we will call A.

Suppose now that Alice receives a separate superposition B, but she doesn't do any measurements on it. Instead, she lets it interact with her copy of A, which she has received from the two-photon source. Then she does some measurements of the result and sends these results (through a classical information channel) to Bob. Bob takes this information and does a series of measurements on his copy of A. It can be shown mathematically (see, e.g., Boschi 1997; Brind 2018) that Bob will now have an exact copy of B. In effect, Alice has sent Bob all the information of qubit B without ever knowing herself its state. This is a type of "teleportation" of information. But of course, no matter has been sent from one place to another, only information.

8.4 Quantum Information Processing

Governments around the world are presently spending billions of dollars to try to make a working *quantum computer*. Such a computer has been proven mathematically to perform some types of calculations millions of times faster than a traditional computer, which we will call a *classical computer*. In particular, a quantum computer could factor large numbers in much less time. This has enormous implications for internet security, code breaking, and electronic currencies like Bitcoin, because the best security protocols involve the factorization of large numbers into prime numbers.

The reason why prime factorization is of such importance in security protocols comes from the fact that it is very time-consuming to take a large number (which we can call N) and find out whether it can be factored into two smaller numbers. In the most basic process, one must take every number smaller than N and do long division to see if one gets an answer with no remainder. The process is speeded up a bit by realizing that one only needs to check numbers up to the square root of N, because for every factor larger than the square root of N, there must be a factor smaller than the square root of N. This still means that finding the factors of a large number can take a very long time. For example, suppose that N has 30 digits, and it takes 100 computer clock cycles to perform one long division process (which is fairly optimistic). For a typical computer clock cycle of 1 nanosecond (one billionth of a second), it will take about three years to check all the possible factors for just this one number. Further mathematical advances have reduced the time to factor large numbers somewhat (see, e.g., Crandall 2005) but not by enough to significantly change the time-consuming nature of the prime factorization problem.

Therefore, for example, one could have a lock that stores the number N, and a user of the system can have one of its factors as a password. When the user enters the password, the device performs a long division of the number N by the password number, and if there is no remainder, access is granted. Even if someone hacks into the device and reads out the number N stored in it, the hacker cannot easily find the password, without doing a massive, time-consuming calculation to factorize N.

In 1998, mathematician Peter Shor showed that a quantum computer could find the prime factors of large numbers much faster than any regular computer (for a review, see Bernhardt 2019). Of course, at the time, a quantum computer did not exist! But that mathematical result was enough to lead many governments and large corporations such as Microsoft and Google to pour money into trying to make quantum computers. The outcome has been that we now have real quantum computers but only with a few bits (the 0s or 1s used in the binary system for computing) instead of the hundreds of bits needed to factor the large numbers used in internet security. But the bottom line is that quantum computers can really be made, and it is just a matter of time and expense to make them larger.

Qubits. In a traditional computer, as discussed in Section 8.1, information is stored in bits, which have two possible values, 0 and 1. In a quantum computer, the basic element of information is the *quantum bit*, or "qubit" (pronounced "cue-bit"). A qubit allows all possible superpositions of the two states, 0 and 1.

This property makes a quantum computer in many ways like an *analog computer*, which stores numbers not as 0s and 1s of the binary mathematical system but as numbers that can have any value on a continuum. The superposition of the two states in a qubit can also vary continuously.

Having continuously variable values might sound good, but it is actually a severe drawback for both analog computers and quantum computers. The reason is that having just two allowed states gives a huge advantage for *error correction*.

In a traditional computer, the information in a bit is stored as a voltage. Any voltage below, say, 2 Volts (V) is counted as a "0." Suppose that the voltage is originally recorded as 0, but then there is a slight error in the circuitry which makes it creep up to 0.1 V. It is a simple matter for another part of the circuit to detect the voltage and force all voltages below 2 down to 0. In the same way, if a level of 5 V was used to represent a "1," and it slid down to 4.7, a later circuit can force all values above 2 back up to 5 V.

In an analog computer, there is no equivalent error correction method. If a voltage level of 4.7 is used to represent the number 4.7, and then an error causes it to fluctuate down to 4.5, there is no easy way for the system to know that this is an error, and not actually supposed to represent the number 4.5. The only way to correct such errors is to keep extra copies of the analog information and hope that all of them do not go wrong at the same time.

There are error correction methods for quantum computers, but these methods take a lot of time and resources. In general, any decoherence in the system (discussed in Section 6.2) will lead to loss of the information of the superposition of states in a qubit.

Entanglement. What is the advantage of quantum computers, then? The important trick is that a quantum system allows massively parallel calculations. To see this, we have to do a little math.

Consider a system with N different states, that is, N different resonances of the quantum field. Each of these resonances can have different quantized amplitudes, which we interpret as particle numbers. For bosons, there is no limit to the number of possible amplitudes,

| 0 | 1 | 1 | 0 | 1 | 0 | 0 | 0 | 1 | 0 | 1 | 1 | 1 | 0 | 0 |

+ | 0 | 1 | 0 | 0 | 1 | 1 | 0 | 0 | 1 | 1 | 1 | 0 | 1 | 0 | 1 |

+ | 0 | 0 | 0 | 0 | 1 | 0 | 1 | 0 | 1 | 0 | 1 | 0 | 1 | 1 | 0 |

+ | 1 | 1 | 1 | 0 | 0 | 1 | 0 | 0 | 1 | 0 | 1 | 1 | 1 | 1 | 0 |

+ . . .

Figure 8.7   Several many-particle states with different numbers of particles in individual single-particle states, in a quantum superposition.

while for fermions there are just two, which we call 0 and 1. We will stick to the fermion case here, and will call these *single-particle* states.

We can now define a unique state of the full system, which we will call the *many-particle* state, as having a definite number 0 or 1 for the amplitude of each single-particle state, as illustrated in Figure 8.7. We then can count the total number of different possible combinations of 0s and 1s to get the total number of possible many-particle states. For example, if there are two single-particle states, each of which can have an amplitude of either 0 or 1, then there are four allowed many-particle states. If there are three single-particle states, then there will be eight possible many-particle states of the full system. The math tells us that the total number of possible unique many-body states of the full system is 2^N, which is a very large number – for example, if there are just 30 single-particle states, there are a billion possible different many-particle states.

The wave properties of quantum mechanics allow for all billions of these many-particle states to exist simultaneously, in a superposition. If there is no significant decoherence of the system (as discussed in Chapter 6), then a quantum computer can operate on all of these states simultaneously. This makes it a type of massively parallel computing, as though one billion traditional computers were all working at the same time.

The math of these parallel quantum states is often described using the concept of *entanglement*. Section 10.3 gives the basic math. It is often stated that only quantum systems can have entanglement, but this is not correct: Chapter 17 gives an example of entanglement in a classical system. But the degree of entanglement possible in a classical system is much less; quantum systems have essentially no upper limit to how many single-particle states can be entangled.

Some of the abilities of quantum computers that have been mathematically deduced have been shown to also be possible with classical analog computing systems with massive parallelism; for example, a set of a million light emitters all sending light in parallel through some region of space can perform the mathematical process known as a Fourier transform in a single step. (Section 11.5.1 gives the math of Fourier transforms.) But Shor's algorithm for quantum computers, mentioned at the beginning of this section, cannot be beaten by any classical system, and, in general, the massive number of possible superpositions in a quantum computer can always beat any practical classical computer.

It has been suggested, for example, by Roger Penrose (1996), that human brains may involve quantum superpositions that allow the same type of massively parallel computation as done by quantum computers. There is good reason to expect that this is not the case, because biological systems at room temperature normally have very strong decoherence, which destroys quantum superpositions. If indeed there are parts of the brain that maintain quantum coherence, it could only be through some type of biological fine-tuning (discussed in Section 8.2).

8.5 Lasers, Superfluids, and Superconductors

For much of this book, we have dealt with systems with spontaneous decoherence, in which coherent, wavelike behavior is dissipated. As discussed in Section 8.4, this is a major problem for quantum computing – the information in a qubit is lost over time due to decoherence. There is a class of systems governed by quantum mechanics, however, in which the opposite happens: systems that are initially incoherent spontaneously become coherent. This type of behavior, which can be called *enphasing*, or *spontaneous coherence*, underlies three types of modern technology that have fascinating behavior: lasers, superfluids, and superconductors.

We've already looked at the basic property that makes these systems special, in Section 2.6: bosons obey the rule that they transition into new states at a rate proportional to the number of bosons already in that state. Photons and phonons are naturally bosons. In some cases, fermions like electrons and protons can also couple together into bound states with an even number of fermions. Because fermions have half-integer spin, when an even number of them couple together, they make up a boson. This new boson obeys the same transition-rate rules as intrinsic bosons like photons.

In many systems, when enough bosons are in one state, there is a runaway amplification effect in which a large fraction of the whole system becomes part of that one wave state. When this happens with photons, it is called a *laser*. When it happens with particles with mass, it is called a *Bose–Einstein condensate*.[4] There are several types of known condensates of this type: superfluid liquid helium, superconducting metals, and super gases of cold atoms.

In these systems, the particle picture breaks down, and the system is best described by its wave properties. To put it another way, not only the *amplitude* of the macroscopic wave becomes important (which could be accounted for by the average number of particles),

---

[4] These two cases are actually not always sharply distinguished. Photons can be given an effective mass, and undergo Bose–Einstein condensation. If the lifetime of the photons in some system is very short, they can be in a state that has some properties of lasing and some properties of condensation (see Snoke 2020, Sections 11.12 and 11.13).

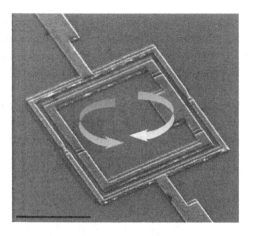

Figure 8.8 A device that can have a macroscopic superposition of electrons circulating both clockwise and counterclockwise. The fact that it is in this superposition can be established by comparing the current through the system with the predictions of quantum mechanics as the magnetic field in the device is varied. From Clarke 2008.

but also its *phase*. Boson systems that start out with lots of random noise and no coherence become spontaneously coherent, with a well-defined phase for the entire large wave. Section 21.1 proves this mathematically.

Therefore, we cannot say that quantum wave behavior is only relevant for microscopic systems. A bucket of superfluid liquid helium, for example, can have a well-defined phase that fills the bucket. In the same way, a large superconducting metal can have wavelike behavior over macroscopic distances.

Superconductors. Superconductors are well known to have zero electrical resistance, which means that they do not dissipate heat when electrical current travels in them. This has been an enormous benefit for the medical world, which uses large magnets made of superconductors in magnetic resonance imaging.

The property of zero electrical resistance is actually a consequence of a more fundamental property of superconductors, namely their phase coherence. Because the electrons in a superconducting metal all act as a single wave, they cannot easily dissipate heat, because dissipation inherently corresponds to loss of phase coherence, as we saw in Section 6.2.

The wave nature of electrons in a superconductor has some important and somewhat bizarre implications. Figure 8.8 shows a small but macroscopic device in which superconducting electron current can be put into a superposition of two wave states: circulating clockwise and also counterclockwise. This is not at all an unusual state for waves, as discussed in Chapter 1, but when we think of this happening for electrons with mass, it is harder to visualize. This is another reason to believe that the particle concept is not essential.

In a sense, we can call this superconducting system a type of Schrödinger's cat, of the type discussed in Section 4.4. A very large number of electrons (trillions and trillions) are in a superposition of going both to the left and to the right. In another sense, however, it is a much simpler system than a cat. The electrons in this system have only one aspect which is in a state of superposition, namely the direction they circulate. A cat, on the other hand, can do many different things, such as eating or not eating, dying or not dying, and so on. A superposition of real cat states has a huge number of different aspects that can be in macroscopic superposition.

The wave nature of the electrons in a superconductor is directly used in a device known as a SQUID (*s*uperconducting *qu*antum *i*nterference *d*evice), discussed in Section 2.6. Because electrons have charge, they react to magnetic field. When magnetic field is varied in a superconducting device like the one shown in Figure 8.8, the two electron currents can have phase shifts relative to each other, which are picked up in the measurement of the current through the device. SQUIDs that use this effect are now used in many cell phone towers to pick up extremely weak electromagnetic fields coming from telephones.

All of the superconducting systems we have at present require low temperature, but there is no intrinsic reason why that must be so. Many scientists are searching for new systems that would allow superconductivity at room temperature, which would revolutionize our technology.

Regular sound and water waves are condensates! In a way, superconductors and superfluids should not seem so strange. We have examples of macroscopic coherent waves all around us, namely sound waves, water waves, and other "classical" waves. The understanding of these waves in modern physics says that they are essentially the same as Bose–Einstein condensates, with one key difference.

Think of a bell, which you can hit with a hammer to make it ring. In quantum physics terms, when you hit it, you create a macroscopic, coherent state of phonons. The same math that predicts Bose–Einstein condensation also predicts that these phonons will tend to remain coherent (see Section 21.1). The coherence of the phonons in the bell transfers to coherent phonons in the air, and this coherent wave is what you hear as the ringing of the bell.

The main difference of this state with a Bose–Einstein condensate is that you used a hammer to directly cause the ringing. In physics language, there was no spontaneous symmetry breaking; you broke the symmetry of the system by the timing of the hit with the hammer. Therefore, the ringing of a bell can be called a "driven" condensate.

The case of a Bose–Einstein condensate is analogous to a bell that you cool down to low temperature, and below some particular temperature, the bell starts to ring on its own! In this case, there would be no hammer hit to start the bell ringing; it would spontaneously start to ring, with timing (i.e., phase) that no one could predict.

This gets us back to deep waters of philosophy. What gives the exact phase of the ringing bell when it cools? One school of thought, the Copenhagen view, says that it remains in a superposition of all possible ringing states until someone looks at it. The many-worlds view says that the entire universe goes into a superposition of all possible phases of the ringing. The spontaneous collapse view says that there is some tiny random term that gives a tiny kick to start the phase coherence. This last view is supported by the calculations given in Section 21.1 that show that the symmetry-breaking term can be very tiny: the property of boson systems to amplify phase coherence means that even an extremely small phase coherent fluctuation will lead to a macroscopic phase symmetry breaking.

Cosmic condensation. Spontaneous symmetry breaking leading to coherent phase is expected to happen not just for small experimental systems on earth but also in outer space; for example, it is widely believed that in neutron stars of sufficient size, the neutrons pair up into Cooper pairs, as in a superconductor, and go into a Bose–Einstein condensate state with definite phase (see, e.g., Pethick 2017). This state of matter can be detected in the radio signal received from these stars.

The implications of this for the Copenhagen view are striking. In the Copenhagen view, the entire physical state of a massive star, trillions of miles away, is waiting for a human to observe it before it can commit to a definite phase!

Some astrophysicists have gone further to argue that the state of the universe itself has the same type of symmetry breaking as a condensate (see Brown 1995; Banik 2017; Dvali 2017). This runs into the same symmetry-breaking problem we explored in Section 7.4. Quantum physics says that spontaneous symmetry breaking can only occur via a cause from "outside" the system; if there is no external perturbation, a system cannot break its own symmetry.

The physics of condensates shows that quantum effects are not just restricted to microscopic effects in the laboratory; even objects as large as stars can have properties that depend crucially on the laws of quantum mechanics. Even the existence of coherent classical waves is fundamentally a quantum effect of coherence of bosons.

References

S. J. Ahn, P. Moon, T.-H. Kim, et al., "Dirac electrons in a dodecagonal graphene quasicrystal," *Science* **361**, 782 (2018).

G. Brown, "Kaon condensation in dense matter," in *Bose–Einstein Condensation*, A. Griffin, D. W. Snoke, and S. Stringari, eds., (Cambridge University Press, 1995).

N. Banik and P. Sikivie, "Cosmic axion Bose–Einstein condensation," in *Universal Themes of Bose–Einstein Condensation*, N. Proukakis, D. W. Snoke, and P. Littlewood, eds., (Cambridge University Press, 2017).

C. Bernhardt, *Quantum Computing for Everyone*, (MIT Press, 2019).

D. Boschi, S. Branca, F. De Martini, L. Hardy, and S. Popescu, "Experimental realization of teleporting an unknown pure quantum state via dual classical and Einstein–Podolski–Rosen channels," *Physical Review Letters* **80**, 1121 (1998).

S. Brind, "The quantum teleportation protocol: How does quantum teleportation work?" https://owlcation.com/stem/Quantum-Teleportation (April 6, 2018).

R. Crandall and C. B. Pomerance, *Prime Numbers: A Computational Perspective*, 2nd ed. (Springer, 2005).

P. C. W. Davies, *The Goldilocks Enigma: Why Is the Universe Just Right for Life?* (Mariner, 2008).

S. De Vries, K. Dörner, M. J. F. Strampraad, and T. Friedrich, "Electron tunneling rates in respiratory complex are tuned for efficient energy conversion," *Angewandte Chemie* (International Edition in English) **54**, 2844 (2015).

G. Dvali and C. Gomez, "Graviton BEC's: A new approach to quantum gravity," in *Universal Themes of Bose–Einstein Condensation*, N. Proukakis, D. W. Snoke, and P. Littlewood, eds., (Cambridge University Press, 2017).

J. Clarke and F. K. Wilhelm, "Superconducting quantum bits," *Nature* **453**, 1031 (2008).

R. Collins, "Modern cosmology and anthropic fine-tuning: Three approaches," in *Georges Lemaitre: Life, Science and Legacy*, R. Holder and S. Mitton, eds., (Springer, 2013).

P. Hosseinzadeh and Y. Lu, "Design and fine-tuning redox potentials of metalloproteins involved in electron transfer in bioenergetics," *Biochimica et Biophysica Acta: Bioenergetics* **1857**, 557 (2016).

F. Hoyle, "The universe: Past and present reflections," *Annual Reviews of Astronomy and Astrophysics* **20**, 1 (1982).

J. K. Jain, "The role of analogy in unraveling the fractional quantum Hall effect mystery," *Physica E* **20**, 79 (2003).

I. Kassal, J. Yuen-Zhou, and S. Rahimi-Keshar, "Does coherence enhance transport in photosynthesis?" *Journal of Physical Chemistry Letters* **4**, 362 (2013).

H. Lee, Y.-C. Cheng, and G. R. Fleming, "Coherence dynamics in photosynthesis: Protein protection of excitonic coherence," *Science* **316**, 1462 (2007).

R. Penrose, *Shadows of the Mind*, (Oxford University Press, 1996).

C. J. Pethick, T. Schäfer, and A. Schwenk, "Bose–Einstein condensates in neutron stars," in *Universal Themes of Bose–Einstein Condensation*, N. P. Proukakis, D. W. Snoke, and P. B. Littlewood, eds., (Cambridge University Press, 2017).

D. W. Snoke, *Electronics: A Physical Approach*, (Pearson, 2015).

D. W. Snoke, *Solid State Physics: Essential Concepts*, (Cambridge University Press, 2020).

Key Points

What are the strong conclusions we can make, and what must we hold more lightly, when it comes to thinking about quantum mechanics?

The following is a summary of the points discussed in Chapters 1–8:

- Common notions of wave-particle duality are mostly misguided. We *know* from quantum field theory where "particles" come from – they are the excitations of the amplitude of the fields, which are as real as the water or the air we breathe. The fact that these excitations have quantum jumps in energy is a simple mathematical result of how waves fit into systems with constraints.
- Nothing in the equations or experiments demands that we view particles intrinsically as localized little objects. The "lumpy" or "jumpy" behavior seen in clicks and tracks in detectors is directly related to the natural size of atoms, which are always involved in the detection process. In other physical systems, there is no natural length scale, in which case a particle size cannot be defined.
- In general, random jumps with abrupt changes are commonly seen in classical, nonlinear systems; so it is not surprising that we also see these in quantum systems. These transitions are not discontinuous in time or space in either quantum or classical systems, although they can be quite abrupt. There is no reason to insist that the randomness seen in quantum systems is "uncaused" randomness; there are plenty of ways to get caused jumps in inhomogeneous quantum systems.
- The strong emphasis on human knowledge often associated with quantum mechanics is part of a specific formulation of quantum mechanics known as the Copenhagen interpretation; that view is not universally accepted among experts, and its treatment of quantum fields as mere expressions of our knowledge runs up against many experiments and experiences that indicate the fields and waves of quantum mechanics have ontological reality. In general, matter waves and other waves such as sound and light, which we commonly take as real, are not sharply distinguished in quantum field theory.

   Philosophies and religions that rely heavily on the mixing of knowledge and physical reality would do well to recognize that the Copenhagen approach is far from proven, and most working physicists simply don't believe that a human brain acts fundamentally differently on the outcome of experiments than a mechanical detector.
- The many-worlds interpretation is rising in popularity because it is in some ways the simplest approach – to take the current form of the equations of quantum mechanics as absolute. But apart from the psychological conundrums it may lead to, it also has issues of math and our experience that make many physicists uncomfortable. The main one is this: all these other worlds are not somewhere else far away, but right with us, all around

us in the same space, according to the laws of quantum mechanics. Why do we not see any evidence of any of them, ever? That requires no reverberations, no tiny echoes, and none of the nonlinear overtones that we are so used to in every other experience of the physical world.

- Many, if not most, working physicists have an intuition that the right interpretation is spontaneous instabilities in macroscopic systems that cause one quantum outcome or another to win out, and many mathematical models of physics, in particular the quantum trajectories method, implicitly assume this. However, such a theory can't be derived from the existing laws of quantum mechanics; a full theory of this would require new mathematical terms in these equations. It is not so easy to come up with such terms; one proposal is given in Chapter 20.

- Every formulation of quantum mechanics involves nonlocal correlations, which is to say, every quantum field seems to act as a whole, and does not always wait for signals to travel to effects that happen later in time. This is hard to wrap our minds around, because we live in a world in which decoherence (loss of wave information) gives irreversibility, which underlies our whole sense of causality. But causality may be a subset of a more comprehensive structure of the universe which allows nonlocal interactions.

- Quantum mechanics doesn't prove that there is a spirit world, create our sense of free will, or explain the origin of the universe. Most of the great questions and arguments of classic philosophy and theology are unchanged by quantum mechanics.

- Quantum mechanics is not just about parlor tricks that we can take or leave. Quantum mechanics underlies an enormous amount of our modern technology and thousands of experiments prove to a high degree of accuracy that it is the proper description of our physical world.

I've often had the experience of telling someone that I study quantum mechanics, and having the reaction that I am something like a magician. But quantum mechanics is not magic. Most of it is like doing hydrodynamics of water waves, or electric fields and radio waves. But it does play with our intuitions, and it tells us that the world is indeed an amazing place.

Part II

Basic Results of Quantum Mechanics

9

Schrödinger Equation Calculations

As discussed throughout this book, quantum field theory is the foundational theory of quantum physics, and the single-particle Schrödinger equation is a simplified model that applies just to special cases. But we can learn a lot about the basic effects of quantum mechanics by studying just this equation, and doing this requires only the math of derivatives and integrals taught in introductory college calculus.

9.1 Wave Equations

As discussed in Section 1.2, the equations for waves in various types of fields are of the same basic form. The equation for a sound wave in air, for example, can be written as

$$\frac{\partial^2 \rho}{\partial t^2} = \alpha \frac{\partial^2 \rho}{\partial x^2}, \tag{9.1.1}$$

where x and t are the position and time, ρ is the density of the air, and α is a constant that depends on the properties of the medium. This is a *scalar* wave equation, since the density ρ is a single number at each point x and time t. For many waves, it is a good approximation to treat α as a constant, equal to v^2, where v is the wave speed in the medium (e.g., the speed of sound in air).

The Maxwell wave equation for electric field in vacuum can be written similarly as

$$\frac{\partial^2 \vec{E}}{\partial t^2} = \alpha \frac{\partial^2 \vec{E}}{\partial x^2}. \tag{9.1.2}$$

The only difference from the case of a sound wave is that $\vec{E}$ is a *vector*, corresponding to three numbers E_x, E_y, and E_z for the strength of the electric field in the three different directions in space in a three-dimensional universe. (E_x is zero in the case of wave motion in the x direction.) The constant α in this case is equal to c^2, where c is the speed of light. This discovery was Maxwell's great "Eureka" moment, because he derived α from entirely different constants of nature for electric and magnetic field but found that the derived value of α agreed with measurements of the speed of light to high precision, thus unifying the phenomenon of light with the phenomena of electricity and magnetism, which had previously been viewed as unrelated to light.

Both of these equations have solutions of the same form,

$$\rho(x,t) = \rho_0 e^{i(kx-\omega t)}, \qquad \vec{E}(x,t) = \vec{E}_0 e^{i(kx-\omega t)}, \tag{9.1.3}$$

where k and ω are constants related to the wavelength and frequency of the wave, given by $k = 2\pi/\lambda$ and $\omega = 2\pi f$, and λ and f are the wavelength and frequency defined in Section 1.2. The value k is called the *wave number* of the wave, and ω is called the *angular frequency*, although often this is shortened to being called just the frequency. The wave solutions of (9.1.3) are written in terms of complex numbers but can be equated to physically measured values by taking the real part,

$$\text{Re } e^{i(kx-\omega t)} = \cos(kx - \omega t). \tag{9.1.4}$$

This form of wave function corresponds to a wave traveling toward positive x. This can be seen by noting that the term in parentheses, $\phi = (kx - \omega t)$, corresponds to the *phase* of the wave. Phase equal to 0 corresponds to a crest of the wave, that is, a maximum of the oscillation. The location of this crest changes; its location as a function of time can be found by solving for the location x:

$$\phi = 0 = kx - \omega t$$
$$\rightarrow x = \frac{\omega}{k} t. \tag{9.1.5}$$

Using the form of solution (9.1.3) in the wave equation (9.1.1) implies $\omega^2 = \alpha k^2$.

A matter wave in vacuum is described by the *Schrödinger equation*,

$$-i\frac{\partial \psi}{\partial t} = \alpha \frac{\partial^2 \psi}{\partial x^2}. \tag{9.1.6}$$

Here, ψ is a complex number, with a real part and an imaginary part associated with each point in space and time. The left side looks different from the previous wave equations, since it has only a first derivative with time, but it can easily be shown that this equation has the same form of solution for the waves,

$$\psi(x, t) = \psi_0 e^{i(kx-\omega t)}. \tag{9.1.7}$$

Because the wave function ψ in this case is complex, consisting of two numbers instead of one number for a scalar field or three for a vector field, some authors have argued that this makes the quantum matter field spooky and utterly unlike sound waves or electromagnetic waves. But in the end, the difference is only of how many numbers to keep track of at each point in space. Scalar fields require one number, complex fields require two, and vector fields require three.

A physical difference of the waves follows from the different forms of the equations for the waves, however. Equations of the form (9.1.2) allow for *standing wave* solutions of the form

$$E_0 e^{i(kx-\omega t)} + E_0 e^{i(kx+\omega t)} = 2E_0 e^{ikx} \cos \omega t. \tag{9.1.8}$$

At certain times t, this entire wave function is equal to zero, at every point in space. The wave function "winks out" twice every oscillation cycle, so to speak. By contrast, for the Schrödinger equation, only solutions proportional to $e^{-i\omega t}$ are allowed, so that a standing wave solution is

$$\psi_0 e^{i(kx-\omega t)} + \psi_0 e^{i(-kx-\omega t)} = 2\psi_0 \cos kx e^{-i\omega t}. \tag{9.1.9}$$

Therefore, it is never the case that both the complex components of the wave function are zero everywhere; if the real part is zero, then the imaginary part is nonzero. This was important for Dirac in his derivation of his wave equation for relativistic matter (see Section 13.2), because he felt it was crucial that for wave excitations associated with mass, the wave could never disappear, because that would violate the principle of conservation of mass. The recently discovered Higgs boson, however, may obey a wave equation like (9.1.2), however, and so violate this principle, since the Higgs boson has mass.

All of the wave equations discussed here are altered in the presence of matter by adding additional terms on the right side. The simplest adjustment to the Schrödinger equation is to add a term proportional to ψ,

$$-i\frac{\partial \psi}{\partial t} = \alpha \frac{\partial^2 \psi}{\partial x^2} + \beta \psi. \tag{9.1.10}$$

If β is a constant, it can be shown that this equation has the same solutions as (9.1.7), with a shift of the value of ω. It is natural to rewrite this equation as an energy equation,

$$i\hbar\frac{\partial \psi}{\partial t} = H\psi, \tag{9.1.11}$$

where H is called the *Hamiltonian*, which gives the total energy of a system, and $\hbar = h/2\pi$ is the reduced Planck's constant (discussed in Section 2.2). In this case, the Hamiltonian is equal to

$$\begin{aligned} H &= -\frac{1}{2m}\frac{\partial^2}{\partial x^2} + U(x) \\ &= \frac{p^2}{2m} + U(x), \end{aligned} \tag{9.1.12}$$

where m is an intrinsic constant of nature treated as the particle mass, $p^2/2m = \frac{1}{2}mv^2$ is the kinetic energy, and $U(x)$ is the potential energy, which can be a function of the location x. This assumes the definition of the momentum p as

$$p \equiv -i\hbar\frac{\partial}{\partial x}, \tag{9.1.13}$$

and the energy as

$$E \equiv i\hbar\frac{\partial}{\partial t}. \tag{9.1.14}$$

These last two equations are known as the *de Broglie relations*. They are generally applicable for all quantum fields, not just matter waves. For wave solutions of the form (9.1.3) and (9.1.7), they imply

$$E = \hbar\omega$$
$$p = \hbar k. \tag{9.1.15}$$

These equations imply that when the frequency of a wave is higher, it has more energy, and also that when the wavelength is shorter, the wave has more energy. We can see the second implication by writing the kinetic energy in terms of momentum and applying it to a traveling wave:

$$\frac{p^2}{2m}\psi = -\frac{\hbar^2}{2m}\frac{\partial^2}{\partial x^2}e^{i(kx-\omega t)}$$

$$= \frac{\hbar^2 k^2}{2m}\psi. \tag{9.1.16}$$

Since $k = 2\pi/\lambda$, when the wavelength λ gets shorter, k gets larger.

9.2 Quantum Confinement Energy: Why Nanometers are Important

For the square well confinement shown in Figure 8.1, reproduced as Figure 9.1 here, the energies are found by setting the allowed wavelengths to $2L/N$, where N is an integer. This gives us

$$E = \frac{\hbar^2 k_z^2}{2m} = \frac{\hbar^2}{2m}\left(\frac{2\pi}{2L/N}\right)^2 = \frac{\hbar^2\pi^2 N^2}{2mL^2}. \tag{9.2.1}$$

The mass m that is used in this formula is in general not the mass of an electron in vacuum but a renormalized mass determined by the material properties of the solid in which electron moves. For the common semiconductor GaAs, the effective mass is around 6% of the vacuum electron mass.

If the energy difference between the $N = 1$ state and the $N = 2$ state in the quantum well is large compared to the typical energy of thermal excitations, $k_B T$, where k_B is Boltzmann's constant, then electrons in the $N = 1$ state will have no degree of freedom to move in the z-direction – they will be stuck in the $N = 1$ state and will be free only to move in the x–y plane. Let us deduce the thickness needed to have this occur at room temperature. At room temperature, $k_B T = 25.6$ meV. We set the energy different between the $N = 1$ and $N = 2$ state to three times this energy:

$$\frac{\hbar^2\pi^2(4 - 1)}{2mL^2} = 3k_B T. \tag{9.2.2}$$

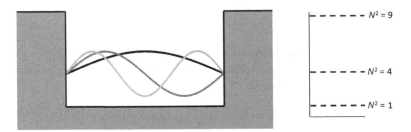

Figure 9.1   A trap for electrons, with several of the waves that fit in this trap. The set of dashed lines on the side indicates the "energy levels" corresponding to these waves.

Solving for L, we obtain

$$L = \sqrt{\frac{\hbar^2 \pi^2}{2mk_B T}}. \tag{9.2.3}$$

To put actual units in this, for convenience, we write the vacuum mass of the electron as the rest energy $m_0 c^2 = 5 \times 10^5$ eV, and $\hbar c = 2 \times 10^{-5}$ eV-cm. For GaAs, we then have

$$L = \sqrt{\frac{\pi^2 (2 \times 10^{-5} \text{ eV-cm})^2}{2(0.06)(5 \times 10^5 \text{ eV})(0.0256 \text{ eV})}}$$

$$= 1.6 \times 10^{-6} \text{ cm} = 16 \text{ nm}. \tag{9.2.4}$$

We see from this analysis why *nanotechnology* is important – when the dimensions of the devices are a few nanometers, the confined wavelengths of the electrons are less than their average wavelengths at room temperature. This means that the effects of wavelength quantization (quantum confinement) will have a significant effect.

9.3 Fermi Pressure in Solids: Why are Solids Solid?

Because the Pauli exclusion principle means that each state can only have one electron, if a system has many electrons (or other fermions), then states of higher and higher energy will be occupied. This leads to a significant energy at typical densities of matter.

We can estimate the Fermi pressure for a solid by assuming that all the electrons in the solid are free waves. This is an approximation, because the positively charged nuclei will alter the wave states of the electrons, but the full calculation cannot be much less than this approximation, because the only thing the positive charge could do would be to compact the electron waves into smaller volumes, which would raise their kinetic energy by making their wavelengths shorter.

We assume that all of the electrons are in the lowest possible states, which means that all of the states are filled up to some energy known as the *Fermi level*. If we count the total number of states below the Fermi level, we have then found the total number of particles.

As we saw in Section 9.2, for a system of length L, standing waves of wavelength $\lambda = 2L/N$ are allowed, where N is any positive integer. These correspond to values of $k = 2\pi/\lambda = \pi N/L$ in the wave solutions of Section 9.1. The number of states per unit range of k is therefore L/π. In three dimensions, we have the same result for each direction. To count the total number of states, we perform an integral over a three-dimensional "k-space," using the number of states per volume in k-space as V/π^3. The total number of states up to a maximum value k_F is then

$$N = 2 \left(\frac{1}{8}\right) \frac{V}{\pi^3} \int_0^{k_F} 4\pi k^2 dk. \tag{9.3.1}$$

The factor of $1/8$ accounts for the fact that we only take positive values of k for standing waves, and so only integrate over one-eighth of the total volume in k-space. We have also multiplied by 2 to account for two spin states of the electrons for every k-state.

We can then change the variable of integration from k to $E = \hbar^2 k^2 / 2m$, the kinetic energy, and integrate up to the Fermi energy E_F that corresponds to k_F. This gives us

$$N = \int_0^{E_F} \frac{V}{\pi^2} \frac{m}{\hbar^3} \sqrt{2mE} dE$$
$$= 2V \frac{\sqrt{2}}{2\pi^2} \frac{m^{3/2}}{\hbar^3} \left(\frac{2}{3} E_F^{3/2} \right). \tag{9.3.2}$$

The density of the electrons is therefore

$$n = \frac{N}{V} = \frac{2\sqrt{2}}{3\pi^2} \frac{m^{3/2}}{\hbar^3} E_F^{3/2}. \tag{9.3.3}$$

We can then solve for E_F as a function of the density,

$$E_F = \left[n \frac{(3\pi^2)}{2\sqrt{2}} \frac{\hbar^3}{m^{3/2}} \right]^{2/3}. \tag{9.3.4}$$

For a density of $n \sim 10^{23}$ cm^{-3} and electron mass equal to the free electron mass in vacuum, this implies $E_F \sim 10$ eV.

Using this number, we can compute the amount of pressure it would take to change the volume of the electrons. The pressure for a closed system is defined as

$$P = -\frac{\partial U}{\partial V}, \tag{9.3.5}$$

where U is the total energy. The integral for U is the same as that for N, with each state weighted by its energy E:

$$U = \int_0^{E_F} \frac{V}{\pi^2} \frac{\sqrt{2}m^{3/2}}{\hbar^3} E^{3/2} dE$$
$$= 2V \frac{\sqrt{2}}{2\pi^2} \frac{m^{3/2}}{\hbar^3} \left(\frac{2}{5} E_F^{5/2} \right)$$
$$= 2V \frac{\sqrt{2}}{5\pi^2} \frac{m^{3/2}}{\hbar^3} \left[\frac{N}{V} \frac{(3\pi^2)}{2\sqrt{2}} \frac{\hbar^3}{m^{3/2}} \right]^{5/3}. \tag{9.3.6}$$

From this, it follows that the pressure is simply

$$P = \frac{2}{3} \frac{U}{V}.$$

The compressibility is given by the change in pressure for a fractional volume change:

$$B = -V \frac{\partial P}{\partial V}$$
$$= \frac{5}{3} P = \frac{2}{3} n E_F. \tag{9.3.7}$$

The pressure needed to change the volume by the fraction $\Delta V/V = 1\%$ is then

$$\Delta P = B(\Delta V/V)$$
$$= \frac{2}{3}nE_F(0.01), \qquad (9.3.8)$$

which for a typical solid density is 8 eV $\times 10^{23}$ cm$^{-3} \times 0.01$, which works out to 10,000 times atmospheric pressure. This is close to real experimental values – typical bulk compressibilities of solids lie in the range of a few times 10^{10} N/m^2. The same type of calculation also leads to the Fermi pressure that gives the stability of stars.

The Coulomb interactions between the charged particles do not change this pressure much at typical solid densities, because there are equal amounts of positive and negative charge. Overall, the Coulomb interactions will lead to a net attraction that tends to make the solid smaller, but this is overcome by the increase of the Fermi energy as the solid contracts.

9.4 Vibration of Atoms: The Simple Harmonic Oscillator Model

Besides the "square well" discussed in Section 9.2, another simple system that can easily be studied with the Schrödinger equation is the *harmonic oscillator*. This is actually the model for all types of springiness in physics, including real springs and the springiness invoked in quantum fields that leads to the quantization into particles, discussed in Section 2.2.

It is no accident that springs can be described by this model, because the springiness of springs arises from the interaction of the atoms they are made of. The interaction of atoms is typically described by the Lennard-Jones, or "6-12" potential, illustrated in Figure 9.2, and familiar to students of chemistry. At long range, atoms attract each other due to covalent bonding, or due to van der Waals attraction. At short distances, the repulsion of the cores of the atoms quickly becomes dominant.

As illustrated in Figure 9.2, the minimum in energy can be approximated as a harmonic potential using the Taylor series approximation,

$$U(x) \simeq U_0 + \frac{1}{2}\frac{\partial^2 U}{\partial x^2}\bigg|_{x_0}(x - x_0)^2$$
$$\equiv U_0 + \frac{1}{2}K(x - x_0)^2, \qquad (9.4.1)$$

where x_0 is the equilibrium atomic spacing. When the amplitude of the motion is small enough, the parabolic approximation is always a good approximation of any minimum. This means that the force between two atoms is well approximated by Hooke's law for springs for low-amplitude oscillations, $F = -K(x - x_0)$. Hooke's law for springs comes from this fact about interatomic forces.

Since we are free to set $x_0 = 0$ and $U_0 = 0$, we can write simply

$$U = \frac{1}{2}Kx^2. \qquad (9.4.2)$$

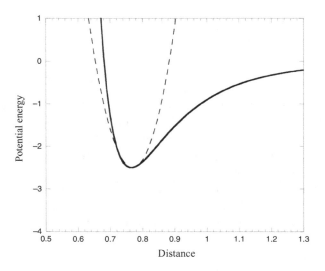

Figure 9.2 Solid line: A typical atomic interaction potential, known as the 6-12, or Lennard-Jones, potential. Dashed line: Approximation as a parabola.

The kinetic energy of an atom with mass M is just $p^2/2M$, where p is the momentum, and therefore we write the Hamiltonian (i.e., the total energy of the system) as

$$H = \frac{p^2}{2M} + \frac{1}{2}Kx^2, \qquad (9.4.3)$$

which can be rewritten as

$$H = \frac{p^2}{2M} + \frac{1}{2}M\omega_0^2 x^2, \qquad (9.4.4)$$

where $\omega_0 = \sqrt{K/M}$ is the *natural frequency* of the oscillator. This is known as the *simple harmonic oscillator*, and is one of the few physical systems that can be solved exactly in quantum physics.

As discussed in Section 2.1, quantized states exist whenever there is a confining potential that restricts how long the wavelength can be. In the case of a wave pinned down at each end, like a guitar string, the wave function must be zero at the ends. As illustrated in Figure 9.1, in a system with fixed walls a distance L apart, the allowed wavelengths are $\lambda = 2L/N$, where N is any positive integer. The standing wave solution in this case has the form of (9.1.9),

$$\psi(x, t) = \psi_0 \cos kx e^{-i\omega t}, \qquad (9.4.5)$$

where $k = 2\pi/\lambda$. The energy is given by the eigenvalue of the kinetic energy operator, which is written in terms of the momentum operator defined in (9.1.13), as

$$\frac{p^2}{2M}\psi = -\frac{\hbar^2}{2M}\frac{\partial^2}{\partial x^2}\psi = \frac{\hbar^2 k^2}{2M}\psi, \qquad (9.4.6)$$

that is, the kinetic energy is

$$E = \frac{\hbar^2 k^2}{2M} = \frac{\pi^2 \hbar^2 N^2}{2ML^2}. \tag{9.4.7}$$

The energies in this case increase proportional to N^2, getting even further and further apart.

A harmonic potential does not have boundaries with a well-defined confinement size L. The size of the confinement region depends on the energy of the particle, because a particle with higher energy moves higher up in potential energy and therefore farther from the center. We can approximate the size of the confinement region felt by the wave, however, by equating the potential energy due to the springy force at maximum distance with the maximum kinetic energy:

$$E = \frac{1}{2}KL^2, \tag{9.4.8}$$

which implies

$$L = \sqrt{2E/K}; \tag{9.4.9}$$

that is, the effective size of the confined region gets larger as the energy in the wave gets larger. If we require the same wave resonance condition $\lambda = 2L/N$, we have

$$E = \frac{\pi^2 \hbar^2 N^2}{2ML^2} = \frac{\pi^2 \hbar^2 N^2}{2M(2E/K)}. \tag{9.4.10}$$

Solving this equation for E gives

$$E = \frac{\pi}{2}N\hbar\sqrt{\frac{K}{M}} = \frac{\pi}{2}N\hbar\omega_0, \tag{9.4.11}$$

which is very close to the exact solution for the eigenstate energies of the harmonic oscillator, namely $E = N\hbar\omega_0$, found in the exact calculation of Section 12.1. The eigenstates of the harmonic oscillator have definite energies for the same reason that the states of a quantum well do – only waves with definite wavelength can fit in the confined geometry.

Note that in this case, however, the energies of the levels are proportional to N instead of N^2. This lends itself to the picture of quantum particles with constant energy being added to the system. The fundamental reason for the quantization, however, is no different from the case of two hard walls, though. It is easy to imagine that for some different potential energy functions, the quantized states would lie somewhere in between the two dependences of N and N^2, for example, $N^{1.001}$. This would be "mostly" particle-like, but not quite.

9.5 Unit Analysis of Atomic States

Unit analysis is a general method of physics that gives intuition about the physical properties of a system without doing a full calculation, just as we have done in Section 9.4. For atoms, we can find approximate numbers for many properties just by doing unit analysis.

The potential energy due to the attraction of two charges is called the *Coulomb energy*, and is given by

$$U(x) = -\frac{e^2}{4\pi \varepsilon_0 x}, \tag{9.5.1}$$

where e is the fundamental constant of charge in nature, and ε_0 is another fundamental constant, known as the vacuum permittivity. To get the total potential energy at some point due to the charge in a wave, one would add up the contribution of this term $U(x)$ for the wave amplitude $\psi(x)$ at every point in space.

In our unit analysis approach, we define a natural length l, which defines the average size over which a wave is localized. This gives a characteristic energy

$$U \sim -\frac{e^2}{4\pi \varepsilon_0 l}. \tag{9.5.2}$$

Since there is a negative sign in this energy, the Coulomb energy will get more negative as the wave gets more compact in space; that is, as l decreases.

Going to lower potential energy is generally favored, for the same reason that objects move downhill. But when l gets smaller, the wavelength of the wave will get shorter, and the kinetic energy of a matter wave gets larger when its wavelength gets shorter. The kinetic energy for a matter wave is written in quantum mechanics as

$$E_K = \frac{p^2}{2m} = \frac{\hbar^2 k^2}{2m} = \frac{(2\pi \hbar)^2}{2m\lambda^2}, \tag{9.5.3}$$

where $\hbar$ is Planck's constant, m is the mass of the relevant particle, and λ is the wavelength of the wave. This equation expresses the physical fact that putting more bends in the wave function requires more energy. In a very real sense, we can view a wave as having stiffness like a spring that resists bending.

In our unit analysis approach, we set λ approximately equal to l, the range of space occupied by the wave. As we reduce l, U gets lower (more negative) while the kinetic energy gets larger. We can find the natural value of l by setting the energy savings of U equal to the energy cost of E_K. This gives us

$$\frac{e^2}{4\pi \varepsilon_0 l} \sim \frac{(2\pi \hbar)^2}{2ml^2}. \tag{9.5.4}$$

Solving this equation for l gives us the natural length

$$l \sim \frac{2\pi (2\pi \hbar)^2 \varepsilon_0}{e^2 m} \simeq 2 \times 10^{-10} \text{cm} = 2\text{Å}. \tag{9.5.5}$$

All atoms have a size on this scale, give or take a small numerical value (which typically could range from 0.1 to 10). For example, the Bohr radius of a hydrogen atom, which consists of a single electron and proton, is found by exact calculations to be equal to

$$l_B = \frac{4\pi (2\pi \hbar)^2 \varepsilon_0}{e^2 m_r}, \tag{9.5.6}$$

where the mass m_r is nearly equal to the electron mass m.[1] This is the same as our unit analysis deduction to within a factor of 2.

We can apply a similar unit analysis method to the energies of the higher, excited states of the atom. If we think of the circumference of the space occupied by the wave function, we can apply the rule that on a circular path around the atom, there must be an integer number of wavelengths of the wave, as illustrated in Figure 2.2. Therefore we write

$$2\pi l \sim N\lambda, \tag{9.5.7}$$

where N is an integer. The energy balance will then be

$$\frac{e^2}{4\pi\varepsilon_0 l} \sim \frac{(2\pi\hbar)^2}{2m(2\pi l/N)^2}. \tag{9.5.8}$$

Solving this for l then implies that the size of the wave l is proportional to N^2. Putting this into the Coulomb energy gives

$$U \propto -\frac{1}{N^2}, \tag{9.5.9}$$

which is exactly the dependence on N of the *Rydberg series* of atomic states.

We thus have a third example of how the form of the potential energy controls the sequence of allowed wave states. For zero potential, or any constant potential energy, we have the energy sequence (9.4.7), proportional to N^2, and for the simple harmonic potential, we have the series (9.4.11), proportional to N. In each of these three cases, the number obtained by unit analysis comes very close to the exact calculation.

Natural timescale for transitions. In the same way that we have deduced a natural length scale, we can deduce a natural timescale for transitions between the electronic states of atoms.

First, the natural length scale we have found implies a natural energy scale, which can be found from putting our natural length into the Coulomb energy (9.5.1), which gives

$$U \sim -7\,\text{eV}. \tag{9.5.10}$$

For transitions between upper states with $N = 3$ or 4 or so, we will have a typical energy of 7 eV divided by an integer of the order of 10, which gives a natural energy scale around 1 eV. From this, we can get the natural timescale

$$\Delta t \sim \frac{E}{\hbar} \sim \frac{1\,\text{eV}}{6.6 \times 10^{-16}\,\text{eV-s}} \sim 10^{-15}\,\text{s} = 1\,\text{fs}. \tag{9.5.11}$$

The natural timescale for electron transitions of atoms is femtoseconds, which is a very short time but not instantaneous. Given how hard it is to measure such short timescales, it is not surprising that many early scientists thought these transitions were indeed instantaneous.

---

[1] The mass m_r, known as the "reduced mass," is a type of average of the proton mass m_P and the electron mass m_e, given by $1/m_r = 1/m_e + 1/m_P$. Since the electron mass is much lighter than the proton mass, the term $1/m_P$ is negligible compared to $1/m_e$, and so the mass m_r is very close to the electron mass.

Some readers may not be comfortable with the level of approximations made in this section, but it is actually a very powerful method, often used by physicists to determine the natural parameters of a system without detailed calculation.

9.6 Universal Conductance in Quantum Wires

A fascinating result of modern physics of solids (known often as *condensed matter* physics) is that many *universal* results can be derived, even though typical solids have all kinds of disorders. This has led some physicists to argue that particle physics is no more "fundamental" than condensed matter physics; there are universal laws at every level of reality we examine (see Laughlin 2005). Here, we examine an example of this.

As discussed in Section 8.1, it is possible to make one-dimensional conductors known as *quantum wires*. There are many ways to make these. One way is to start with a quantum well and etch away two sides until a narrow strip only a few nanometers wide is left. After this etching, new barrier material can be deposited over the entire structure. It is also possible to use chain-type molecules made by chemistry as one-dimensional conductors. Metals such as gold can also be drawn into very thin whiskers that are only a few nanometers across. In all of these cases, the electronic states will be quantized in two dimensions while continuous in the other dimension, along the wire. At low temperature, that means the electrons can only move in one direction.

A surprising property of quantum wires is quantization of current; that is, the current through a quantum wire depends only on universal constants of nature, independent of the details of the geometry of the wire. We can prove this in the following calculation.

Assuming that the electrons move without scattering, the current carried by an electron is given simply by the charge of the electron divided by the transit time across the wire. The transit time is just the length of the wire L divided by the velocity v of the electron, and therefore for a given value of k, the current is

$$I_k = \frac{e}{(L/v)} = \frac{e}{L}\left(\frac{p}{m}\right) = \frac{e\hbar k}{Lm}, \tag{9.6.1}$$

where we have used the de Broglie relation $p = mv = \hbar k$ from Section 9.1.

For a wire of length L, the allowed wavelengths are $\lambda = 2L/N$, as discussed in Section 9.2. We can write this as a set of allowed wave numbers $k = 2\pi/\lambda = \pi N/L$. We can then say that the spacing between different k-states is a constant number π/L. The total number of k-states within a range Δk is then $\Delta k/(\pi/L) = (L/\pi)\Delta k$.

For electrons with mass m, the kinetic energy is $E = \frac{1}{2}mv^2 = p^2/2m$, since the momentum $p = mv$. We can then write the amount of energy change per k-state as the derivative

$$\Delta E = \left(\frac{\partial E}{\partial k}\right)\Delta k = \frac{\partial}{\partial k}\frac{(\hbar k)^2}{2m}\Delta k$$
$$= \frac{\hbar^2 k}{m}\Delta k. \tag{9.6.2}$$

The number of quantum states of the electrons in an energy range ΔE is then

$$2\frac{L}{\pi}\Delta k = 2\frac{L}{\pi}\frac{\Delta E}{\hbar^2 k/m},\tag{9.6.3}$$

where we have multiplied by 2 to account for two allowed electron spin states.

Each of these quantum states can carry current as an electron moves in that state down the wire. The total current carried in an energy range ΔE is therefore equal to the number of quantum states in that range times the current per state:

$$I = \sum_k I_k = \left(2\frac{mL}{\pi\hbar^2 k}\Delta E\right)\frac{e\hbar k}{Lm} = \frac{2}{\pi}\frac{e}{\hbar}\Delta E.\tag{9.6.4}$$

The dependence on k has dropped out. Essentially, although states with higher k carry more current, since the electrons move faster, this is canceled out by the fact that there are more states with low k in a given energy range.

To convert this result to a formula for conductivity, we suppose that the wire is connected between two conductors. For there to be a steady current, the two ends of the wire must be at different potential energies. The difference in energy is given by $\Delta U = e|\Delta V|$, where ΔV is the electric potential difference. The range of energy of electrons flowing in the wire is then found by equating the $\Delta E = \Delta U$. Formula (9.6.4) can then be written as

$$I = \frac{2}{\pi}\left(\frac{e^2}{\hbar}\right)\Delta V.\tag{9.6.5}$$

This has the same form as Ohm's law, $I = \Delta V/R$, where R is the resistance, in units of ohms. This is one version of the *Landauer* formula for one-dimensional conductivity.

Surprisingly, the intrinsic conductance is just a function of universal constants, namely the electron charge e and Planck's constant $\hbar$. This is just one example of behavior in solids that has universal properties despite the many complicated interactions of the atoms and the presence of disorder. Another related example is the current carried by electrons in a two-dimensional sheet (which could be created using a quantum well, as discussed in Section 8.1) in the presence of a strong magnetic field. The same universal conductivity $e^2/\hbar$ arises, and this is not an accident: under a strong magnetic field, the electrons move in circular orbits, and the only electrons that can conduct electricity are those that follow a one-dimensional path bouncing along the edges of the system. The universal properties of such a system are known as the *quantum Hall effect*, and the work on this topic has been recognized by several Nobel Prizes.

More generally, we can take into account that some electrons may be reflected from the entrance of the wire, and we can account for conduction in higher energy states of the electrons due to lateral confinement in the sideways directions of the wire, as discussed in Section 9.2. In that case, this formula is adjusted to

$$I = \frac{2}{\pi}\sum_i t_i \frac{e^2}{\hbar}\Delta V,\tag{9.6.6}$$

where the sum is over the quantized states that contribute to the current, and t_i is the transmission coefficient for each state, which depends on the reflection of the electron waves from the contacts to the wire. These depend on the wave properties of the electrons just like the reflection of coefficient of light at a boundary.

References

R. Laughlin, *A Different Universe: Reinventing Physics from the Bottom Down*, (Basic Books, 2005).

Comparing Classical and Quantum Systems

Because much of the philosophy of quantum mechanics was developed while the math and experiments were still being worked out, many overstatements about quantum mechanics have persisted in the literature to this day. In this chapter, we look at the modern understanding of several common effects.

10.1 Derivation of the Planck Spectrum

A standard example in textbooks on the history of quantum mechanics is the failure of the Rayleigh–Jeans radiation law, and its correction by Planck on the basis of the assumption of the existence of particles. This indeed played an important role in the thinking at the time, but the explanation in modern quantum field theory goes along completely different lines.

The failure of the Rayleigh–Jeans radiation law comes down to this, that the assumption was made in classical thermodynamics that every mode of oscillation in a system carries on average the same amount of energy, equal to $\bar{E} = k_B T$, where k_B is Boltzmann's constant (which sets the temperature scale) and T is the temperature relative to absolute zero. This has the unphysical implication that in a system with no upper limit to the number of degrees of freedom, there will be an infinite amount of energy – the so-called "ultraviolet catastrophe."

The resolution of this was Planck's distribution of the amount of energy carried by different wave states. This was

$$\bar{E} = \hbar\omega N(\hbar\omega) = \hbar\omega \frac{1}{e^{\hbar\omega/k_B T} - 1}, \tag{10.1.1}$$

where $\hbar$ is Planck's constant and ω is the oscillation frequency of a given wave mode, defined in Section 9.1. This is equivalent to saying that the average energy of a wave state is given by the energy $\hbar\omega$ times a weight factor, $N(\hbar\omega)$, which decreases exponentially as ω becomes large. In the limit of either high temperature or low frequency, this distribution just becomes the classical value $k_B T$, as can be seen by writing the exponential factor by a Taylor series:

$$\hbar\omega \frac{1}{e^{\hbar\omega/k_B T} - 1} = \hbar\omega \frac{1}{1 + \hbar\omega/k_B T + \cdots - 1}$$

$$\simeq \hbar\omega \frac{k_B T}{\hbar\omega} = k_B T. \tag{10.1.2}$$

This solves the problem of the ultraviolet catastrophe. The question is then how to justify Planck's formula.

10.1.1 Planck's Derivation in Terms of Particle Statistics

Planck derived his equation by a statistical argument, assuming that the system consisted of billiard-ball-like photons moving around randomly. The math is the following.

One writes the energy of a state with frequency ω in terms of the Boltzmann probability factor e^{-E/k_BT}, which was already known at the time of Planck and derived from classical statistical mechanics; Planck simply made the identification that the energy of a single photon is $\hbar\omega$. Allowing for the number of particles in a state to be any number N, with no upper bound, gives

$$\bar{E} = \frac{1}{Z}\sum_{N=0}^{\infty} N\hbar\omega e^{-\beta N\hbar\omega}, \tag{10.1.3}$$

where $\beta = 1/k_BT$, and Z is the *partition function*, equal to the sum over all the Boltzmann factors of the different allowed numbers of particles:

$$Z = \sum_{N=0}^{\infty} e^{-\beta N\hbar\omega}. \tag{10.1.4}$$

This is a geometric series, which can be solved using the standard math method for summing an infinite geometric series, to obtain

$$Ze^{-\beta\hbar\omega} = \sum_{N=1}^{\infty} e^{-\beta N\hbar\omega}$$
$$= Z - 1$$
$$\Rightarrow Z = \frac{1}{1 - e^{-\beta\hbar\omega}} = \frac{e^{\beta\hbar\omega}}{e^{\beta\hbar\omega} - 1}. \tag{10.1.5}$$

The sum in (10.1.3) is given by the first derivative of Z with respect to β, and so we can write

$$\bar{E} = -\frac{1}{Z}\frac{\partial Z}{\partial\beta}$$
$$= -\frac{e^{\beta\hbar\omega} - 1}{e^{\beta\hbar\omega}}\left(\frac{\hbar\omega^{\beta\hbar\omega}}{e^{\beta\hbar\omega} - 1} - e^{\beta\hbar\omega}\frac{\hbar\omega^{\beta\hbar\omega}}{(e^{\beta\hbar\omega} - 1)^2}\right)$$
$$= -\hbar\omega\left(1 - \frac{e^{\beta\hbar\omega}}{e^{\beta\hbar\omega} - 1}\right)$$
$$= \hbar\omega\frac{1}{e^{\hbar\omega/k_BT} - 1}. \tag{10.1.6}$$

This is the formula given in (10.1.1). A different calculation must be done for the equilibrium of fermions, assuming that they have a maximum in any state of $N = 1$.

10.1.2 Derivation Using Quantum Field Theory

The Planck result can also be obtained from the quantum operator properties discussed in Section 13.1.1. As proven in Section 14.1.2, the rate of transition into any state is proportional to $(1+N)$ for bosons and proportional to $(1-N)$ for fermions, where N is the number of particles in the final state. It is also proportional to the number of particles in the initial state moving toward that final state. For example, for a transition in which two particles collide and move into two different states, the rate of this transition is proportional to

$$N_1 N_2 (1 \pm N_3)(1 \pm N_4), \qquad (10.1.7)$$

where N_1 and N_2 are the numbers in the original states, N_3 and N_4 are the numbers already present in the final states, and the $+$ signs are for bosons and the $-$ signs are for fermions.

The principle of *detailed balance* states that in equilibrium, the reverse process should have the same rate. Therefore, we have

$$N_1 N_2 (1 \pm N_3)(1 \pm N_4) = N_3 N_4 (1 \pm N_1)(1 \pm N_2). \qquad (10.1.8)$$

Assuming that each number depends only on the energy of that state, then the only function $N(E)$ that satisfies this equation for all energies, assuming energy conservation $E_1 + E_2 = E_3 + E_4$, is

$$N(E) = \frac{1}{e^{\beta E + \alpha} \mp 1}, \qquad (10.1.9)$$

where the $-$ is for bosons and the $+$ is for fermions, and α and β are numbers that must be found based on the total number of particles and the temperature; doing this gives the same value $\beta = 1/k_B T$ as used in (10.1.1). This result is exactly the same as Planck's $N(E)$, when $\alpha = 0$, which is always the case for photons in equilibrium; we have the additional benefit that we have derived the equilibrium distribution here for both bosons and fermions. The detailed balance argument used here is the same that Boltzmann used to derive his Boltzmann factor, except that he did not include the quantum final-states factors, which then gave him the single solution $N(E) = e^{-(\beta E + \alpha)}$.

We have used particle language for the scattering rates in (10.1.7) and (10.1.8), but we didn't need to. The above is a specific case of the more general formalism of the *quantum Boltzmann equation*, which is entirely formulated in terms of the time dependence of wave functions and never involves any random variables at all (see Section 18.2). In that formulation, the values of N are just the squares of the amplitudes of the wave in each state (equal to $\langle \psi | a_k^\dagger)(a_k | \psi \rangle$ in field theory notation), and the transitions are just due to the nonlinear coupling of waves in different states. Without ever assuming there is a definite number of particles in any state, all of the standard results of thermodynamics can be deduced from the quantum Boltzmann equation, including Boltzmann's H-theorem, which says that systems always move toward equilibrium, and therefore proves the second law of thermodynamics.

10.2 Classical Chaos Theory

Often the distinction has been made that classical systems are deterministic, while quantum systems are indeterministic, that is, random. While some may insist on the second statement, we now know that classical systems can be, to all intents and purposes, indeterministic; that is, the outcomes of classical systems can be completely unpredictable unless one has absolutely perfect knowledge of the initial state. This is a mathematical result of the field known as *chaos theory*. (For a review, see Gleck 2011.) In principle, these systems could have been studied even in the 1800s based on classical physics, but the necessary math and the computer resources had not been developed yet.

We all have experience with things that are effectively unpredictable. For example, imagine trying to roll a marble down the edge of a knife. In principle, it is possible to have the marble roll all the way down the blade. Most likely, however, it will fall off on one side or the other. Tiny little changes in the initial placement of the marble will lead to big changes in its later behavior. Yet, in principle, a person could measure the initial state of the marble accurately enough to predict its motion. Chaos theory says that some classical systems can be unpredictable even with nearly perfect knowledge of the initial state.

One interesting consequence of chaos theory is that, if the laws of gravity are calculated for millions of years into the future, it is predicted that at least one of the planets of our solar system will eventually leave the sun and shoot off into space (see, e.g., Lithwick 2014, and references therein). When the attraction of each planet to all the other planets is taken into account, the calculations become chaotic on timescales of millions of years. Thus, the driving force in many of the early years of the Scientific Revolution, that the orbits of the heavenly spheres are absolutely certain, is found to be only an approximation based on looking at the paths of the planets over short times compared to the timescale for chaos to be important.

Anyone can create a chaotic system using a very simple mathematical model. As an example, we can create a series of x's, labeled $x_1, x_2, x_3, \ldots$. In general, x_n is the nth value in the series. This series might represent the population of different generations of rabbits, or the velocity of an object at different times, or some other physical quantity as it changes in time. Now consider the following rule for going from one x to the next:

$$x_{n+1} = cx_n(1 - x_n). \tag{10.2.1}$$

This rule is plotted in Figure 10.1. It says that there is a growth in x that is proportional to x, and also a reduction of x that is proportional to x^2. For instance, suppose x is the population of rabbits in a field. The growth of the population is proportional to the number of rabbits. If there are too many rabbits, however, then they will not have enough food, and the competition for food will tend to decrease their numbers. The number c is a constant of proportionality which says how fast these effects occur.

This series can easily be plotted using a spreadsheet program with the "fill down" feature. The recipe is the following:

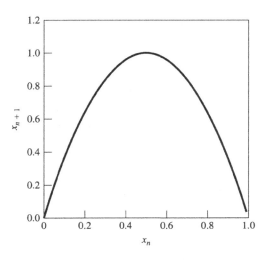

Figure 10.1 A rule for deterministic time-dependent steps that will produce chaotic behavior.

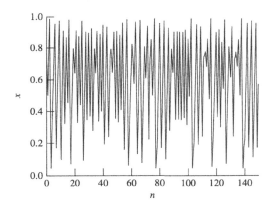

Figure 10.2 Chaos produced by a deterministic computer program (Microsoft Excel) using the rule (10.2.1) with $c = 3.95$.

- Enter a number between 0 and 1 for the first cell.

- Enter rule (10.2.1) for the second cell. (You must refer to the first cell in the formula for the second one, which in Excel notation for column A would be this: =2*A1*(1-A1).)

- Use the fill-down feature to copy this formula for the next 100 cells or so. Make sure to use relative addressing, so that each cell refers to the one just above it.

This has meaningful results for values of c between 0 and 4. For values of c less than 3, the series of x's will either converge to a single value or oscillate between two values. Both of these types of behavior are very predictable. But as c increases above 3, the behavior becomes steadily more complicated, until as $c \to 4$, the behavior of x becomes completely *chaotic*. The plot looks random, as in Figure 10.2.

To see how unpredictable this system is, try the following. For c close to 4, change the first value of x very slightly. For example, try plotting first $x_1 = 0.4$, and then $x_1 = 0.400001$. When the plot is updated (modern versions of Excel do this automatically), you will see that the pattern changes completely. Even for tiny changes in x_1, the final value, for example, x_{100}, is completely different. The whole pattern looks similar ("noisy"), but you cannot predict how the final value will change if you change the initial value.

Yet this system is completely deterministic. There is a definite rule for how each value depends on the previous value. In the same way, for a chaotic physical system, if we had perfect knowledge of the initial value, and we had a perfect computer, we could run the computer to get the exact final value. In practice, since any real measurement always has some uncertainty, we cannot have a perfectly correct initial value for any real system. Not only that, but even if we had a perfectly certain initial input, no computer is perfectly accurate. Computers only calculate to a certain number of digits (typically more than 10 digits, which seems very accurate, but it is not perfectly accurate). In the process of computing hundreds of numbers, little errors will build up in this chaotic calculation, which make the final outcome uncertain.

The proposals for spontaneous collapse in quantum mechanics, as discussed in Chapter 6, implicitly assume that something like classical chaos happens to quantum fields, leading to unpredictable behavior described by the Born rule. This cannot be generated by a unitary theory; it requires some additional, nonunitary part. At present, this nonunitary part is supplied by "measurement," which is ill-defined.

10.3 Quantum and Classical Entanglement

In Section 8.4, we discussed how quantum systems can have a superposition of a large number of distinguishable quantum states. This property is often described in terms of *entanglement*.

Consider the state

$$\Psi(x, y) = \frac{1}{2}[\psi_0(x)\psi_0(y) + \psi_0(x)\psi_1(y) + \psi_1(x)\psi_0(y) + \psi_1(x)\psi_1(y)], \qquad (10.3.1)$$

where x and y are two different spatial positions, which may be taken as the location of two different particles, or more generally as two different degrees of freedom of a physical system, and ψ_0 and ψ_1 are two different functions. This may at first look like a superposition of four quantum states. But it can be factorized into

$$\Psi(x, y) = \frac{1}{2}[\psi_0(x) + \psi_1(x)][\psi_0(y) + \psi_1(y)]. \qquad (10.3.2)$$

We are then free to define new quantum states for each degree of freedom,

$$\psi_0'(x) = \frac{1}{\sqrt{2}}[\psi_0(x) + \psi_1(x)]$$

$$\psi_0'(y) = \frac{1}{\sqrt{2}}[\psi_0(y) + \psi_1(y)], \qquad (10.3.3)$$

in which case the state (10.3.1) becomes

$$\Psi(x,y) = \psi_0'(x)\psi_0'(y). \tag{10.3.4}$$

This state is the simple product of two functions of x and y and does not need to be written as a superposition.

On the other hand, consider the state

$$\Psi(x,y) = \frac{1}{\sqrt{2}}[\psi_0(x)\psi_1(y) + \psi_1(x)\psi_0(y)]. \tag{10.3.5}$$

This state is not factorizable into a product of two states that depend separately on just x and y. The degrees of freedom of x and y are therefore *entangled*.

Generalized Born rule for entangled states. Physically, these two states have very different characteristics when we do a measurement – the entangled state implies the possibility of nonlocal correlations.

To see this, we write the Born rule in its most general form. If the system is in a state $\Psi(x)$, we can *project* it onto some other state $\psi_n(x)$ by doing the overlap integral over all x,

$$P_n = \int dx\ \psi_n^*(x)\Psi(x). \tag{10.3.6}$$

For some states $\psi_n(x)$, this integral may be zero, which means that the state $\Psi(x)$ does not contain any part that is the same as $\psi_n(x)$. The prescription of quantum measurement using the Born rule is that the probability of detecting the system in state ψ_n is given by the square of this number, and after such a measurement, the wave function $\Psi(x)$ of the system is replaced by $\psi_n(x)$. It is assumed that each wave function is *normalized*, so that

$$\int dx\ \psi_n^*(x)\psi_n(x) = 1. \tag{10.3.7}$$

This means that, if the initial state is ψ_n, there is 100% probability of finding it in that state by a measurement, as one would expect.

If the measurement is done by a small detector at location x_0, then the final state of interest is highly localized around position x_0. We can approximate this mathematically as the *Dirac delta function*, $\delta(x - x_0)$, which is peaked sharply at $x = x_0$. This function has the property that

$$\int dx\ \delta(x - x_0)\Psi(x) = \Psi(x_0). \tag{10.3.8}$$

The probability of a measurement at the location x_0 is therefore given by the square of the projection, which is $|\Psi(x_0)|^2$, in agreement with the simple version of the Born rule given in Section 3.4.

If the state of the system has two or more degrees of freedom, and a measurement is done for a state dependent only on x, the prescription is to project the full wave function onto just the x degree of freedom and leave the remaining part of the wave function alone. Thus, for the entangled state (10.3.5), a projection onto the state ψ_0 corresponds to doing the following integral:

$$P_n(y) = \int dx \; \psi_0^*(x)\Psi(x,y)$$

$$= \int dx \; \psi_0^*(x)\frac{1}{\sqrt{2}}[\psi_0(x)\psi_1(y) + \psi_1(x)\psi_0(y)]$$

$$= \frac{1}{\sqrt{2}}\left(\int dx \; \psi_0^*(x)\psi_0(x)\right)\psi_1(y) + \frac{1}{\sqrt{2}}\left(\int dx \; \psi_0^*(x)\psi_1(x)\right)\psi_0(y).$$

$$(10.3.9)$$

If we assume that ψ_0 and ψ_1 represent two nonoverlapping states, for example, two detectors at different locations, then the second integral in parentheses is equal to zero. Because the wave function ψ_0 is normalized, the first integral is equal to 1. This means that the final state is $\psi_0(x)\psi_1(y)$, according to the prescription of the Born rule. Even though we have only measured the wave function ψ_0 at position x, we have automatically also forced the system into a definite state for y. If the detector for y is very far away, it doesn't matter – the entire wave function responds to what happens with the x measurement.

Classical entanglement. Sometimes statements are made in the literature that no classical system can have a wave function of the form (10.3.5). This is not true – Chapter 17 gives an example of a classical wave system (an optical resonator made of two mirrors), which can have states described by an electromagnetic wave function with exactly this form. However, there are key differences. One is that the number of degrees of freedom that can be entangled in a classical system is small, limited by the number of dimensions of free space, while in quantum systems, there is essentially an infinite number of possible degrees of freedom that can become entangled. Second, systems with classical entanglement do not have nonlocality. The different degrees of freedom that are entangled occur at the same location in space.

References

J. Gleck, *Chaos: Making a New Science*, (Open Road Media, 2011).

Y. Lithwick and Y. Wu, "Secular chaos and its application to Mercury, hot Jupiters, and the organization of planetary systems," *Proceedings of the National Academy of Sciences* (USA) **111**, 12610 (2014).

Part III

A Short Course in Quantum Field Theory

11

Preliminary Mathematics

As we have discussed throughout this book, quantum field theory is the canonical theory of quantum mechanics. Yet many philosophers and scientists are untutored in it. This is often because quantum field theory is not taught until later years of graduate school, and even in that case, it is often embedded in a larger program of relativistic particle physics.

Quantum field theory does require some effort, but not as much as one might think. It follows naturally from the undergraduate quantum mechanics many students learn, and its elements could be taught at much lower levels. Many of the methods of quantum field theory are used in the theory of solids and fluids, without the need for relativity.[1]

This short introduction gives the main concepts of quantum field theory. Much of the discussion will involve phonons, that is, quanta of sound. While some philosophers treat phonons as somehow less ontologically real than other particles such as electrons and photons, there is no reason to do this; the prescription of quantum field theory is that any field can be quantized, and the particles that arise from this quantization, whether phonons, electrons, or quarks, are treated the same way in every field.

11.1 Dirac Wave Notation

As discussed in Chapter 1, waves in fields are described by continuous functions that ascribe some number or set of numbers to each point in space. A continuous function can be written as $f(x)$, which means that a number x for the location in space is fed into a calculation, and another number, f, is obtained as the result of this calculation, whatever it may be. In the context of the quantum matter field, one typically writes $\psi(x)$ for the *wave function*, which has the same form, giving the complex value, with real and imaginary components, of the matter wave at any location in space x.

In advanced quantum mechanics, another approach is used, known as Dirac's "bra–ket" notation (because it writes wave functions in brackets). This notation can be understood in terms of linear algebra. We write any vector as a *ket*, which is defined as a vector arranged vertically. In standard vector notation, we would write $\vec{v}$, while in bra–ket notation, we write

---

[1] The field theory methods of this book are a simplified version of the introduction to field theory presented in Snoke 2020. For greater depth, see Fetter and Walecka 1971.

$$|v\rangle = \begin{pmatrix} v_1 \\ v_2 \\ \cdot \\ \cdot \\ \cdot \\ v_n \end{pmatrix}, \tag{11.1.1}$$

where n is the number of vector components. A *bra* is the same vector transposed, and with the complex conjugate taken:

$$\langle v| = \begin{pmatrix} v_1^*, & v_2^*, & \ldots, & v_n^* \end{pmatrix}. \tag{11.1.2}$$

The *inner product* of two vectors is defined as the product of a bra and a ket,

$$\begin{aligned} \langle v|u\rangle &= \begin{pmatrix} v_1^*, & v_2^*, & \ldots, & v_n^* \end{pmatrix} \begin{pmatrix} u_1 \\ u_2 \\ \cdot \\ \cdot \\ \cdot \\ u_n \end{pmatrix} \\ &= v_1^* u_1 + v_2^* u_2 + \cdots + v_n^* u_n. \end{aligned} \tag{11.1.3}$$

For this to make sense, the two vectors must have the same dimension n, or in other words, span the same number of dimensions. We see that any term in a bracket $\langle \ldots \rangle$ (i.e., a "bra–ket") is a simple number.

The *outer product* of the same two vectors is formed by writing a bra and a ket in reverse order. When we use a ket and its corresponding bra, we obtain

$$\begin{aligned} |v\rangle\langle v| &= \begin{pmatrix} v_1 \\ v_2 \\ \cdot \\ \cdot \\ \cdot \\ v_n \end{pmatrix} \begin{pmatrix} v_1^*, & v_2^*, & \ldots, & v_n^* \end{pmatrix} \\ &= \begin{pmatrix} v_1^* v_1 & v_2^* v_1 & \cdot & \cdot & \cdot & v_n^* v_1 \\ v_1^* v_2 & v_2^* v_2 & & & & v_n^* v_2 \\ \cdot & & \cdot & & & \cdot \\ \cdot & & & \cdot & & \cdot \\ \cdot & & & & \cdot & \cdot \\ v_1^* v_n & v_2^* v_n & \cdot & \cdot & \cdot & v_n^* v_n \end{pmatrix}. \end{aligned} \tag{11.1.4}$$

This forms a square matrix. Any square matrix is known as an *operator* because when it multiplies a vector, it transforms that vector into another vector with the same number of components. The matrix in (11.1.4) is known as the *projection operator* because it projects one vector onto another vector, giving a result that has length equal to the inner product of the two vectors, and that points in the direction of $|v\rangle$. This is written succinctly in bra–ket notation as

$$P_v|u\rangle \equiv \big(|v\rangle\langle v|\big)|u\rangle = |v\rangle\langle v|u\rangle. \tag{11.1.5}$$

So far, we have looked at vectors with a known number of rows and columns. We can go further to treat any continuous function $f(x)$ as a vector with an infinite number of components. This is equivalent to writing a vector

$$|f\rangle = \begin{pmatrix} f(x_1) \\ f(x_2) = f(x_1 + dx) \\ f(x_3) = f(x_1 + 2dx) \\ . \\ . \\ . \\ f(x_n) \end{pmatrix}. \tag{11.1.6}$$

Note that the dimensions we are talking about here do *not* correspond to any real dimensions in space. They are simply a mathematical convenience to represent the infinite number of points in a continuous function, which allows us to use vector concepts.

The inner product in the continuous case will be an integral instead of a simple sum. The bra in this case must be adjusted by multiplying by dx:

$$\langle f| = \big(f^*(x_1), \quad f^*(x_2), \quad f^*(x_3), \quad \ldots, \quad f^*(x_n) \big)dx. \tag{11.1.7}$$

The inner product for any functions $f(x)$ and $g(x)$ is then

$$\langle f|g\rangle = \lim_{dx \to 0} \sum_i f^*(x_i)g(x_i)dx$$

$$= \int f^*(x)g(x)dx, \tag{11.1.8}$$

which is the standard definition of the inner product of two continuous functions in Schrödinger wave mechanics.

11.2 General Properties of Operators

As discussed in Section 11.1, any square matrix is an operator because it transforms a vector into another vector of the same length. Operators acting on continuous functions can still be thought of as square matrices that act on one vector and give another vector with the same number of components. Any operation that transforms one function into another function is therefore a type of operator. For example, a derivative transforms the value of a function at every point to a different value at the same point, and therefore is a type of operator.

The *adjoint* of an operator is defined as the matrix that would have the same effect acting to the left on a bra that the original operator has acting to the right on a ket. Thus, for example, if $A|v\rangle = |u\rangle$, then the adjoint of A is written as $A^\dagger$, and has the property that

$$\langle u| = \langle v|A^\dagger. \tag{11.2.1}$$

An *eigenvector* of a square matrix is defined as any vector that has the property that multiplication by the matrix is the same as multiplying by a simple number, that is, $|\psi\rangle$ is an eigenvector of the square matrix A if and only if

$$A|v\rangle = a|v\rangle, \tag{11.2.2}$$

where a is a simple number. The number of eigenvectors of a square matrix equals its dimension, which is equal to the number of components of the vectors it operates on.

In general, real-world measurements are represented by operators that have eigenvalues that are real numbers (in other words, not complex numbers). Mathematically, this is the same as saying that the operator is *Hermitian*, which means that $A = A^\dagger$. Any non-Hermitian operator can be turned into a Hermitian operator, however, by adding it to its adjoint, since the adjoint of the adjoint is the original matrix:

$$(A + A^\dagger)^\dagger = A^\dagger + A. \tag{11.2.3}$$

Projection onto complete sets of states. Both the discrete vector version of states and the continuous function version can be expressed in terms of the very general mathematical concept known as a *complete set of states*. A set of states $\{|n_1\rangle, |n_2\rangle, \ldots\}$ is complete if, for every state $|s\rangle$ in the physical system of interest, there is at least one nonzero inner product $\langle s|n_i\rangle$.

A complete set of states is particularly useful if all the states in the set are *orthonormal*, which means that $\langle n_j|n_i\rangle = \delta_{ij}$. Here, δ_{ij} is the *Kronecker delta* function, which is equal to 0 if $i \neq j$ and equal to 1 if $i = j$. With such a set, we can write

$$\sum_j |n_j\rangle\langle n_j|n_i\rangle = \sum_j \delta_{ij}|n_j\rangle = |n_i\rangle, \tag{11.2.4}$$

which implies

$$\sum_j |n_j\rangle\langle n_j| = 1 \tag{11.2.5}$$

for any orthonormal complete set of states. This means that we can insert the sum $\sum_j |n_j\rangle\langle n_j|$ anywhere in an equation, whenever we wish.

This implies the useful result that any state of the system can be written in terms of an orthonormal complete set of states according to the rule

$$|s\rangle = \sum_j |n_i\rangle\langle n_j|s\rangle, \tag{11.2.6}$$

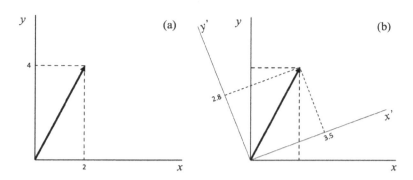

Figure 11.1 (a) A vector seen as its projection onto two axes. (b) The same vector in two different coordinate systems.

where $\langle n_j | s \rangle$ is a complex number that gives the inner product of $|s\rangle$ with each state. The states $|n_j\rangle$ are often called *basis states* of the system.

This result is obvious in the case when the states $|s\rangle$ are simple vectors. Consider a vector like that shown in Figure 11.1(a). The complete set of orthonormal basis vectors in this case can be written simply as unit vectors along the axes $|x\rangle$, $|y\rangle$, and $|z\rangle$, which we can label in shorthand as $|x_1\rangle$, $|x_2\rangle$, and $|x_3\rangle$. To write the two-dimensional vector shown in Figure 11.1(a) in terms of these axes, we expand it using the rule given in (11.2.14),

$$|v\rangle = \left(\sum_i |x_i\rangle \langle x_i| \right) |v\rangle \tag{11.2.7}$$
$$= |x_1\rangle \langle x_1 | v\rangle + |x_2\rangle \langle x_2 | v\rangle.$$

The term $\langle x_i | v \rangle$ is the "projection" of the vector onto the axis vector $|x_i\rangle$, which we can think of quite literally as the length of the shadow formed by projecting light across the vector $|v\rangle$ onto that axis. For the axes shown, the projections onto the two axes have lengths 2 and 4, respectively, which means that we can write the vector compactly as $|v\rangle = (2, 4)$.

The benefit of Dirac notation. This shows us why Dirac notation is useful, instead of just using standard vector or function notation. Dirac notation allows us to consider a state as a "thing," independent of any particular choice of coordinate system.

Suppose we adopted a different coordinate system, as shown in Figure 11.1(b), which we are always free to do, since coordinate systems are not printed on the natural world; they are chosen by humans. This will give us new basis vectors $|x_1'\rangle$ and x_2'. In the new coordinate system, projection onto these new basis vectors will tell us to write the vector as $(3.5, 2.8)$.

But the vector itself has not changed; it is a "thing" that we have simply described in two different ways. In terms of Dirac notation, we can define $|v\rangle$ as the "thing," and write it in terms of its projections on various different axis systems as we choose. The Dirac notation therefore allows us to treat each "ket" as a physical thing independent of any choice of mathematical representation.

Complete sets for continuous functions. The exact same approach can be used for continuous functions, which, as we have seen, can be thought of as vectors with an infinite number of components. In this case, instead of a sum of discrete vectors, we have an integral of the same form as (11.2.5),

$$\int dx\, |x\rangle \langle x| = 1, \tag{11.2.8}$$

where each ket $|x\rangle$ corresponds to a different point in the space that the function occupies. (The integration is assumed to be over all of the space that the wave function occupies.) Any function can then be written as

$$|f\rangle = \int dx\, |x\rangle \langle x|f\rangle. \tag{11.2.9}$$

The inner product $\langle x|f\rangle$ gives the projection of the function onto each point in space. This is typically what we write as the value of the function at each point, $f(x)$. If we set $|f\rangle = |x'\rangle$, we obtain

$$
\begin{aligned}
|x'\rangle &= \int dx\, |x\rangle \langle x|x'\rangle \\
&= \left(\int dx\, \langle x|x'\rangle \right) |x\rangle, \tag{11.2.10}
\end{aligned}
$$

which implies $\langle x|x'\rangle = \delta(x-x')$, where $\delta(x)$ is the Dirac delta function, equal to 0 if $x \neq x'$, and with a total integrated *area* equal to 1.

Writing functions this way allows many powerful mathematical methods, in particular Fourier transforms, discussed at length in Section 11.5. The Fourier transform method, as well as many other transform methods, is based on the fact that just as we can change the basis states for vectors, as seen in Figure 11.1, we can also change the basis states of continuous functions to any other orthonormal complete set of states. If we can find any such complete set of states, we can rewrite any function in terms of those states.

Let us pick the set of states $|k\rangle$ defined by

$$\langle x|k\rangle = \frac{1}{\sqrt{2\pi}} e^{ikx} \tag{11.2.11}$$

for all possible values of k. The analogous orthonormality condition to (11.2.10) is

$$\langle k|k'\rangle = \delta(k - k'). \tag{11.2.12}$$

This is consistent with the result we obtain using the rule (11.2.8),

$$\int dx\, \langle k|x\rangle \langle x|k'\rangle = \frac{1}{2\pi} \int dx\, e^{i(k'-k)x} = \delta(k - k'), \tag{11.2.13}$$

where on the right side, we have used a mathematical formula for the Dirac delta function. It is easy to see that, for $k \neq k'$, the integrand will give a sum over many oscillating functions that cancel out to a total of 0, and the result will be nonzero only if $k = k'$.

This then implies that, if $k = k'$, the integral over k is given by

$$\int dk \int dx\, \langle k|x\rangle \langle x|k\rangle = \int dx \int dk\, \langle x|k\rangle \langle k|x\rangle = 1. \tag{11.2.14}$$

Since the integrals can be done in any order, we must then have a projection relation analogous to (11.2.8),

$$\int dk \, |k\rangle\langle k| = 1. \qquad (11.2.15)$$

The set of states $|k\rangle$ can therefore be used as a set of basis functions in just the same way as the $|x\rangle$ states. We can write, for any function $|f\rangle$,

$$\langle x|f\rangle = \int dk \, \langle x|k\rangle\langle k|f\rangle. \qquad (11.2.16)$$

The function $\langle k|f\rangle$ is only dependent on k and is known as the *Fourier transform* of f. Since $|f\rangle$ is a "thing" independent of which basis functions we use, $\langle k|f\rangle$ and $\langle x|f\rangle$ are just two different ways of writing down the same information.

11.3 Operators and Measurements

So far, we have defined operators simply as mathematical actions that can be done to states (which could be either vectors or continuous functions) that produce some new state. This is a very general concept, which can include the action of derivatives on functions, or simple queries that do not change the state, but just return the same state times some number that is derived from that state.

For each physical, measurable property of a system, we can define an operator that takes the information of the state of the system and gives us a number for that property. As we have seen, every operator corresponds to a square matrix with eigenstates. Quantum mechanics says that, if a system is in an eigenstate of some operator that measures a physical property, then when a measurement of that property is made, the state of the system will not change; it will just be multiplied by the value of the property of interest. The value of the property returned is the eigenvalue of that eigenstate.

If the system is in a superposition of more than one eigenstate of an operator, the action of that operator will be to return each eigenstate multiplied by its own eigenvalue. In most formulations of quantum mechanics, notably the Copenhagen interpretation discussed in Chapter 4, an additional step is taken to truncate the resulting state to just one of the eigenstates in the superposition. This action is not done by the operator representing the measured property but by an additional projection operator that is applied randomly, with a probability given by the Born rule. In mathematical language, the Born rule is that the probability of ending up in a state after a measurement is proportional to the square of the inner product of the original wave function with that state.

Momentum and position operators. As with any measurable property, momentum p and position x have corresponding operators acting on the wave functions. We will generally signify operators by "hats," so that the operators corresponding to position and momentum are $\hat{x}$ and $\hat{p}$, respectively. Sometimes the hats are left off when it is obvious that the term is an operator.

When the wave function is represented as a function of x, that is, $\psi(x)$, the $\hat{x}$ operator has the action of simply multiplying the wave function by the number x. As discussed in Section 9.1, for the same wave function $\psi(x)$, the momentum operator corresponds to a derivative acting on the matter wave function,

$$\hat{p} \equiv -i\hbar \frac{\partial}{\partial x}, \tag{11.3.1}$$

where $\hbar$ is Planck's constant.

In operator language, we say that the value x is the eigenvalue corresponding to the state $|x\rangle$, which is an eigenvalue of the operator $\hat{x}$. The state $|x\rangle$ is not an eigenstate of the operator $\hat{p}$. But as discussed in Section 11.2, we don't have to stay in the basis of the states $|x\rangle$. We could instead switch to the basis composed of the eigenstates of the $\hat{p}$ operator. It will come as no surprise that these are the same as the $|k\rangle$ states we introduced in Section 11.2. We can see this in the following calculation.

We start by assuming that we don't know what the eigenstates of $\hat{p}$ are; we just know that they have eigenvalues equal to $p = \hbar k$. Using the projection relation (11.2.15), we write

$$\langle x|\hat{p}|k\rangle = \int dk' \, \langle x|\hat{p}|k'\rangle \langle k'|k\rangle$$

$$= \int dk' \, \hbar k' \langle x|k'\rangle \delta(k - k')$$

$$= \hbar k \langle x|k\rangle$$

$$= (\hbar k)\frac{1}{\sqrt{2\pi}}e^{ikx}, \tag{11.3.2}$$

where, in the last line, we have used our knowledge of $\langle x|k\rangle$ from Section 11.2. We now observe that the derivative of the exponential function just gives us multiplication by ik, which means we can write

$$\langle x|\hat{p}|k\rangle = -i\hbar \frac{\partial}{\partial x}\langle x|k\rangle, \tag{11.3.3}$$

thus showing that the $\hat{p}$ operator has the expected action.

Commutation relations. In general, operators do not always *commute*; that is, it is not always the case that $AB = BA$ for two operators A and B, When operators are used in a different order, the final result can be different. We write the *commutator* $[A, B] = AB - BA$. If the commutator is zero, then the order of the operators doesn't matter.

The definition of the $\hat{p}$ operator implies the commutation relation

$$[x, p] \equiv xp - px = i\hbar. \tag{11.3.4}$$

This can be seen by applying this term to any function $f(x)$, and applying the rule for the derivative of a product to the first term:

$$[x, -i\frac{\partial}{\partial x}]f(x) = -ix\frac{\partial}{\partial x}f(x) - \left(-ix\frac{\partial}{\partial x}f(x) - if(x)\right)$$

$$= if(x). \tag{11.3.5}$$

This commutation relation will be used to deduce an uncertainty relation in Section 11.5.

A similar commutation relation exists for time and energy measurements. As discussed in Section 9.1, in the Schrödinger equation, we equate

$$E = i\hbar \frac{\partial}{\partial t}. \tag{11.3.6}$$

By the same argument used in (11.3.29), there will be a commutation relation between the energy and time operators. We apply this term to any function $f(t)$, and applying the rule for the derivative of a product:

$$[t, i\hbar \frac{\partial}{\partial t}]f(t) = i\hbar t \frac{\partial}{\partial t}f(t) - \left(i\hbar t \frac{\partial}{\partial t}f(t) + i\hbar f(t)\right)$$
$$= -i\hbar f(t). \tag{11.3.7}$$

This implies the commutation relation $[E, t] = i\hbar$.

11.4 The Schrödinger Equation

In bra–ket notation, the Schrödinger equation is written as

$$i\hbar \frac{\partial}{\partial t}|\psi\rangle = H|\psi\rangle, \tag{11.4.1}$$

where H is the Hamiltonian operator, which we view as a square matrix acting on the vector $|\psi\rangle$ (which can have an infinite number of dimensions, to account for continuous wave functions). The Hamiltonian operator acts on the wave function and gives its total energy. In other words, the eigenvalues of H are

$$H|\psi\rangle = E|\psi\rangle, \tag{11.4.2}$$

where E gives the energy (e.g., in units of Joules).

If an eigenstate is used in the Schrödinger equation, we then have

$$i\hbar \frac{\partial}{\partial t}|\psi\rangle = E|\psi\rangle, \tag{11.4.3}$$

which is equivalent to the equation in function form,

$$i\hbar \frac{\partial}{\partial t}\psi(x, t) = E\psi(x, t), \tag{11.4.4}$$

which has the solution

$$\psi(x, t) = \psi(x, 0)e^{-iEt/\hbar}. \tag{11.4.5}$$

The eigenstates of the Hamiltonian operator correspond to wave states that keep the same amplitude as a function of x but rotate in the complex plane over time. This property ensures that the evolution of the wave function is unitary. For the wave solutions used in Section 9.1, we can equate $E = \hbar\omega$.

It can be proved that the set of all the eigenstates of the Hamiltonian make up a complete set of states, which, as we saw in Section 11.2, means any physical state of the system can be written as a sum of those eigenstates, each weighted by some numerical factor.

Separable Hilbert spaces. Suppose that the Hamiltonian of a system can be written as a sum of two separate terms,

$$H = H_1 + H_2, \tag{11.4.6}$$

each of which has its own set of eigenstates. It is then easy to show that the eigenstates of the total Hamiltonian are products of the eigenstates of the two individual terms:

$$H|\psi_1\rangle|\psi_2\rangle = (H_1|\psi_1\rangle)|\psi_2\rangle + |\psi_1\rangle(H_2|\psi_2\rangle)$$
$$= (E_1 + E_2)|\psi_1\rangle|\psi_2\rangle. \tag{11.4.7}$$

Since H_1 acts only on states composed of eigenstates of H_1, and H_2 acts only on its own eigenstates, we can view the eigenstates of these different operators as existing in different "spaces." Again, for the case of simple vectors, this is easy to visualize. Suppose that H_1 depends only on x, and H_2 depends only on y. In this case, we can write the eigenstates of the system as product functions $\psi_1(x)\psi_2(y)$. The variables x and y exist in orthogonal dimensions, measured by the x- and y-axes. In mathematical language, we say that the x and y states exist in different *Hilbert spaces*. What happens in one does not affect what happens in the other.

The same concept can be used for Hamiltonians that depend on different types of fields. For example, suppose that H_1 represents the energy of the electromagnetic field, quantized into photons, while H_2 represents the energy of a fermionic electron field. We could apply the same approach to say that the energies of these fields exist in different Hilbert spaces.

Now suppose that there is an additional term that depends on both x and y, that is,

$$H = H_1(x) + H_2(y) + \hat{V}(x,y). \tag{11.4.8}$$

In this case, we cannot write the eigenstates as pure products of states in different Hilbert spaces, because the additional term mixes them together. However, it is often useful, if $\hat{V}$ is weak enough, to write the states of the system in terms of the eigenstates of just H_1 and H_2. Since $\hat{V}$ does not introduce any new degrees of freedom, we know that the eigenstates of H_1 and H_2 are an adequate complete set of states that can be used to write any states of the whole system. We can think of $\hat{V}$ as an "interaction" term that couples the eigenstates of the original Hamiltonian to each other.

Interaction representation. It is often useful to rewrite the Schrödinger equation in a different form to account for interactions. Suppose that the Hamiltonian is

$$H = H_0 + \hat{V}, \tag{11.4.9}$$

where $H_0 = H_1 + H_2$, which, for example, could be the energy of just photons and electrons separately, and $\hat{V}$ is a term for the energy associated with interactions between the photons and electrons.

Let the state of the system at time $t = 0$ be $|\psi_0\rangle$. At a later time t, the state is $|\psi_t\rangle$. The Schrödinger equation gives the time evolution of the system as

$$i\hbar \frac{\partial}{\partial t} |\psi_t\rangle = (H_0 + \hat{V})|\psi_t\rangle. \qquad (11.4.10)$$

In the *interaction representation*, we define a new state $|\psi(t)\rangle$ (written with the t in parentheses rather than subscript), given by

$$|\psi(t)\rangle = e^{iH_0t/\hbar}|\psi_t\rangle, \qquad (11.4.11)$$

and a new, time-dependent operator,

$$\hat{V}(t) = e^{iH_0t/\hbar}\hat{V}e^{-iH_0t/\hbar}. \qquad (11.4.12)$$

In this representation, the Schrödinger equation is rewritten as

$$i\hbar \frac{\partial}{\partial t}|\psi(t)\rangle = \hat{V}(t)|\psi(t)\rangle, \qquad (11.4.13)$$

which has the advantage of not depending on H_0.

This represents the physical reality that in many systems, there is a very high-frequency phase rotation that depends on the energies of H_1 and H_2 (e.g., the individual energies of electrons and photons), and in addition, a much lower-frequency evolution that gives the interactions between the particles. The high-frequency phase rotation is effectively subtracted off by moving to the interaction representation, so that we can focus on the interaction dynamics alone.

Note that the operator $\hat{V}(t)$ now has an explicit time dependence. As discussed in Section 12.2.3, it is also possible to give operators an explicit spatial dependence, and to write down an equation that looks like a Schrödinger equation for the evolution of these operators in space and time. As discussed in Section 12.2.3, operators by themselves hold no information content about the physical state of the system; the information about the physical state of the system is held in the wave function. Operators, even ones that evolve in time and space, are only prescriptions for actions to be done to the wave function.

11.5 The Uncertainty Principle

The uncertainty principle in quantum mechanics is often presented as a deep mystery. However, as we will see, it is a straightforward deduction of the math of continuous functions. It is only when we try to apply it to particles with definite, localized properties that it has strange implications.

11.5.1 Fourier Analysis

We have already justified the Fourier transform theorem in Section 11.2 based on the mathematics of complete sets of states. This theorem says that any function $f(x)$ can be written

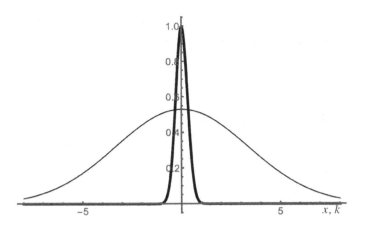

Figure 11.2   A Gaussian function (heavy line) and its Fourier transform (thin line), which is also a Gaussian, with width inversely proportional to the original Gaussian peak width.

as an integral of the oscillating function e^{ikx} times another function $F(k)$, known as the *Fourier transform* of $f(x)$, as follows:

$$f(x) = \frac{1}{2\pi} \int F(k)e^{-ikx}dk. \tag{11.5.1}$$

The way to think of this is to view the integral as a sum, that is, a superposition, of many waves of the form e^{ikx}, all with different values of k, each weighted by a factor $F(k)$, giving how much of each wave to include in the sum.

The function $F(k)$ can be proved to equal

$$F(k) = \int_{-\infty}^{\infty} f(x)e^{ikx}dx. \tag{11.5.2}$$

Since $F(k)$ is computed from $f(x)$, and $f(x)$ is computed from $F(k)$, both functions are just two different ways of representing the same information. Physicists talk of functions $f(x)$ giving the *real-space* description of a wave function, and $F(k)$ giving the *k-space* description of the same wave function (also sometimes called *reciprocal space*, or *momentum space*).

One particular choice for $f(x)$ is known as a *Gaussian wave packet*, which has a peak in one place, as illustrated in Figure 11.2. This function and its Fourier transform are given by

$$f(x) = e^{-(x-x_0)^2/2\sigma^2}$$
$$F(k) = \sqrt{\pi\sigma^2}e^{-\sigma^2k^2/2}e^{-ikx_0}. \tag{11.5.3}$$

Note that σ^2, which gives the width of the peak, appears in the denominator of the exponent in the original function $f(x)$ but appears in the numerator of the exponent in the Fourier transform. This means that a narrow peak in real space corresponds to a wide peak in k-space, and vice versa.

So far, all of this has been pure math. The quantum physics comes in when we identify the *momentum* of a wave with its first derivative, according to the definition

$$p = -i\hbar\frac{\partial}{\partial x}. \tag{11.5.4}$$

The oscillating functions used in the Fourier transform are eigenstates of this operator (defined in Section 11.3):

$$-i\hbar\frac{\partial}{\partial x}(e^{ikx}) = \hbar k(e^{ikx}), \tag{11.5.5}$$

with eigenvalue $\hbar k$, which we identify as the momentum.

This allows us to express the uncertainty principle in terms of waves. The width of the Gaussian peak in real space, known as the standard deviation, is $\Delta x = \sigma$, and the width of the k-space peak is the variance of that peak. The product of the two widths for the Gaussian wave packet is given by

$$\Delta x \Delta k = (\sigma)(1/\sigma) = 1. \tag{11.5.6}$$

This is a pure wave mathematics result, which is quite general: the narrower a function is in real space, the broader it will be in k-space. If we equate momentum as $\hbar k$, then the above relation becomes

$$\Delta x \Delta p = \hbar. \tag{11.5.7}$$

The term "uncertainty" comes from the Born rule, which treats a quantum wave function $\psi(x)$ as giving the probability $P(x)$ of measuring a particle at location x, according to the rule

$$P(x) \propto |\psi(x)|^2. \tag{11.5.8}$$

That means that we interpret the width of the peak in real space as the range of typical measurements of x, and the range of the peak in k-space as giving the range of possible measurements of p. The Fourier result means that the narrower the peak is in real space, the greater the range of possible measurements of p will be. There is a tradeoff between the two uncertainties. As we have seen, this follows simply from the properties of any wave, plus the assignment of the momentum p to be proportional to the first-derivative operator.

Uncertainty principle in diffraction. As discussed in Section 3.5, the uncertainty principle for waves can be seen directly in the observation of *diffraction*, in which a constriction of the size of a wave source leads to spreading out of the wave as it propagates away from the source. When the wave gets very far from the source, its behavior at a point $\vec{r}$ is described by the *Fraunhofer diffraction* formula (see, e.g., Hecht 2015),

$$\psi(\vec{r}) = \int_S d^2x \, e^{i(\vec{k}\cdot(\vec{r}-\vec{x})-\omega t)}, \tag{11.5.9}$$

in which the integration is over all emitting points $\vec{x}$ on the source. This can be seen as the sum of all the plane waves propagating from the source.

More generally, instead of assuming that the source has constant intensity, we can write down a source function $f(\vec{x})$, which gives the amplitude of the wave emitted from the source at each point. The Fraunhofer diffraction integral is then

$$\psi(\vec{r}) = e^{i(\vec{k}\cdot\vec{r}-\omega t)} \int d^2x\, f(\vec{x})e^{-i\vec{k}\cdot\vec{x}}. \tag{11.5.10}$$

This is exactly proportional to the two-dimensional Fourier transform of the function $f(\vec{x})$. The physical process of diffraction therefore automatically performs Fourier transforms.

Time-energy uncertainty. The results of this section follow from the commutation relation of the operators x and p, as shown in Section 11.5.2. It is therefore common to refer to a commutation relation as an uncertainty relation, or an "incompatibility" relation.

When operators do not commute, this implies a tradeoff relationship in measurements that those operators represent. Just as a wave that is squeezed to a small region will have a wide range of wavelengths, and therefore large uncertainty in momentum, in the same way, a wave that has a very short time duration will have a large range of frequencies, and therefore large uncertainty in energy.

As discussed in Section 11.3, a similar relationship exists for energy and time. This means that the total energy of a system is not always strictly conserved. In the long run, it will be, but on short timescales, a system can pass through a temporary state in which energy conservation is violated to an arbitrarily high degree. One well-known effect of this is the energy of vacuum, discussed in Section 15.3, which can be viewed as arising from the interactions of extremely short-lived particles that can appear and disappear.

11.5.2 Derivation of the Uncertainty Relationship

The uncertainty relationship of the previous section applies not only to the special case of a Gaussian wave packet but also to any wave function. The following analysis shows that a Gaussian wave packet has the minimum value of the product $\Delta x \Delta p$ of any function.

The mean measured value of x is the weighted average

$$\langle x \rangle = \alpha \int dx\, x|\psi(x)|^2, \tag{11.5.11}$$

where α is the *normalization* factor, giving the total probability in all locations, determined by

$$\alpha^{-1} = \int dx|\psi(x)|^2. \tag{11.5.12}$$

If the wave represents just one particle, then the total probability will be $\alpha = 1$.

The *uncertainty* in x is the average distance it deviates from its mean value in any given measurement. To quantify this, we define the average of the *root-mean-squared* deviation of x from its average value $\bar{x}$:

$$\Delta x = \sqrt{(x - x_0)^2}. \tag{11.5.13}$$

To find this, we can do the quantum mechanical average of $|\Delta x|^2$, and afterwards take the square root. This gives us

$$\langle|\Delta x|^2\rangle = \alpha \int dx\,(x - \langle x\rangle)^2|\psi(x)|^2$$

$$= \alpha \int dx\,(x^2 - 2x\langle x\rangle + \langle x\rangle^2)|\psi(x)|^2$$

$$= \alpha \left[\int dx\,x^2|\psi(x)|^2 - 2\langle x\rangle \int dx\,x|\psi(x)|^2 + \langle x\rangle^2\right]$$

$$= \langle x^2\rangle - \langle x\rangle^2. \tag{11.5.14}$$

Since we are free to set $x = 0$ at any point, we can define $\langle x\rangle = 0$, so that we have simply

$$\langle|\Delta x|^2\rangle = \langle x^2\rangle. \tag{11.5.15}$$

The same analysis applies to the p operator. The product of the uncertainties of x and p can therefore be written as

$$\langle|\Delta x|^2\rangle\langle|\Delta p|^2\rangle = \langle x^2\rangle\langle p^2\rangle. \tag{11.5.16}$$

The Cauchy–Schwarz inequality of mathematics says that, for any two functions $f(x)$ and $g(x)$,

$$\left(\int dx\,|f(x)|^2\right)\left(\int dx\,|g(x)|^2\right) \geq \left|\int dx\,f^*(x)g(x)\right|^2. \tag{11.5.17}$$

We also have the result of the math of complex numbers that the square of the magnitude of any number is at least as great as the square of the magnitude of either of its real or imaginary parts. That is, for any number A,

$$|A|^2 \geq |\mathrm{Im}A|^2. \tag{11.5.18}$$

Assigning $f(x) = x\psi(x)$ and $g(x) = p\psi(x)$, we then have

$$\langle|\Delta x|^2\rangle\langle|\Delta p|^2\rangle \geq \left|\int dx\,(f^*(x)g(x) - g^*(x)f(x))\right|^2$$

$$= \left|\int dx\left[x\psi^*\left(-i\hbar\frac{\partial\psi}{\partial x}\right) - \left(i\hbar\frac{\partial\psi^*}{\partial x}\right)x\psi\right]\right|^2. \tag{11.5.19}$$

Using integration by parts, assuming f and g are negligible very far away, we can move the derivative in the second term to ψ:

$$\int dx\left[x\psi^*\left(-i\hbar\frac{\partial\psi}{\partial x}\right) - \left(i\hbar\frac{\partial\psi^*}{\partial x}\right)x\psi\right]$$

$$= \int dx\left[-i\hbar\psi^*x\frac{\partial\psi}{\partial x} + i\hbar\psi^*x\frac{\partial\psi}{\partial x} + i\hbar\psi^*\psi\right]$$

$$= \int dx\,\psi^*\left(-i\hbar x\frac{\partial}{\partial x} + i\hbar\frac{\partial}{\partial x}x\right)\psi. \tag{11.5.20}$$

We have already seen in Section 11.3 that a simple consequence of treating x and p as operators acting on continuous functions is the *commutation relation*,

$$[x, p] \equiv xp - px = i\hbar. \tag{11.5.21}$$

Equation (11.5.19) therefore implies that, for our definitions of x and p,

$$\langle |\Delta x| \rangle \langle |\Delta p| \rangle \geq |\langle [x, p] \rangle| = \hbar. \tag{11.5.22}$$

The uncertainty relation between two measurements is directly related to the commutator of the operators that represent those measurements, in this case x and p.

References

A. L. Fetter and J. D. Walecka, *Quantum Theory of Many-Particle Systems*, (McGraw-Hill, 1971).

E. Hecht, *Optics*, 5th ed., (Pearson, 2015).

D. W. Snoke, *Solid State Physics: Essential Concepts*, 2nd ed., (Cambridge University Press, 2020).

Boson Quantization

12.1 The Harmonic Oscillator

All of the basic properties of field theory can be seen starting with the physical model of the harmonic oscillator. As discussed in Sections 2.2 and 2.3, the properties of particles are derived from the underlying wave mechanics of any system, assuming that the stretchiness of the waves is like a harmonic oscillator.

12.1.1 Derivation of Harmonic Oscillator States

Here we give the detailed math to find the exact eigenstates of the simple harmonic oscillator introduced in Section 9.4. A summary of the results of these calculations is given in Section 12.1.2.

We start by defining the new operators

$$\hat{x} = \sqrt{\frac{M\omega_0}{\hbar}}x$$

$$\hat{p} = \sqrt{\frac{1}{M\hbar\omega_0}}p. \tag{12.1.1}$$

The commutation relation $[x, p] = i\hbar$ implies the commutation relation between $\hat{x}$ and $\hat{p}$

$$[\hat{x}, \hat{p}] = i. \tag{12.1.2}$$

The Hamiltonian (9.4.4) then takes the simple form

$$H = \frac{1}{2}(\hat{x}^2 + \hat{p}^2)\hbar\omega_0. \tag{12.1.3}$$

The Hamiltonian is further simplified by defining the new operators

$$a = \frac{1}{\sqrt{2}}(\hat{x} + i\hat{p})$$

$$a^\dagger = \frac{1}{\sqrt{2}}(\hat{x} - i\hat{p}). \tag{12.1.4}$$

Then we have

$$H = \left(a^\dagger a + \tfrac{1}{2}\right)\hbar\omega_0$$

$$= \left(\hat{N} + \tfrac{1}{2}\right)\hbar\omega_0, \tag{12.1.5}$$

where we have defined a new operator $\hat{N} = a^\dagger a$. Note that a and $a^\dagger$ are operators but are commonly are written without "hats."

Using (12.1.1), it is easy to show that the operators a and $a^\dagger$ have the commutation relation

$$[a, a^\dagger] = aa^\dagger - a^\dagger a = 1. \tag{12.1.6}$$

This implies the following commutation relations:

$$\begin{aligned}
[\hat{N}, a] &= a^\dagger a a - a a^\dagger a \\
&= a^\dagger a a - (a^\dagger a + 1)a \\
&= -a, \tag{12.1.7}
\end{aligned}$$

and

$$\begin{aligned}
[\hat{N}, a^\dagger] &= a^\dagger a a^\dagger - a^\dagger a^\dagger a \\
&= a^\dagger(a^\dagger a + 1) - a^\dagger a^\dagger a \\
&= a^\dagger. \tag{12.1.8}
\end{aligned}$$

We now define $|\phi_N\rangle$ as an eigenstate of the operator $\hat{N}$ with eigenvalue N. Any state has a norm that is real and greater than or equal to zero:

$$\left(\langle\phi_N|a^\dagger\right)(a|\phi_N\rangle) \geq 0. \tag{12.1.9}$$

This implies

$$\langle\phi_N|a^\dagger a|\phi_N\rangle = \langle\phi_N|\hat{N}|\phi_N\rangle = N \geq 0. \tag{12.1.10}$$

Since the norm is real and nonnegative, the eigenvalues of $\hat{N}$ must be real, and the lowest eigenvalue is 0. We also have, using (12.1.8),

$$\begin{aligned}
a^\dagger|\phi_N\rangle &= (\hat{N}a^\dagger - a^\dagger\hat{N})|\phi_N\rangle \\
&= (\hat{N} - N)a^\dagger|\phi_N\rangle \\
\Rightarrow \hat{N}a^\dagger|\phi_N\rangle &= (N + 1)a^\dagger|\phi_N\rangle. \tag{12.1.11}
\end{aligned}$$

Thus, $a^\dagger|\phi_N\rangle$ is an eigenstate of $\hat{N}$ with eigenvalue $N + 1$, which we write as $\beta_N|\phi_{N+1}\rangle$, where β_N is a complex number. Assuming each of the eigenstates is normalized so that $\langle\phi_N|\phi_N\rangle = 1$, we obtain

$$\begin{aligned}
\langle\phi_N|aa^\dagger|\phi_N\rangle &= \langle\phi_N|(a^\dagger a + 1)|\phi_N\rangle \\
\langle\phi_{N+1}|\beta_N^*\beta_N|\phi_{N+1}\rangle &= \langle\phi_N|(N + 1)|\phi_N\rangle = N + 1, \tag{12.1.12}
\end{aligned}$$

which implies $|\beta_N|^2 = N + 1$. Since an eigenstate multiplied by any phase factor $e^{i\theta}$ is also an eigenstate, we can always write

$$\begin{aligned}
a^\dagger|\phi_N\rangle &= \beta_N|\phi_N + 1\rangle = e^{i\theta_N}\sqrt{N+1}|\phi_{N+1}\rangle \\
&= \sqrt{N+1}\left(e^{i\theta_N}|\phi_{N+1}\rangle\right) = \sqrt{N+1}|\phi'_{N+1}\rangle, \tag{12.1.13}
\end{aligned}$$

where $|\phi'_{N+1}\rangle$ is a new definition of the eigenstate. Choosing this phase convention for every state, we can therefore write

$$a^\dagger|\phi_N\rangle = \sqrt{N+1}|\phi_{N+1}\rangle. \tag{12.1.14}$$

Similarly,

$$\begin{aligned}
a|\phi_N\rangle &= (a\hat{N} - \hat{N}a)|\phi_N\rangle \\
&= (N - \hat{N})a|\phi_N\rangle \\
\Rightarrow \hat{N}a|\phi_N\rangle &= (N-1)a|\phi_N\rangle.
\end{aligned} \tag{12.1.15}$$

Defining $a|\phi_N\rangle = \gamma_N|\phi_{N-1}\rangle$, we have

$$\begin{aligned}
\langle\phi_N|a^\dagger a|\phi_N\rangle &= \langle\phi_N|\hat{N}|\phi_N\rangle \\
\langle\phi_{N-1}|\gamma_N^*\gamma_N|\phi_{N-1}\rangle &= \langle\phi_N|N|\phi_N\rangle = N,
\end{aligned} \tag{12.1.16}$$

which implies $|\gamma_N|^2 = N$, and by the same phase convention,

$$a|\phi_N\rangle = \sqrt{N}|\phi_{N-1}\rangle. \tag{12.1.17}$$

If N is a fraction between 0 and 1, then $\hat{N}|\phi_{N-1}\rangle = (N-1)|\phi_{N-1}\rangle$, which gives an eigenvalue less than zero, which is not allowed according to (12.1.10). Since such a state could be generated by applying a successively to any state with fractional N greater than 1, all states with fractional N are forbidden, and n must be an integer.

We therefore have a ladder of eigenstates of $\hat{N}$ equal to all the nonnegative integers. The $a^\dagger$ and a operators act as "creation" and "destruction" operators, or "raising" and "lowering" operators, which take one eigenstate to another. Note, however, that definition (12.1.4) implies that these are *amplitude* operators, which measure the amount of excursion of the oscillator.[1] These operators are the fundamental tools of "second quantization," which will be the basis of quantum field theory.

12.1.2 Basic Rules for Particle Operators

The quantum mechanics of the harmonic oscillator revolve around the definition of the new operators $a^\dagger$ and a. As we will see in later mathematical sections, these operators will become fundamental in quantum field theories. These operators have the commutation relation

$$[a, a^\dagger] \equiv aa^\dagger - a^\dagger a = 1. \tag{12.1.18}$$

Using these, the Hamiltonian of a harmonic oscillator can be written as

$$H = \left(a^\dagger a + \tfrac{1}{2}\right)\hbar\omega_0, \tag{12.1.19}$$

with the energies of its eigenstates equal to

$$E = \left(N + \tfrac{1}{2}\right)\hbar\omega_0, \tag{12.1.20}$$

---

[1] Note that the amplitude measured by the a operators is the amplitude of the motion of the oscillator, not the amplitude of the quantum wave function of the mass M that oscillates. As noted before Equation (12.1.12), that wave function is always normalized.

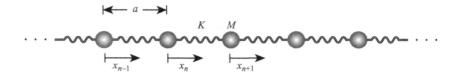

Linear chain of identical masses and springs.

where N is a nonnegative integer. We can write these eigenstates as $|N\rangle$. These are called *Fock states* or *number states*.

The operators a and $a^\dagger$ are known as *destruction* and *creation* operators, respectively, because they have the following action on the eigenstates of the Hamiltonian:

$$a|N\rangle = \sqrt{N}|N-1\rangle$$
$$a^\dagger|N\rangle = \sqrt{1+N}|N+1\rangle. \qquad (12.1.21)$$

The operator $a^\dagger$ "creates" a quantum of energy in the system, while a "destroys" one quantum of energy. Since energy is associated with the amplitude of a wave, these operators jump the wave from one well-defined amplitude to another.

Equation (12.1.20) implies that the wave function of the harmonic oscillator is *quantized* into states with definite energies. Because the jumps in energy, or "excitations," have equal energy equal to $\hbar\omega_0$, it is natural to define N as the number of "particles." This is the mathematical origin of particles in quantum field theory.

12.2 Phonon Quantization

We can take the formalism of the single harmonic oscillator from Section 12.1 and apply it to a linear chain of identical atoms. Since the force between two atoms can be modeled as a spring force, as discussed in Section 9.4, we treat the atoms as connected by springs, as illustrated in Figure 12.1. Section 12.2.2 gives a summary of the results of this calculation.

12.2.1 Derivation of Phonon Properties

For simplicity, we assume that all the atoms are identical. The Hamiltonian for this system now depends on the distance between each of the two neighboring atoms, plus the kinetic energy of each; it is given by

$$H = \sum_n \left(\frac{1}{2}\frac{p_n^2}{M} + \frac{1}{2}K(x_n - x_{n-1})^2 \right), \qquad (12.2.1)$$

where the index n labels the individual atoms.

We define the following *normal variables* which are sums of the quantum mechanical operators:

$$x_k = \frac{1}{\sqrt{N}} \sum_n x_n e^{-ikan}$$

$$p_k = \frac{1}{\sqrt{N}} \sum_n p_n e^{-ikan}, \qquad (12.2.2)$$

where N is the total number of atoms.

We can rewrite the Hamiltonian (12.2.1) in terms of these new operators by using a few mathematical tricks. First, we expand the sum over the x_n operators in (12.2.1) as

$$\frac{1}{2} K \sum_n (x_n - x_{n-1})^2 = \frac{1}{2} K \sum_n (x_n^2 - 2x_n x_{n-1} + x_{n-1}^2). \qquad (12.2.3)$$

Since the sum is over all n, the index $(n-1)$ is just a dummy variable that can be replaced by n or $(n+1)$, if the chain is infinite or finite with periodic boundary conditions. We therefore can write

$$\frac{1}{2} K \sum_n (x_n - x_{n-1})^2 = \frac{1}{2} K \sum_n (2x_n^2 - 2x_n x_{n-1})$$

$$= K \sum_n \left(x_n^2 - \frac{1}{2}(x_n x_{n-1} + x_n x_{n+1}) \right). \qquad (12.2.4)$$

Next, we can write

$$\sum_n x_n^2 = \sum_{n,n'} x_n x_{n'} \delta_{n,n'}$$

$$\sum_n x_n x_{n\pm1} = \sum_{n,n'} x_n x_{n'} \delta_{n\pm1,n'}, \qquad (12.2.5)$$

where $\delta_{n,n'}$ is the Kronecker delta, equal to 1 when $n = n'$ and zero otherwise. In the limit $N \to \infty$, the Kronecker delta can be written in terms of the identity

$$\delta_{n,n'} = \frac{1}{N} \sum_k e^{ika(n-n')}, \qquad (12.2.6)$$

where $k = (v/N)(2\pi/a)$, with v being an integer ranging from $-N/2$ to $N/2$. Substituting this for the Kronecker delta in the two terms in (12.2.5), we have

$$\sum_n x_n^2 = \sum_k \left(\frac{1}{\sqrt{N}} \sum_n x_n e^{-ikan} \right)^\dagger \left(\frac{1}{\sqrt{N}} \sum_{n'} x_n' e^{-ikan'} \right)$$

$$\sum_n x_n x_{n\pm1} = \sum_k \left(\frac{1}{\sqrt{N}} \sum_n x_n e^{-ika(n\pm1)} \right)^\dagger \left(\frac{1}{\sqrt{N}} \sum_{n'} x_{n'} e^{-ikan'} \right), \qquad (12.2.7)$$

or

$$\sum_n x_n^2 = \sum_k x_{-k} x_k$$

$$\sum_n x_n x_{n\pm1} = \sum_k |x_k|^2 e^{\pm ika}, \tag{12.2.8}$$

and therefore,

$$\frac{1}{2}K\sum_n (x_n - x_{n-1})^2 = K\sum_k |x_k|^2 (1 - \cos ka)$$

$$= 2K\sum_k |x_k|^2 \sin^2(ka/2)$$

$$= 2K\sum_k |x_k|^2 \left(\frac{\omega_k}{\omega_0}\right)^2$$

$$= \frac{1}{2}M\sum_k |x_k|^2 \omega_k^2, \tag{12.2.9}$$

where we have defined

$$\omega_k = \omega_0 \sin(ka/2). \tag{12.2.10}$$

The identities (12.2.8) can be applied to the sum over p_n operators, to finally give

$$H = \sum_k \left(\frac{|p_k|^2}{2M} + \frac{1}{2}M\omega_k^2 |x_k|^2\right). \tag{12.2.11}$$

Comparing this to the Hamiltonian (9.4.4), we see that the Hamiltonian for the linear chain is a sum of independent Hamiltonians for *single* harmonic oscillators. Each value of k corresponds to a different harmonic oscillator, with a different natural frequency given by the classical eigenfrequency ω_k.

Equation (12.2.11) allows us to treat each vibrational mode of the chain of atoms as a separate, independent, simple harmonic oscillator. It therefore makes sense to quantize each of these modes just as we did the single harmonic oscillator. We can then follow the same logic as in Section 12.1 to write

$$H = \sum_k \left(a_k^\dagger a_k + \tfrac{1}{2}\right)\hbar\omega_k \tag{12.2.12}$$

with

$$a_k = \frac{1}{\sqrt{2}}(\hat{x}_k + i\hat{p}_k)$$

$$a_k^\dagger = \frac{1}{\sqrt{2}}(\hat{x}_k^\dagger - i\hat{p}_k^\dagger) \tag{12.2.13}$$

and

$$\hat{x}_k = \sqrt{\frac{M\omega_k}{\hbar}}x_k$$

$$\hat{p}_k = \sqrt{\frac{1}{M\hbar\omega_k}}p_k. \tag{12.2.14}$$

The commutation relation is now

$$[a_k, a_{k'}^\dagger] = \delta_{k,k'}, \tag{12.2.15}$$

where $\delta_{k,k'}$ is the Kronecker delta-function, equal to 1 if $k = k'$ and zero otherwise. The new operators have the action

$$a_k|N_k\rangle = \sqrt{N_k}|N_k - 1\rangle,$$
$$a_k^\dagger|N_k\rangle = \sqrt{1 + N_k}|N_k + 1\rangle. \tag{12.2.16}$$

In this formulation, we call the energy quanta of the sound field *phonons*. These particles appear simply as the eigenstates of the Hamiltonian of the field.

Continuum limit. The fact that our model of the system consisted of discrete atoms does not affect whether or not there are phonons. We can equally well quantize the sound field in the continuum limit. The mass M can be rewritten in terms of the mass density ρ,

$$M = \rho V_{\text{cell}} = \rho(ahw), \tag{12.2.17}$$

where V_{cell} is the volume of a small region, or "cell," of the medium, and a, h, and w are length, height, and width, respectively, of this cell. The spring constant can be rewritten in terms of an effective elastic constant $C = Ka/hw$, which gives the force per area F/A, as a function of the fractional change of the unit cell. Substituting these definitions into (12.2.10), and taking the limit of the cell size going to zero, we have

$$\omega_k = \lim_{a \to 0} 2\sqrt{\frac{K}{M}} \sin(ka/2) = 2\sqrt{\frac{C/a}{\rho a}}\frac{ka}{2} = \sqrt{\frac{C}{\rho}}k. \tag{12.2.18}$$

The value of $\sqrt{C/\rho}$ just turns out to be the speed of sound.

12.2.2 Basic Rules for Phonons

In Section 12.1, we wrote down operators a and $a^\dagger$ for the destruction and creation of quanta of excitation ("particles") of a single harmonic oscillator. For a system with many particles, such as a solid composed of many atoms, we can write the Hamiltonian of the system in the same way but with many possible vibration states, each treated as a separate harmonic oscillator. Each vibration state is identified by its wave number $k = 2\pi/\lambda$, where λ is the wavelength. We now write the operators with a subscript, as a_k and $a_k^\dagger$, and write the Hamiltonian for the total energy as

$$H = \sum_k \left(a_k^\dagger a_k + \tfrac{1}{2}\right)\hbar\omega_k, \tag{12.2.19}$$

where ω_k is the frequency of the wave with wave number k. The same rules apply to these operators as for a single harmonic oscillator, namely

$$[a_k, a_{k'}^\dagger] = \delta_{k,k'}, \tag{12.2.20}$$

where $\delta_{k,k'}$ is the Kronecker delta-function, equal to 1 if $k = k'$ and zero otherwise, and

$$a_k|N_k\rangle = \sqrt{N_k}|N_k - 1\rangle,$$

$$a_k^\dagger|N_k\rangle = \sqrt{1 + N_k}|N_k + 1\rangle. \tag{12.2.21}$$

To fully describe the sound field, we now need Fock states with a separate value of N_k defined for each possible k, that is, $|\ldots N_{k_1}, N_{k_2}, N_{k_3}, \ldots\rangle$. The vacuum state corresponds to each of these numbers set to zero. The creation and destruction operators for one k do not affect the number of particles in a state with different k.

Note that each mode has the energy $\hbar\omega_k/2$ even when N_k equals zero. If there is an infinite number of atoms, the sum over all k is then infinite. Some people make a big story about this *zero-point energy*, but it really is not a problem, and it does not provide some new energy source that could be tapped into. Because it is a constant, we are always free to change our definition of potential energy to add or subtract a constant, and so we can just subtract off this constant value of all the zero-point energies.

12.2.3 Spatial Field Operators

Once we have defined the operators for pure k-states, the Fourier transform theorem says that we can write any wave function as a sum of k-states. Therefore, we can make a creation operator for a wave in any spatial location by writing a sum of k-state creation and destruction operators. We write, for a one-dimensional system,

$$\hat{\psi}^\dagger(r) = \sum_k \frac{e^{-ikr}}{\sqrt{L}}a_k^\dagger$$

$$\hat{\psi}(r) = \sum_k \frac{e^{ikr}}{\sqrt{L}}a_k, \tag{12.2.22}$$

where L is the size of the system. (The limit $L \to \infty$ can be taken, in general; see Section 12.3.) These operators have a commutation relation similar to the operators for k-states:

$$[\hat{\psi}(r), \hat{\psi}^\dagger(r')] = \delta(r - r'), \tag{12.2.23}$$

where $\delta(r)$ is the Dirac delta function, equal to zero everywhere that $r \neq r'$, and with a total area integral equal to 1. The spatial operator $\psi^\dagger(r)$ can be viewed as creating a particle at exactly the point r, but more properly, it is a weighting function for the wave amplitude in space.

In general, a particle can be created with any wave function $\phi(r)$ by writing the superposition

$$|\phi\rangle = \int dr\, \phi(r)\hat{\psi}^\dagger(r)|0\rangle, \tag{12.2.24}$$

where $|0\rangle$ is the vacuum state. The k-state $\phi(r) = (1/\sqrt{L})e^{ikr}$ can also be created this way. Using the definition of $\hat{\psi}^\dagger(r)$, this resolves to

$$|\phi\rangle = \int dr \, \frac{1}{\sqrt{L}} e^{ikr} \left(\frac{1}{\sqrt{L}} \sum_{k'} e^{-ik'r} a_k^\dagger \right) |0\rangle$$

$$= \frac{1}{L} \sum_{k'} \left(\int dr \, e^{i(k-k')r} \right) a_k^\dagger |0\rangle. \tag{12.2.25}$$

Using the identity

$$\frac{1}{L} \int dr \, e^{i(k-k')r} = \delta_{k,k'}, \tag{12.2.26}$$

(12.2.25) becomes simply

$$|\psi\rangle = a_{\bar{k}}^\dagger |0\rangle, \tag{12.2.27}$$

as one would expect.

Note the difference between a plane-wave state created by the creation operator $a_{\bar{k}}^\dagger$ and a coherent state with amplitude $A_k^2 = 1$, which we will discuss in Section 12.5. The former has exactly one particle, and indefinite phase (though it has a definite wavelength). The latter has definite phase, and is a superposition of different number states, with average number equal to 1.

Local atomic motion. The operators (12.2.22) are different from the operators we use if we want to know the motion of the atoms in a sound wave. In that case, we solve the equations in (12.2.13) for $\hat{x}_k$ and $\hat{p}_k$ to get

$$\hat{x}_k = \frac{1}{\sqrt{2}} \left(a_k + a_{-k}^\dagger \right),$$

$$\hat{p}_k = \frac{-i}{\sqrt{2}} \left(a_k - a_{-k}^\dagger \right). \tag{12.2.28}$$

Definition (12.2.2) implies

$$x_n = \frac{1}{\sqrt{N}} \sum_k x_k e^{ikan}, \tag{12.2.29}$$

and therefore, by substitution, the position x of an atom at site n is given by the operator

$$x_n = \frac{1}{\sqrt{N}} \sum_k \sqrt{\frac{\hbar}{2M\omega_k}} \left(a_k + a_{-k}^\dagger \right) e^{ikan}, \tag{12.2.30}$$

or, in the continuum limit,

$$x(r) = \sum_k \sqrt{\frac{\hbar}{2\rho V \omega_k}} \left(a_k e^{ikr} + a_k^\dagger e^{-ikr} \right), \tag{12.2.31}$$

where we have defined the continuous variable $r = na$, and $V = NV_{\text{cell}}$ is the total volume. Since the summation is over the k running from $-\infty$ to ∞, we have switched $k \to -k$ in the second term (ω_k depends only on the magnitude of k). Comparison to (12.2.22) shows that this consists of the unitless phonon spatial amplitude operators, times a multiplicative constant, added to its complex conjugate to ensure that the value for the position is a real number. In the same way, we can write for the velocity of the medium in the continuum limit,

$$\dot{x}(r) = \frac{p(r)}{M} = -i \sum_k \sqrt{\frac{\hbar \omega_k}{2\rho V}} \left(a_k e^{ikr} - a_k^\dagger e^{-ikr} \right). \tag{12.2.32}$$

Spatial operators as a "field." Because the spatial field operators $\hat{\psi}^\dagger(r)$ and $\hat{\psi}(r)$ are defined for a continuum of locations r, one can define a new type of field consisting of just these operators (see, e.g., Teller 1995). This is allowable if we define a field as any set of mathematical objects assigned to a continuous range of spatial positions. Furthermore, one can define field operators with both spatial and temporal dependence, of the form

$$\hat{\psi}^\dagger(r, t) = \sum_k \frac{1}{\sqrt{L}} e^{-i(kr - \omega_k t)} a_k^\dagger$$

$$\hat{\psi}(r, t) = \sum_k \frac{1}{\sqrt{L}} e^{i(kr - \omega_k t)} a_k. \tag{12.2.33}$$

These operators obey a time-evolution equation entirely analogous to the Schrödinger equation.

It is important to understand, however, that all operators in quantum theory represent queries of the physical properties of a system, and/or prescriptions for changes to that system, and do not hold in themselves any information content about the physical state of the system. That information is held in the many-body wave function upon which the operators act. This many-body wave function is typically represented by a superposition of many Fock states, defined in Section 12.1. Even when the operators are given time evolution, their time dependence does not represent the time evolution of any real state of affairs; it just defines the variation of the operators over time, which can act upon any number of different physical wave functions.

Therefore, the oft-repeated statement, that quantum fields are fields of operators, while classical fields are fields of numbers, is very misleading. First, a field of operators can also be defined for a classical field; for example, one can apply the gradient operator to a scalar field at every point in space, as is done when deriving the force field from a potential-energy function, $\vec{F} = -\nabla U(\vec{r})$. Some might argue that in this case, the same operator applied is the same at every point in space, while quantum field operators have an explicit r-dependence, but, of course, we can always add an explicit r-dependence to obtain a location-dependent classical operator, for example, $\vec{G}(\vec{r}) = f(\vec{r})\nabla$. It would make no sense to say that force fields are somehow fundamentally different from other classical fields because they can be discussed in terms of "operators." As in the fully quantum case, the operator field $\vec{G}(\vec{r})$ contains no information about the real physical state of affairs; it is merely a prescription for what to do with the information of that physical state.

Also, in quantum field theory, the many-body wave function is formally a *vector*, consisting of the set of all the inner products of the state with each of the definable Fock states of the system. It is no different mathematically from the vector field of electromagnetism, except that it has many more components, that is, degrees of freedom. This is the fundamental difference between classical and quantum fields: quantum fields have a vastly greater number of degrees of freedom.

In quantum field theory, we may say that for electromagnetic waves and matter waves, there is a continuous, real "thing" that plays the same role as the water in a water wave.[2] Throughout this book, this "thing" is what we call the "quantum field;" it is a physical entity, and waves in it are mathematically described by the many-body wave function. By contrast, the "operator field," consisting of the set of all possible spatial field operators, carries no physical information and represents only a set of potential queries or actions performed on the physical state of the system.

Note that the water waves discussed here are not just an *analogy* for quantum waves. Real water waves are, of course, fully quantum in a universe governed by quantum mechanics. Water waves are coherent states (defined in Section 12.5) in phonon fields, and phonon fields are quantized by the same math as used for other fields, such as the electromagnetic field, as we will see in Section 12.4. Phonon and electromagnetic fields are, in turn, the same as matter fields except for one sign change, as we will see in Section 13.1. Thus, water waves are an easily accessible *example* of a quantum field. If we ascribe reality to the water field we see, there is no mathematical justification to treat other types of quantum fields differently.[3]

12.3 The Thermodynamic Limit in Quantum Field Theory

As discussed in Section 12.2.3, we can write the spatial field operator for creation and destruction of a particle at position r as

$$\hat{\psi}^{\dagger}(r) = \sum_k \frac{e^{-ikr}}{\sqrt{L}} a_k^{\dagger}$$

$$\hat{\psi}(r) = \sum_k \frac{e^{ikr}}{\sqrt{L}} a_k, \tag{12.3.1}$$

where L is the size of the system.

---

[2] As shown in Section 12.2.1, the fact that a water wave is made of atoms is irrelevant; the same math can be worked out for a continuous fluid. Therefore, phonon waves, of which water waves are one example, are not fundamentally different from electromagnetic waves or matter waves in this respect.

[3] Teller (1995) points out that, formally, one can write down all of quantum field theory starting with just Fock states, without ascribing any physical meaning to these states. However, for the cases of phonons and photons, we *do* know what these Fock states represent, physically – they represent spatially extended wave states with various quantized amplitudes. As we will discuss in Section 13.1.1, we don't have a similar physical intuition for fermion fields. However, as discussed in that section, the only difference between photon and phonon fields and matter fields is a single sign change from + to −. Therefore, it stands to reason that what is real in one case is real in the other case, and that there is a similar underlying physical entity for fermions that plays the same role as the water in water waves.

This can be generalized to a three-dimensional volume V as

$$\hat{\psi}^{\dagger}(\vec{r}) = \sum_{k} \frac{e^{-i\vec{k}\cdot\vec{r}}}{\sqrt{V}} a_{\vec{k}}^{\dagger}$$

$$\hat{\psi}(\vec{r}) = \sum_{\vec{k}} \frac{e^{i\vec{k}\cdot\vec{r}}}{\sqrt{V}} a_{\vec{k}}, \tag{12.3.2}$$

where $\vec{r}$ and $\vec{k}$ are now explicitly vectors.

Suppose that we create a many-particle state with N particles in various k-states by a product of creation operators acting on the vacuum state:

$$|\psi_N\rangle = \prod_{i=1}^{N} a_{\vec{k}_i}^{\dagger} |0\rangle. \tag{12.3.3}$$

The density of particles at position $\vec{r}$ is given by the square of spatial field amplitude:

$$\rho = \langle \psi_N | \psi^{\dagger}(\vec{r})\psi(\vec{r}) | \psi_N \rangle. \tag{12.3.4}$$

Substituting in our definitions (12.3.2) for the operators and (12.3.3) for the state, this is

$$\rho = \langle 0 | \prod_{i=1}^{N} a_{\vec{k}_i} \frac{1}{V} \sum_{\vec{k},\vec{k}'} e^{-i(\vec{k}-\vec{k}')\cdot\vec{r}} a_{\vec{k}}^{\dagger} a_{\vec{k}'} \prod_{j=1}^{N} a_{\vec{k}_j}^{\dagger} |0\rangle. \tag{12.3.5}$$

The destruction operator $a_{\vec{k}'}$ acting to the right will only give a nonzero result if $\vec{k}'$ equals one of the $\vec{k}_j$, and the same happens for the $a_{\vec{k}}^{\dagger}$ operator acting to the left. We then have

$$\rho = \frac{1}{V} \sum_{i=1}^{N} \sum_{j=1}^{N} e^{-i(\vec{k}_i - \vec{k}_j)\cdot\vec{r}}. \tag{12.3.6}$$

For a large number of different k-states, the phase factors will, in general, cancel unless $\vec{k}_i = \vec{k}_j$. We then have

$$\rho = \frac{1}{V} \sum_{i=1}^{N} = \frac{N}{V}. \tag{12.3.7}$$

The *thermodynamic limit* is defined as the case when we let both N and V go to infinity, with N proportional to V. This gives us a finite density ρ even in an infinite system. The set of plane-wave states we have created gives the same average particle density everywhere in space.

Note that, if we had first taken the limit $V \to \infty$, then the spatial operators would have been set to 0, and we would have gotten the incorrect result $\rho = 0$. The thermodynamic limit is a standard method in physics for treating quantities that are independent of the volume in the large-volume limit. It requires that we first compute the properties of a finite system, and then take the limit of that finite system going to infinity. Nonsensical results can be obtained if the infinite-volume limit is not taken properly.

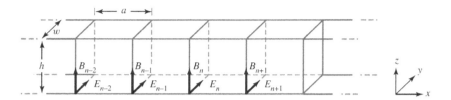

Figure 12.2 A linear chain of vacuum cells.

12.4 Photon Quantization

Because we have generated phonons from quantized sound waves, some people view them as not really real, that is, an "epiphenomenon" in philosophical terms, while they view photons as truly fundamental particles. The approach of field theory in generating photons is exactly the same as that for generating phonons, however.

12.4.1 Derivation of Photon Properties

To start, we consider a model of space consisting of discrete elements of length a, height h, and width w, as shown in Figure 12.2. We assume that the electric and magnetic fields are nearly constant within each element, that the electric field points in the $\hat{y}$ direction, and that the magnetic field points in the $\hat{z}$ direction, as shown in the figure.

We write down the Maxwell Hamiltonian for the electromagnetic field as

$$H = V_{\text{cell}} \sum_n \left(\frac{1}{2}\varepsilon_0 E_n^2 + \frac{1}{2}\frac{1}{\mu_0}B_n^2 \right), \tag{12.4.1}$$

where $V_{\text{cell}} = ahw$ is the volume of each element, and ε_0 is a constant of nature known as the *vacuum permittivity*. We can write $\vec{E}$ and $\vec{B}$ in terms of the vector potential $\vec{A}$ according to the definitions

$$\vec{E} = -\frac{\partial \vec{A}}{\partial t}$$

$$\int_S \vec{B} \cdot d\vec{\sigma} = \oint \vec{A} \cdot d\vec{l}. \tag{12.4.2}$$

The first equation implies that $\vec{A}$ is parallel to $\vec{E}$, which means for a volume element shown in Figure 12.2, the latter equation can be written as

$$B_n wa = -(A_n - A_{n-1})w. \tag{12.4.3}$$

Substituting into the Hamiltonian (12.4.1), we have

$$H = V_{\text{cell}} \sum_n \left(\frac{1}{2}\varepsilon_0 (\dot{A}_n)^2 + \frac{1}{2}\frac{1}{\mu_0 a^2}(A_n - A_{n-1})^2 \right). \tag{12.4.4}$$

Recall that, for the sound wave, we had the Hamiltonian (12.2.1),

$$H = \sum_n \left(\frac{1}{2} M \dot{x}_n^2 + \frac{1}{2} K (x_n - x_{n-1})^2 \right). \tag{12.4.5}$$

The one-to-one correspondence of these two suggests that we should treat them the same way. Not only that, we know that the vector potential A acts as a momentum. In classical electrodynamics, the momentum of a charged object moving in a magnetic field is equal to

$$p = M\dot{x} + qA. \tag{12.4.6}$$

If there is a commutation relation of the classical momentum p and x, then there must also be the same commutation relation between A and its conjugate, the electric field E.[4] We can therefore apply the same quantization procedure as in Section 12.2, replacing M with $V_{cell}\varepsilon_0$, K with $V_{cell}/\mu_0 a^2$, A with x, and E with $-\dot{x}$. We write

$$H = \sum_k \left(a_k^\dagger a_k + \frac{1}{2} \right) \hbar \omega_k \tag{12.4.7}$$

with

$$a_k = \frac{1}{\sqrt{2}} \left(\sqrt{\frac{\varepsilon_0 V_{cell} \omega_k}{\hbar}} A_k - i \sqrt{\frac{\varepsilon_0 V_{cell}}{\hbar \omega_k}} E_k \right)$$

$$a_k^\dagger = \frac{1}{\sqrt{2}} \left(\sqrt{\frac{\varepsilon_0 V_{cell} \omega_k}{\hbar}} A_k^* + i \sqrt{\frac{\varepsilon_0 V_{cell}}{\hbar \omega_k}} E_k^* \right), \tag{12.4.8}$$

and the normal variables

$$A_k = \frac{1}{\sqrt{N}} \sum_n A_n e^{-ikan}$$

$$E_k = \frac{1}{\sqrt{N}} \sum_n E_n e^{-ikan} \tag{12.4.9}$$

and the mode frequency

$$\omega = 2\frac{c}{a} \sin(ka/2). \tag{12.4.10}$$

In the continuum limit $a \to 0$, this mode frequency becomes, naturally, $\omega = ck$, where c is the speed of light. Although we used a fictional cell volume V_{cell} for the photon Hamiltonian, when we compute real amplitudes in the continuum limit, we always encounter $V = V_{cell}N$, so the cell size drops out.

---

[4] This analysis assumes transverse electromagnetic waves in the Coulomb gauge. In general, the commutation relation is $[A_i(\vec{r}), \dot{A}_j(\vec{r}')] = (i\hbar/\varepsilon_0)\delta_{ij}\delta^T(\vec{r} - \vec{r}')$, where $\delta^T(\vec{r})$ is the "transverse" δ-function, which picks out only the transverse part of a wave (see Loudon 1973, pp. 145–147). For the finite cell method used here, $\delta^T(0) = 1/V_{cell}$.

Just as we did for phonons for x and p, we can write spatial field operators to give the local values of the fields A and E. In the continuum limit, these are

$$A(r) = \sum_k \sqrt{\frac{\hbar}{2\varepsilon_0 V \omega}} \left(a_k e^{ikr} + a_k^\dagger e^{-ikr} \right),$$ (12.4.11)

and the electric field is

$$E(r) = i \sum_k \sqrt{\frac{\hbar \omega}{2\varepsilon_0 V}} \left(a_k e^{ikr} - a_k^\dagger e^{-ikr} \right).$$ (12.4.12)

12.4.2 Basic Rules for Photons

By treating the vacuum as a set of finite cells and then taking the limit when these cells go to infinitesimal size, we have found that the exact same quantization method applies to electromagnetic field, with the resulting *photons*, as did for a sound wave with *phonons*. Although we used an imaginary cell volume V_{cell} for the photon Hamiltonian, when we compute real amplitudes in the continuum limit, we always encounter $V = V_{cell} N$, so the cell size drops out.

In fact, we do not know that the vacuum is really continuous – it may be that there is a smallest "cell size" as in the case of a solid material or water carrying a sound wave. Some people have proposed that space is granular on length scales of the *Planck length*, which is very tiny, of the order of 10^{-35} m. We also don't know that the electromagnetic field is strictly *linear*, that is, that its Hamiltonian is exactly given by (12.4.1), which allows us to have eigenstates perfectly equally spaced. In fact, we know that it is not, when interactions with charged matter are allowed.

The one-to-one correspondence with phonons allows us to write destruction and creation operators in the same form, namely we write the Hamiltonian for the total energy as

$$H = \sum_k \left(a_k^\dagger a_k + \tfrac{1}{2} \right) \hbar \omega_k,$$ (12.4.13)

where $\omega_k = ck$ is the frequency of the wave, with the action of a_k and $a_k^\dagger$ governed by the commutation relation The same rules apply to these operators as for a single harmonic oscillator, namely

$$[a_k, a_{k'}^\dagger] = \delta_{k,k'},$$ (12.4.14)

and

$$a_k |N_k\rangle = \sqrt{N_k} |N_k - 1\rangle,$$
$$a_k^\dagger |N_k\rangle = \sqrt{1 + N_k} |N_k + 1\rangle.$$ (12.4.15)

As with the sound field, we write Fock states for the photons with a separate value of N_k defined for each possible k. The vacuum state corresponds to each of these numbers set to zero. As with phonons, each mode has a zero-point energy $\hbar \omega_k / 2$.

To all intents and purposes, photons have no more "reality" in quantum field theory than phonons. Both are the quantized energy of the excitations of the systems that they apply to.

12.5 Coherent States of Bosons

The operators a_k and $a_k^\dagger$ introduced in Section 12.2 are called creation and destruction operators because of their roles in the algebra to create and destroy excitations of the field. These operators correspond physically, however, to measurements of the *complex amplitude* of a wave. The eigenstates of a_k correspond to states with definite phase and amplitude, that is,

$$a_k|\alpha_k\rangle = \alpha_k|\alpha_k\rangle, \tag{12.5.1}$$

where $\alpha_k = A_k e^{i\theta_k}$ is a complex number that gives the complex amplitude of the wave.

By itself, the operator a_k does not have real numbers for its eigenvalues, which means it cannot correspond to experimental observations. The sum $(a_k + a_k^\dagger)$ does, however, and corresponds to

$$\langle\alpha_k|(a_k + a_k^\dagger)|\alpha_k\rangle = 2A_k\cos\theta_k|\alpha_k\rangle. \tag{12.5.2}$$

This is the real amplitude of a field, which can be measured. The state $|\alpha_k\rangle$, which is an eigenstate of a_k, is called a *coherent state*, because it has a definite phase and amplitude.

By definition (12.5.1), the product $a_k^\dagger a_k$ therefore gives

$$\langle\alpha_k|a_k^\dagger a_k|\alpha_k\rangle = A_k e^{-i\theta_k} A_k e^{i\theta_k}\langle\alpha_k|\alpha_k\rangle = A_k^2. \tag{12.5.3}$$

Recall from (12.2.16) that product $a_k^\dagger a_k$ acting on a Fock number state gives the value

$$\langle N_k|a_k^\dagger a_k|N_k\rangle = \sqrt{N_k}\sqrt{N_k}\langle N_k|N_k\rangle = N_k. \tag{12.5.4}$$

In other words, the product $a_k^\dagger a_k$ gives the number of particles in a state. We write this as the *number operator* $\hat{N}_k \equiv a_k^\dagger a_k$ (the hat distinguishes this from the simple occupation number, N_k). As we see in (12.5.3), the same measurement gives the amplitude squared of a coherent state. In other words, a measurement of the square of the amplitude is a measurement of the total number of particles.

Clearly, a coherent state is not also an eigenstate of the number operator. In terms of the Fock number states, a coherent state is equal to the superposition

$$|\alpha_k\rangle = e^{-|\alpha_k|^2/2}\sum_{N_k=0}^{\infty}\frac{\alpha_k^{N_k}}{\sqrt{N_k!}}|N_k\rangle. \tag{12.5.5}$$

12.5.1 Time Dependence of a Coherent State

The time evolution of the complex amplitude can be found using Schrödinger's equation:

$$\begin{aligned}
\frac{\partial}{\partial t}\alpha_k &= \frac{\partial}{\partial t}\langle\alpha_k|a_k|\alpha_k\rangle = \left(\frac{\partial}{\partial t}\langle\alpha_k|\right)a_k|\alpha_k\rangle + \langle\alpha_k|a_k\left(\frac{\partial}{\partial t}|\alpha_k\rangle\right) \\
&= \left(-\frac{1}{i\hbar}\langle\alpha_k|H\right)a_k|\alpha_k\rangle + \langle\alpha_k|a_k\left(\frac{1}{i\hbar}H|\alpha_k\rangle\right) \\
&= \frac{1}{i\hbar}\langle\alpha_k|[a_k,H]|\alpha_k\rangle.
\end{aligned} \tag{12.5.6}$$

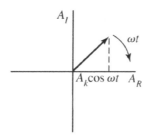

Figure 12.3 The field oscillation represented by a phasor.

The Hamiltonian has the form in (12.2.12). Only the term in H with $\hat{N}_k = a_k^\dagger a_k$ has a nonzero commutator with a_k, so that we have

$$\frac{\partial}{\partial t}\alpha_k = -i\omega_k \langle\alpha_k|[a_k, \hat{N}_k]|\alpha_k\rangle$$
$$= -i\omega_k \langle\alpha_k|a_k|\alpha_k\rangle$$
$$= -i\omega_k \alpha_k, \tag{12.5.7}$$

which implies

$$\alpha_k(t) = A_k e^{-i\omega_k t}. \tag{12.5.8}$$

Therefore, in a coherent state, the real and imaginary components of the amplitude are

$$A_R = \frac{1}{2}A_k \left(e^{-i\omega_k t} + e^{i\omega_k t}\right) = A_k \cos\omega_k t$$

$$A_I = \frac{1}{2i}A_k \left(e^{-i\omega_k t} - e^{i\omega_k t}\right) = -A_k \sin\omega_k t, \tag{12.5.9}$$

as illustrated in Figure 12.3. We can represent the state of the system as a vector (A_R, A_I) in the complex plane, known as a *phasor*, as shown in Figure 12.3. This vector rotates clockwise with angular frequency ω_k.

More generally, we can write operators for the real and imaginary components of the phasor, given by

$$\hat{A}_R = \text{Re } a_k = \frac{a_k + a_k^\dagger}{2}$$

$$\hat{A}_I = \text{Im } a_k = \frac{a_k - a_k^\dagger}{2i}. \tag{12.5.10}$$

Classical correspondence. The amplitude of a coherent state has no upper limit, while as shown in Section 12.5.2, the uncertainty in the amplitude is a constant. Therefore, in coherent states with very large amplitude, the uncertainty in the amplitude as a fraction of the amplitude becomes negligible. The solution for A_R in (12.5.9) can then be taken as a classical wave, that is, as a wave with continuous range of the amplitude A, with no quantization. This is the modern view of *all* classical waves, whether water waves or sound waves or light or radio waves – they are quantum coherent states with a macroscopic amplitude, so that the energy difference due to adding or subtracting single particles is negligibly small.

12.5.2 Number-Phase Uncertainty and Coherent States

There is no rule that a system must always be in a Fock state, that is, that it must have a definite number of particles. States can range anywhere from Fock states with a definite number of particles, to coherent states with definite phase (defined in Section 12.5.1), and anywhere in between these.

Just as there is an uncertainty relationship between x and p, there is also *number-phase incompatibility* due to the fact that the number operator $\hat{N}_k$ and the complex amplitude operator a_k do not commute. From (12.1.7), we have

$$[a_k, \hat{N}_k] = a_k N_k - N_k a_k = a_k. \tag{12.5.11}$$

Since the phase θ_k is found by a measurement of complex amplitude $\langle a_k \rangle = A_k e^{i\theta_k}$, we cannot determine the phase exactly if we know the number of particles exactly. A measurement of definite phase is a common physical measurement, though, which implies that physical states exist that *cannot* have a definite number of particles.

For coherent states, we can write an uncertainty relationship between the real and imaginary parts of the complex amplitude analogous to the relationship between x and p. We use the time evolution equation (12.5.9) to obtain the uncertainty in the real component:

$$\begin{aligned}
\langle (\Delta A_R)^2 \rangle &= \langle \alpha_k | (\hat{A}_R - \bar{A}_R)^2 | \alpha_k \rangle \\
&= \langle \alpha_k | (\hat{A}_R^2 - 2\hat{A}_R \bar{A}_R + \bar{A}_R^2) | \alpha_k \rangle \\
&= \langle \alpha_k | \left(\tfrac{1}{4}(a_k^\dagger a_k^\dagger + a_k a_k + 2 a_k^\dagger a_k + 1) \right. \\
&\quad \left. - (a_k^\dagger + a_k) A_k \cos \omega_k t + A_k^2 \cos^2 \omega_k t \right) | \alpha_k \rangle \\
&= \frac{A_k^2}{4} \left(e^{2i\omega_k t} + e^{-2i\omega_k t} + 2 \right) + \tfrac{1}{4} \\
&\quad - A_k^2 \left(e^{i\omega_k t} + e^{-i\omega_k t} \right) \cos \omega_k t + A_k^2 \cos^2 \omega_k t \\
&= \tfrac{1}{4}, \tag{12.5.12}
\end{aligned}$$

that is, $\Delta A_R = \tfrac{1}{2}$. The same result is found for ΔA_I. Thus, although a coherent state is not an eigenstate of the Hamiltonian, the uncertainty in the complex amplitude does not increase in time.

We can represent the total uncertainty as an area in the complex plane equal to $\Delta A_R \Delta A_I = \tfrac{1}{4}$, as shown in Figure 12.4(a). This is the minimum possible total uncertainty. All states other than the coherent state have larger total uncertainty. The uncertainty principle along one axis can be reduced, however, if the uncertainty in the perpendicular direction is increased. For example, as shown in Figure 12.4(b), the uncertainty in the phase can be reduced by increasing the uncertainty in the number, keeping the total area the same.

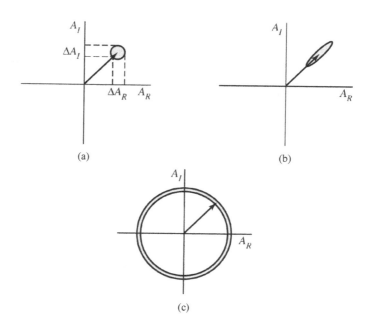

Figure 12.4 (a) Uncertainty of a coherent state in the phasor picture. (b) A squeezed wave in the phasor picture. (c) A Fock state in the phasor picture.

Figure 12.4(c) shows a Fock state, which has minimal amplitude uncertainty and maximal phase uncertainty. States that have one type of uncertainty traded off against another are called *squeezed* states. Many experiments have demonstrated the effect of reducing the uncertainty of one component below the value of $1/2$ by increasing the other.

References

R. Loudon, *The Quantum Theory of Light*, (Oxford University Press, 1973).

P. Teller, *An Interpretive Introduction to Quantum Field Theory*, (Princeton University Press, 1995).

13 Fermion Quantization

13.1 Fermion Field Operators

We have seen in Sections 12.2 and 12.4 that phonons and photons arise as the energy of excitation of vibrational and electromagnetic fields. It is therefore natural to suppose that *all* particles have a similar origin, from the quantization of an underlying field. In particular, it is natural to view electrons as having the same status.

13.1.1 Quantum Field Hamiltonians

As we have seen, both phonons and photons have the form of Hamiltonian:

$$H = \sum_k \left(a_k^\dagger a_k + \tfrac{1}{2} \right) \hbar \omega_k, \tag{13.1.1}$$

with the commutation relation

$$[a_k, a_{k'}^\dagger] = a_k a_{k'}^\dagger - a_{k'}^\dagger a_k = \delta_{k,k'}, \tag{13.1.2}$$

where $\delta_{k,k'}$ is the Kronecker delta-function, equal to 1 if $k = k'$ and zero otherwise, and the action of the operators is given by

$$a_k |N_k\rangle = \sqrt{N_k} |N_k - 1\rangle,$$
$$a_k^\dagger |N_k\rangle = \sqrt{1 + N_k} |N_k + 1\rangle. \tag{13.1.3}$$

These are the characteristic properties of *bosonic* fields, and the particles associated with these fields are called *bosons*.

Following the work of Dirac and others, it was found that the properties of electrons, protons, quarks, and a vast number of other types of particles can be described by the same formalism but with a single sign change from + to −. One can write

$$H = \sum_k a_k^\dagger a_k \hbar \omega_k, \tag{13.1.4}$$

where we ignore (for now) the constant 1/2, and instead of a commutation relation for the operators, we use the *anticommutation* relation

$$\{a_k, a_{k'}^\dagger\} \equiv a_k a_{k'}^\dagger + a_{k'}^\dagger a_k = \delta_{k,k'}, \tag{13.1.5}$$

which implies that the action of the operators is given by

$$a_k|N_k\rangle = \sqrt{N_k}|N_k - 1\rangle,$$
$$a_k^\dagger|N_k\rangle = \sqrt{1 - N_k}|N_k + 1\rangle. \qquad (13.1.6)$$

This is known as a *fermionic* field, and the associated particles are called *fermions*. Although we have left off the constant 1/2 in (13.1.4), one can argue that there is a similar zero-point energy, as we will see in Section 13.1.2.

Since there are many possible k-states, we can define Fock states and field operators for fermions just as for boson particles like phonons and photons. The many-body Fock number states are written as $|\ldots N_{k_1}, N_{k_2}, N_{k_3}, \ldots\rangle$ just as for phonons and photons but with the constraint that all N_k must have values only of either 0 or 1.

Although the fermion formalism involves a simple sign change in going from the commutation relation of bosons to the anticommutation relation (13.1.5), it has enormous physical consequences. Relation (13.1.6) ensures the law of Pauli exclusion, that each quantum state can only have one of two amplitudes, either 0 or 1. This means, among other things, that we cannot construct a coherent state, as discussed in Section 12.5, from fermion operators. This is the primary reason why phonons and photons are associated with classical waves while electrons and other fermions are not.

Which version is the fundamental version? In Sections 12.2 and 12.4, we started with the familiar properties of classical fields, namely sound and light, and showed that when they are treated quantum mechanically, one naturally gets the particle picture as excitations of the underlying fields, which can then be described by the particle operators defined by (13.1.1)–(13.1.3). Could we do the same thing with a fermion field?

Section 13.1.2 gives an analogous calculation, but it's not actually necessary to find such equivalence. Instead, we can make yet one more mental shift, to view the commutator (13.1.2) and anticommutator (13.1.5) of *both* types of fields given above as the fundamental properties of the fields. Then it is easy to show that one can start with the properties of a_k and $a_k^\dagger$ operators of boson fields and go backwards to derive the properties of the x and p operators used for classical fields. In other words, the "fundamental" field formalism for both types of field is given by the commutation and anticommutation relations, and because of the physical implications of the sign change, one type of field (bosons) allows classical fields to be constructed from the excitations, while the other does not. The things we observe as x and p, in this approach, are actually epiphenomena of the more basic field quantization rules for bosons.

The fact that the boson and fermion fields have the same mathematical structure except for a positive or negative sign seems to indicate that they have the same underlying basis in reality. Hanging an entire philosophy of ontology (i.e., saying that fermion fields are not real) on a simple sign change seems unwarranted. Why should we say that fields with $+$ sign are "real" and fields with $-$ sign are not real? (Or vice versa, for that matter.)

13.1.2 Visualizing the Fermion Field

The fact that there are only two allowed Fock states for fermionic fields, $|0\rangle$ and $|1\rangle$, allows us to write the fermion creation and destruction operators explicitly in terms of 2×2

matrices. We define creation and destruction operators for fermions analogous to those we used for the bosonic field:

$$a = \frac{1}{2}(\sigma_x - i\sigma_y)$$

$$a^\dagger = \frac{1}{2}(\sigma_x + i\sigma_y), \tag{13.1.7}$$

where the σ operators are Pauli operators (discussed further in Section 13.2), acting on two states $|0\rangle$ and $|1\rangle$. Note that the Pauli operators used here do not act on the spin degree of freedom, introduced in Section 13.2, which interacts with magnetic field. Instead, these Pauli operators act on a different degree of freedom which also has two states, which we have called $|0\rangle$ and $|1\rangle$. In the common picture, we identify $|0\rangle$ as "no particle" and $|1\rangle$ as "one particle."

Written explicitly, these operators are

$$a = \begin{pmatrix} 0 & 0 \\ 1 & 0 \end{pmatrix}, \qquad a^\dagger = \begin{pmatrix} 0 & 1 \\ 0 & 0 \end{pmatrix}. \tag{13.1.8}$$

It is easy to see that $a^\dagger$ has the action of taking the ground state of the system and converting it to the upper state, and vice versa:

$$a^\dagger \begin{pmatrix} 0 \\ 1 \end{pmatrix} = \begin{pmatrix} 1 \\ 0 \end{pmatrix}, \qquad a \begin{pmatrix} 1 \\ 0 \end{pmatrix} = \begin{pmatrix} 0 \\ 1 \end{pmatrix}. \tag{13.1.9}$$

As with the bosonic operators, we can define the number operator

$$\hat{N} = a^\dagger a = \frac{1}{4}(\sigma_x + i\sigma_y)(\sigma_x - i\sigma_y)$$

$$= \frac{1}{4}(\sigma_x^2 + \sigma_y^2 - i[\sigma_x, \sigma_y])$$

$$= \frac{1}{2}(1 + \sigma_z)$$

$$= \begin{pmatrix} 1 & 0 \\ 0 & 0 \end{pmatrix}, \tag{13.1.10}$$

where we have used the properties of the Pauli matrices,

$$\sigma_x^2 = \sigma_y^2 = \sigma_z^2 = 1$$

$$[\sigma_x, \sigma_y] = 2i\sigma_z. \tag{13.1.11}$$

The diagonal elements of the matrix operator for $\hat{N}$ in the last line of (13.1.10) correspond to the allowed fermion occupation numbers, 0 and 1.

These creation and destruction operators also satisfy the fermion anticommutation relation:

$$\{a, a^\dagger\} = \frac{1}{4}(\sigma_x + i\sigma_y)(\sigma_x - i\sigma_y) + (\sigma_x - i\sigma_y)(\sigma_x + i\sigma_y)$$

$$= \frac{1}{4}(2\sigma_x^2 + 2\sigma_y^2 - i[\sigma_x, \sigma_y] + i[\sigma_x, \sigma_y])$$

$$= 1, \tag{13.1.12}$$

which is also easy to see by multiplying the explicit forms of the operators.

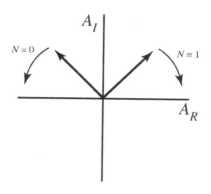

Figure 13.1   The number states of fermions seen as opposite rotations in a complex plane.

As will be shown in Section 13.2, the fermion Hamiltonian can be written in terms of four relativistic spin-eigenstates as

$$H = \sum_{\vec{k},s}(\hbar\omega_{\vec{k},s}\hat{N}_{\vec{k},s,+} - \hbar\omega_{\vec{k},s}\hat{N}_{\vec{k},s,-}), \tag{13.1.13}$$

where the index s refers to one of two spin states, and the final subscript $\pm$ refers to one of two "bands," with either positive or negative energy. It is suggestive to rewrite this in terms of two zero point energies:

$$H = \sum_{\vec{k},s}\left[\hbar\omega_{\vec{k},s}\left(\frac{1}{2}\sigma_z^{(\vec{k},s,+)} + \frac{1}{2}\right) - \hbar\omega_{\vec{k},s}\left(\frac{1}{2}\sigma_z^{(\vec{k},s,-)} + \frac{1}{2}\right)\right]$$

$$= \sum_{\vec{k},s}\frac{1}{2}\hbar\omega_{\vec{k},s}\left(\sigma_z^{(\vec{k},s,+)} - \sigma_z^{(\vec{k},s,-)}\right), \tag{13.1.14}$$

where $\sigma_z^{(\vec{k},s,\pm)}$ is the Pauli spin matrix defined in (13.1.11), for an eigenstate with a particular choice of $\vec{k}$, spin s, and positive-or negative-energy band. This can be explicitly written as

$$\sigma_z^{(\vec{k},s,\pm)} = \begin{pmatrix} 1 & 0 \\ 0 & -1 \end{pmatrix}, \tag{13.1.15}$$

acting on the two states $|0\rangle$ and $|1\rangle$ in each band. If we now switch to the x basis states of the Pauli matrices, the z eigenstates correspond to rotations between these two states. This then leads to the natural interpretation that the state $|1\rangle$ corresponds to clockwise rotation in a complex plane with frequency $\omega_{\vec{k},s}/2$, and the state $|0\rangle$ corresponds to counterclockwise motion, that is, frequency $-\omega_{\vec{k},s}/2$, for the positive-energy band, as illustrated in Figure 13.1. The rotations are the opposite for the negative-energy band.

We thus see that it is no accident that there are two allowed occupation numbers 0 and 1. These can be seen as just two different expressions of having an intrinsic degree of freedom of the field with two allowed states. In other words, *the existence of the states $|0\rangle$ and $|1\rangle$ can be seen as a consequence of rotation of an element of the field with constant amplitude.*

Instead of having an oscillation corresponding to a springiness that stretches from zero to large amplitude, as in the case of bosons, the oscillation in a fermion wave corresponds to a rotation with constant amplitude, with two choices of clockwise or counterclockwise rotation.

The calculation of this section also can be viewed as a *spin-statistics argument*; that is, the anticommutation relation (13.1.12), which gives rise to Pauli exclusion, can be seen as a direct consequence of the assumption that fermion fields correspond to rotations between two states, which in turn follows from the assumption of Paul Dirac, discussed in Section 13.2.1, that the amplitude of the wave must always be constant and never go to zero.

This approach also gives a natural way of thinking about the fact that the Schrödinger wave equation always gives clockwise rotation in the complex plane and never the opposite rotation. It is natural to take the two axes of the rotation plane as the real and imaginary parts of the wave function. The ground state of the system corresponds to all the states in the upper band rotating counterclockwise, and all the states in the negative-energy band rotating clockwise. An excitation out of this ground state flips one rotation in the positive-energy band to the clockwise direction, and one rotation in the negative-energy band to the counterclockwise direction.

The form of the Hamiltonian (13.1.14) means that this ground state has infinite negative total energy, as also found in the analysis of the Dirac equation in terms of a negative-energy "Dirac sea," as we will see in Section 13.2. As discussed in Section 12.2.2, an infinite ground-state energy is not a fundamental problem, because it is a constant like any other constant, which can be subtracted from the total energy. One might argue that the infinite positive zero-point energy of bosons cancels the infinite negative zero-point energy of fermions, but there is no need to do so; the zero-point energy is a constant in any case.

13.1.3 Fermion Spatial Field Operators

As we did for bosonic fields in Section 12.2.3, we can define fermionic spatial field operators that give the amplitude at a point r in space:

$$\hat{\psi}^{\dagger}(r) = \sum_k \frac{e^{-ikr}}{\sqrt{L}} a_k^{\dagger}$$

$$\hat{\psi}(r) = \sum_k \frac{e^{ikr}}{\sqrt{L}} a_k. \tag{13.1.16}$$

The anticommutation relation for the electron spatial field operators is then

$$\{\hat{\psi}(r), \hat{\psi}^{\dagger}(r')\} = \delta(r - r'). \tag{13.1.17}$$

For both fermions and bosons, one can create a particle in a general state ϕ using

$$|\phi\rangle = \int dr\, \phi(r)\hat{\psi}^{\dagger}(r)|0\rangle. \tag{13.1.18}$$

From this, one can define the general creation operator for an electron in state n,

$$a_n^\dagger = \int dr\, \phi_n(r)\hat{\psi}^\dagger(r). \tag{13.1.19}$$

We can multiply by $\phi_n^*(r')$ and sum over a set of states n, to obtain

$$\sum_n \phi^*(r')a_n^\dagger = \int dr \left[\sum_n \phi_n^*(r')\phi_n(r)\right]\hat{\psi}^\dagger(r). \tag{13.1.20}$$

By the same math as discussed in Section 11.2, for a complete set of states, the term in parentheses is equal to $\delta(r - r')$. We therefore can resolve the integral and find

$$\hat{\psi}^\dagger(r') = \sum_n \phi_n^*(r')a_n^\dagger, \tag{13.1.21}$$

for any complete set of state n. Definition (13.1.16) is just a special case of this, since the set of all waves of the form $e^{ikr}/\sqrt{L}$ is a complete set.

13.2 The Dirac Fermion Field

The theory of fermions is one of the great triumphs of twentieth-century physics. Most of the credit belongs to Paul Dirac, who started, like Einstein, with some simple assumptions and laid the foundations for the Pauli exclusion principle of chemistry, Fermi statistics in solids, and antimatter in particle physics.

13.2.1 Derivation of the Dirac Equation

In this section, we reproduce Dirac's (1947) simple but elegant argument to deduce the relativistic wave equation for particles with mass. The basic problem is that the Schrödinger equation

$$i\hbar\frac{\partial}{\partial t}\psi = H\psi \tag{13.2.1}$$

is not relativistically invariant if H is given by the standard kinetic energy $p^2/2m = -\hbar^2\nabla^2/2m$. In relativity, the conserved quantity is $E^2 = (mc^2)^2 + (cp)^2$. The energy is squared in this relativistic invariant, however, while Dirac believed strongly that the equations for the particles should be linear in the time dependence. His argument was that, if it is not so, then the square of the amplitude of the wave function is not constant over time; as discussed in Section 9.1, it "winks out" during every oscillation. This is the case for boson fields, but that did not bother Dirac, because those did not have mass, for any fields he knew.

To obtain a linear equation, we can factor the relativistically invariant term $E^2 - (mc^2)^2 - (cp)^2$ into two linear terms as follows:

$$E^2 - (mc^2)^2 - |c\vec{p}|^2 = (E + \alpha_0 mc^2 + c\vec{\alpha}\cdot\vec{p})(E - \alpha_0 mc^2 - c\vec{\alpha}\cdot\vec{p}) = 0, \tag{13.2.2}$$

where α_i are four new operators that commute with $\vec{p}$. Clearly, if either of the two factors is zero, then the full relativistic term is zero, satisfying relativistic invariance. Since the components of α_i all commute with the components of $\vec{p}$, they must describe some extra degree of freedom.

It turns out that this factorization is only possible if the operator $\vec{\alpha}$ has the anticommutation property

$$\{\alpha_i, \alpha_j\} = \alpha_i\alpha_j + \alpha_j\alpha_i = 2\delta_{ij}. \tag{13.2.3}$$

In order to have four linearly independent operators that anticommute, the new operator must be represented by a matrix with at least four rows and columns. There is not one unique choice for these matrices, and various theories have been developed for different representations. The standard choice, following Dirac, is the following:

$$\alpha_0 = \left(\begin{array}{c|c} E & 0 \\ \hline 0 & -E \end{array}\right), \qquad \alpha_i = \left(\begin{array}{c|c} 0 & \sigma_i \\ \hline \sigma_i & 0 \end{array}\right), \tag{13.2.4}$$

where σ_i are the standard 2×2 Pauli spin matrices

$$\sigma_x = \left(\begin{array}{cc} 0 & 1 \\ 1 & 0 \end{array}\right), \qquad \sigma_y = \left(\begin{array}{cc} 0 & -i \\ i & 0 \end{array}\right), \qquad \sigma_z = \left(\begin{array}{cc} 1 & 0 \\ 0 & -1 \end{array}\right),$$

$$E = \left(\begin{array}{cc} 1 & 0 \\ 0 & 1 \end{array}\right). \tag{13.2.5}$$

These are the standard spin matrices used in quantum mechanics textbooks.

We can therefore write a relativistically invariant wave equation for particles with mass as follows:

$$i\hbar\frac{\partial}{\partial t}|\psi\rangle = H|\psi\rangle = (\alpha_0 mc^2 + c\vec{\alpha} \cdot \vec{p})|\psi\rangle, \tag{13.2.6}$$

where $|\psi\rangle$ has four components. This is the *Dirac equation*. Note that even when the momentum $p_i = -i\hbar\nabla$ gives zero contribution, there are both positive- and negative-energy solutions, or bands, with $\langle H \rangle = \pm mc^2$. This symmetry means that it doesn't matter that we used (13.2.6) instead of $H = -\alpha_0 mc^2 - c\vec{\alpha} \cdot \vec{p}$, though both are equally valid according to (13.2.2). It does raise the interesting equation of what we mean by negative-energy solutions, though. Dirac hypothesized that the negative-energy solutions are all filled with electrons, one per state according to the Pauli exclusion principle. Therefore, unless they are given enough energy to jump up to the positive-energy band (at least $2mc^2$, which is a million electron volts), they will have no effect on electrons in positive states. The energy of this *Dirac sea* of negative-energy electrons does not matter because it is a constant; as discussed in Section 12.2, we are free to define the vacuum as the ground state of the system and treat its energy as a constant, even if that constant is infinite.

The existence of the negative-energy sea led Dirac to predict the existence of *positrons* and *antimatter* in general. This was a tremendous success of theoretical physics driving experiment, with a successful prediction. When energy is given to an electron in the negative-energy band, it can jump to the positive-energy band, making a standard, free electron. It will leave behind an empty state, sometimes called a *hole*, in the negative-energy states. This hole is like a bubble of air in a glass of water: it can move around, and it reacts in the opposite direction as an electron to forces. To all intents and purposes, it acts like a particle with positive charge. It is therefore called a *positron*. An electron in the positive-energy states can fall back down into a hole like this in the negative-energy states, at which point both the free electron and the positron would disappear, releasing energy of at least $2mc^2$. This is known as *recombination* or *annihilation* of electrons and holes. In other words, matter and antimatter annihilate each other.[1]

Although we may think of electrons as elementary particles, we can see from this analysis that they are excitations out of the vacuum state, which can be created and annihilated, just like photons and phonons. The same is true of all fermions.

The symmetry of the Dirac equations raises a philosophical problem. If the ground state of the universe is a filled Dirac sea with no excitations (i.e., no free electrons or holes), we would expect equal amounts of matter and antimatter, because the excitation of one electron to the positive-energy states leaves behind exactly one hole. If this were the case, however, we would expect all the free electrons to recombine with holes eventually, and we could not endure the energy released by all the recombination. Some cosmologists therefore believe that there is some small, additional term that must be added to the equations to give an extra amount of matter even in the ground state, but this would break the aesthetically appealing symmetry of the Dirac equations.

13.2.2 The Dirac Equation and Spin

Dirac's formalism requires four states, two in a positive-energy band and two in a negative-energy band. The two states in each band are known as the two "spin" states. We can see the connection of this spin degree of freedom with angular momentum by looking at the effect of a magnetic field on the particles.

In the presence of a magnetic field, we modify the relativistically invariant term by the standard substitution $\vec{p} \to \vec{p} - q\vec{A}$, where q is the charge, and $\vec{A}$ is the vector potential. The invariant term (13.2.2) thus becomes

$$E^2 - (mc^2)^2 - |c\vec{\alpha} \cdot (\vec{p} - q\vec{A})|^2 = 0. \qquad (13.2.7)$$

We would like to find the Hamiltonian that includes magnetic field in the nonrelativistic limit. To do this, we can adopt the strategy used in standard relativistic mechanics, in which we assume that the momentum terms are small compared to mc, and expand in a Taylor series, which allows us to write $E = \sqrt{(mc^2)^2 + (cp)^2} \simeq mc^2 + p^2/2m$. The last term of (13.2.7) can be simplified by using the properties of the α_i matrices, namely the anticommutation rule (13.2.3) and $\alpha_x \alpha_y = i\sigma_z$, for cyclic permutations, as well as the definition $p = -i\hbar\nabla$, so that we obtain

$$E^2 = (mc^2)^2 + c^2|\vec{p} - q\vec{A}|^2 + \hbar c^2 \vec{\sigma} \cdot (\nabla \times \vec{A}). \qquad (13.2.8)$$

---

[1] The concept of free electrons and holes was actually first worked out in the context of solid state physics, for electron bands in solids, and adapted to Dirac's theory. Conceptually, there is no difference between the two cases: in each case, the ground state of the system is found, with an energy gap to the first set of excited states. For an extended discussion of hole states in solids, see Snoke 2020, Chapter 2.

Writing E as a Taylor series of the square root of the right side, and recalling $\vec{B} = \nabla \times \vec{A}$, we then have

$$E \simeq mc^2 + \frac{1}{2m}|\vec{p} - q\vec{A}|^2 + \frac{q\hbar}{2m}\vec{\sigma} \cdot \vec{B}. \qquad (13.2.9)$$

The first two terms correspond to the standard nonrelativistic Hamiltonian for a particle in a magnetic field, while the last term is a new term that arises from the fact that we needed to introduce the α matrices for the relativistic wave equation. This term is the Zeeman spin splitting in magnetic field, which corresponds to particles with magnetic moment of $\pm\hbar/2$. Thus, we obtain the standard picture of fermions as having half-integer spin.

Recalling the discussion of Section 13.1.2, we see that there is a natural connection of the fermion statistics to the existence of half-integer spin. As we have seen, the spin of fermions follows from the assumption of Dirac that the amplitude of the electron wave in any state is never zero; instead, it rotates in the complex plane with constant amplitude. Getting this requirement to work with relativity led to all of the formalism of spin and negative-energy states given in Section 13.2.1.

As seen in Section 13.1.2, a rotation with constant amplitude in the complex plane, with two components equal to the real and imaginary parts, can be rewritten in the basis of the two states $|0\rangle$ and $|1\rangle$ that correspond to "unoccupied" and "occupied." As shown in that section, to make this work out, the amplitude operators must anticommute, which in turn implies fermion statistics.

References

P. A. M. Dirac, *The Principles of Quantum Mechanics*, 3rd ed., (Oxford University Press, 1947).

D. W. Snoke, *Solid State Physics: Essential Concepts*, (Cambridge University Press, 2020).

14

Transition Rules

The equations of quantum mechanics give predictions for the deterministic evolution of the wave function of many-particle systems. Schrödinger's equation (introduced in Section 9.1 and presented in Dirac notation in Section 11.4) is the overall controlling equation for time dependence. Another rule, known as *Fermi's golden rule*, can be derived from it. Fermi's golden rule is useful when we can treat a system as a set of distinct energy states ("eigenstates") with weak interactions between them. Then we can use Fermi's golden rule to find the relative weight of each eigenstate in the system after some time.

Fermi's golden rule implicitly assumes the existence of *decoherence*, which is introduced by coupling the system of interest to a huge number of other states. Decoherence leads to irreversible behavior, and is a major topic of study, which is discussed in Part V of this book. In the second half of this chapter, we introduce a different model, known as the *Bloch equations*, which allows decoherence to be either left out or included in small amounts.

14.1 Fermi's Golden Rule

Fermi's golden rule is purely a wave result but has a natural interpretation in terms of particle transition probabilities when we write it for the many-particle Fock states introduced in Sections 12.2 and 13.1.

14.1.1 Derivation of Fermi's Golden Rule

We begin by assuming that the Hamiltonian of a quantum mechanical system is given by

$$H = H_0 + \hat{V}, \tag{14.1.1}$$

where H_0 is the energy of just the photons and electrons separately, and $\hat{V}$ is a term for the energy associated with interactions between the particles. We assume that $\hat{V}$ gives small alterations of the eigenstates of H_0. We then talk of transitions between the eigenstates (i.e., the particle states) of H_0 due to $\hat{V}$.

We now adopt the interaction representation, introduced in Section 11.4, and write the state of the system at time $t = 0$ as $|\psi(0)\rangle$. In the interaction representation, we define the state $|\psi(t)\rangle$, which obeys the Schrödinger equation

$$i\hbar \frac{\partial}{\partial t}|\psi(t)\rangle = \hat{V}(t)|\psi(t)\rangle. \tag{14.1.2}$$

Integrating both sides of this equation, we have

$$|\psi(t)\rangle = |\psi(0)\rangle + \frac{1}{i\hbar} \int_0^t \hat{V}(t')|\psi(t')\rangle dt'. \tag{14.1.3}$$

If $\hat{V}$ is sufficiently weak, and the time elapsed is sufficiently short, then to first order in t, we can approximate $|\psi(t')\rangle = |\psi(0)\rangle$, which implies

$$|\psi(t)\rangle = \left(1 + \frac{1}{i\hbar} \int_0^t \hat{V}(t')dt'\right)|\psi(0)\rangle. \tag{14.1.4}$$

To find $|\psi(t)\rangle$ to second order in t, one can substitute this approximation for $|\psi(t)\rangle$ into (14.1.3). By repeated substitution, one can show that

$$|\psi(t)\rangle = \left(1 + (1/i\hbar) \int_0^t dt'\, \hat{V}(t') + (1/i\hbar)^2 \int_0^t dt' \int_0^{t'} dt''\, \hat{V}(t')\hat{V}(t'') + \dots\right)|\psi(0)\rangle, \tag{14.1.5}$$

which we can write in shorthand as

$$|\psi(t)\rangle = e^{-(i/\hbar)\int_0^t \hat{V}(t')dt'}|\psi(0)\rangle \equiv S(t,0)|\psi(0)\rangle. \tag{14.1.6}$$

The operator $S(t, 0)$ is unitary, which means that the wave function is normalized at all times so that $\langle\psi(t)|\psi(t)\rangle = 1$.

Suppose that the system at time $t = 0$ is in the initial state $|i\rangle$, which is an eigenstate of H_0. Using the definition of the full state,

$$|\psi_t\rangle = e^{-iH_0 t/\hbar}|\psi(t)\rangle, \tag{14.1.7}$$

the first-order expansion (14.1.4) gives the overlap of the system being in a final, different eigenstate $|f\rangle$ at time t,

$$\begin{aligned}
\langle f|\psi_t\rangle &= e^{-(i/\hbar)E_f t}\langle f|\psi(t)\rangle \\
&= e^{-(i/\hbar)E_f t}\frac{1}{i\hbar} \int_0^t \langle f|\hat{V}(t')|i\rangle\, dt' \\
&= e^{-(i/\hbar)E_f t}\frac{1}{i\hbar}\langle f|\hat{V}|i\rangle \int_0^t e^{(i/\hbar)(E_f - E_i)t'}\, dt',
\end{aligned} \tag{14.1.8}$$

where we have used the fact that at time $t = 0$, the initial state is the same in both representations, $|\psi(0)\rangle = |\psi_0\rangle = |i\rangle$, and we have assumed that $\langle f|i\rangle = 0$.

The total weight of state $|f\rangle$ at time t is therefore

$$\begin{aligned}
|\langle f|\psi_t\rangle|^2 &= \frac{1}{\hbar^2}|\langle f|\hat{V}|i\rangle|^2 \left|\int_0^t e^{(i/\hbar)(E_f - E_i)t'}\, dt'\right|^2 \\
&= \frac{1}{\hbar^2}|\langle f|\hat{V}|i\rangle|^2 \left|\frac{e^{(i/\hbar)(E_f - E_i)t} - 1}{(i/\hbar)(E_f - E_i)}\right|^2 \\
&= |\langle f|\hat{V}|i\rangle|^2 \frac{\sin^2[(E_f - E_i)t/2\hbar]}{[(E_f - E_i)/2]^2}.
\end{aligned} \tag{14.1.9}$$

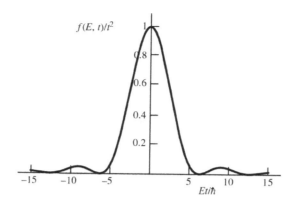

$f(E, t)/t^2$

The peak function $f(E, t) = \sin^2(Et/2\hbar)/(E/2\hbar)^2$, which appears in the calculation of Fermi's golden rule.

The function $\sin^2(Et/2\hbar)/(E/2)^2$ is a peak with oscillating wings, as shown in Figure 14.1. As time increases, the zeroes of the oscillations get closer together in energy, so that the width of the central peak decreases, and the height of the peak increases, so that it looks more and more like a δ-function, which is equal to infinite when $E_i = E_f$ and zero otherwise. For times t long compared to the oscillation time, we can take the mathematical limit,

$$\lim_{t \to \infty} \frac{|e^{ixt} - 1|^2}{x^2} = \lim_{t \to \infty} \frac{\sin^2(xt/2)}{(x/2)^2}$$
$$= \delta(x)2\pi t. \tag{14.1.10}$$

We can therefore write

$$\frac{\partial}{\partial t}|\langle f|\psi_t\rangle|^2 = \frac{2\pi}{\hbar}|\langle f|\hat{V}|i\rangle|^2\delta(E_f - E_i). \tag{14.1.11}$$

This is one form of *Fermi's golden rule* for transitions from one quantum state to another. The term $\langle f|\hat{V}|i\rangle$ is called the *matrix element*, and gives the strength of the interaction, which controls the speed of how fast transitions occur.

If we interpret the square of the wave function as a probability, according to the Born rule (defined in Section 11.3), Fermi's golden rule gives us the probability of a transition from state $|i\rangle$ to state $|f\rangle$ per unit time. More generally, it tells us that an initial, prepared state will tend to dissipate as it seeps into many other states, and gives us the relative fraction of the weight given to each of these states.

Although the δ-function has value equal to infinity when $E_f = E_i$, it will not cause a problem, because in any real calculation, a sum is done over many final states $|f\rangle$. When this sum is converted to an integral for a continuous range of states, the δ-function inside the integral will give a finite number.[1]

---

[1] In deriving Fermi's golden rule, we made two assumptions about the time scale. In (14.1.8), we assumed that t was small, so that we could ignore higher-order terms. On the other hand, in using (14.1.10), we assumed that t was long enough to treat the oscillating function as a δ-function. Both of these conditions can be true if

We see here that the general principle of *conservation of energy* arises naturally in quantum mechanics as a wave property; it is seen here in the fact that the δ-function ensures that the initial and final state must have the same energy. This comes from the interference of the initial and final wave states; if they have different frequencies, they will not oscillate in phase with each other and will tend to cancel out. Although we have used an approximation here, keeping only the lowest-order term in (14.1.4), it can be shown that, when higher-order terms are kept, there will always be the same interference term between the initial and final states which enforces energy conservation.

14.1.2 Fermi's Golden Rule and Quantum Statistics

Suppose now that the eigenstates of H_0 are many-body Fock states. If there are interactions between these states, then the Hamiltonian will have terms that do not only contain the number operator $N = a_k^\dagger a_k$ but also have unbalanced creation and destruction operators that do not conserve the number in a k-state. In Section 14.2, we will look at some typical interactions. As a generic example, consider the interaction term for bosons

$$\hat{V} = \sum_{k,k'} A_{kk'}\, a_{k'}^\dagger a_k, \tag{14.1.12}$$

where $A_{k,k'}$ is a number that gives the strength of the interaction between states k and k'. Each term in the sum corresponds to destroying a particle in state k and recreating it in state k'.

If the initial state is $|i\rangle = |\ldots, N_i, \ldots, N_f, \ldots\rangle$ and the final state is $|f\rangle = |\ldots, N_i - 1, \ldots, N_f + 1, \ldots\rangle$, Fermi's golden rule gives us

$$\frac{\partial}{\partial t}|\langle f|\psi_t\rangle|^2 = \frac{2\pi}{\hbar} \left| \langle f| \sum_{k,k'} A_{k,k'}\, a_{k'}^\dagger a_k\, |i\rangle \right|^2 \delta(E_f - E_i)$$

$$= \frac{2\pi}{\hbar} \left| A_{i,f}\sqrt{1 + N_f}\sqrt{N_i} \right|^2 \delta(E_f - E_i)$$

$$= \frac{2\pi}{\hbar} |A_{i,f}|^2 (1 + N_f)N_i\, \delta(E_f - E_i). \tag{14.1.13}$$

The Fock states were eliminated in the second step because we assume that the pair of operators $a_{k'}^\dagger a_k$ turned the initial state $|i\rangle$ into the final state $|f\rangle$, and we assume that the eigenstates are normalized, that is, $\langle f|f\rangle = 1$. If the pair of operators does not turn $|i\rangle$ into $|f\rangle$, then we use the property that the eigenstates are orthogonal, that is, $\langle f|m\rangle = 0$ if the state m is not equal to the state f.

Treating the numbers N_i and N_f as particle numbers, the factor N_i in this formula expresses the simple fact that the rate of scattering out of a state is proportional to the number of particles in that state in the first place. The $(1 + N_f)$ term depends on the number already in the final state. This expresses the law of *stimulated transitions* of bosons,

---

t falls in a proper intermediate time range. The field of study known as *quantum kinetics* deals with situations when Fermi's golden rule breaks down, for example, when a laser pulse excites the electrons in a solid on femtosecond time scales. (For a review, see Haug 2004.)

as discussed in Section 2.4; namely, the rate of transition to a final state f depends on the number of particles in the final state.

If we had used fermion states instead of bosons, then we would have a factor $(1 - N_f)$ instead, which expresses the law of Pauli exclusion, that two fermions cannot occupy the same state. We can therefore summarize that the rate of a transition per particle is given by

$$\Gamma_{i \to f} = \frac{1}{N_i} \frac{\partial}{\partial t} |\langle \psi_f | \psi_t \rangle|^2 \propto \begin{cases} (1 + N_f) & \text{bosons} \\ (1 - N_f) & \text{fermions.} \end{cases} \tag{14.1.14}$$

A tremendous amount of physics is included in these two statements. All of chemistry is dependent on the principle of Pauli exclusion, while all of the physics of Planck radiation, lasers, and superfluids are implied by the principle of stimulated scattering.

14.2 Interaction Terms

In Chapters 12 and 13, we derived the form of the Hamiltonian for boson and fermion fields by themselves. Interesting things happen, however, when different fields interact with each other. In this section, we will derive some typical interaction terms.

The general method of deriving these interactions starts with knowing the classical limit of these interactions and then substituting quantized fields for the classical fields in these interactions.

14.2.1 Electron–Phonon Interactions

Let us start with the simple case of electrons in a crystal subjected to a strain. In general, it is not hard to imagine that the energy of the electron will depend on the strain in the crystal, because as discussed in Section 9.3, there is an associated energy of moving electrons near to each other. In this section, we will derive the electron–phonon interaction for a simple longitudinal sound wave.

The simplest assumption, which can be viewed as the lowest-order approximation, is that the change in electron energy is proportional to the strain. This is known as the *linear response* assumption. It is generally a good approximation for weak enough strains. We can therefore write the energy of the interaction as

$$V_{\text{int}} = D \left(\frac{\partial u_x}{\partial x} + \frac{\partial u_y}{\partial y} + \frac{\partial u_z}{\partial z} \right), \tag{14.2.1}$$

where D is a constant with units of energy known as the *deformation potential*, and u_x, u_y, and u_z are displacement distances for the atoms in a given region of the crystal. The energy caused by the strain depends on the derivatives of these, not on their absolute values, because a rigid shift of the crystal to a new position will not change the energy of an electron inside it; electron energies are only affected by distortion of the crystal, which gives gradients of the positions of the atoms.

The displacement u can be related to the quantized phonon amplitude by formula (12.2.31) from Section 12.2. We write (using u for displacement instead of x)

$$\vec{u}(\vec{x}) = \sum_{\vec{k}} \hat{k} \sqrt{\frac{\hbar}{2\rho V \omega_k}} \left(a_{\vec{k}} e^{i\vec{k}\cdot\vec{x}} + a_{\vec{k}}^{\dagger} e^{-i\vec{k}\cdot\vec{x}} \right), \tag{14.2.2}$$

where we have used vectors to account for the possibility of motion in all directions. The displacement of the atoms is assumed to be in the same direction as the motion of the wave, which is given by a unit vector $\hat{k} = \vec{k}/k$, where k is the magnitude of the vector $\vec{k}$, that is,

$$k = \sqrt{k_x^2 + k_y^2 + k_z^2}. \tag{14.2.3}$$

Substituting the quantized displacement term into (14.2.2), we obtain

$$V_{\text{int}} = \sum_{\vec{k}} D \sqrt{\frac{\hbar}{2\rho V \omega_k}} \, ik \left(a_{\vec{k}} e^{i\vec{k}\cdot\vec{x}} - a_{\vec{k}}^{\dagger} e^{-i\vec{k}\cdot\vec{x}} \right). \tag{14.2.4}$$

The derivatives in (14.2.1) acting on the exponential factors have led to multiplication by ik.

The interaction energy given by (14.2.4) is the energy per electron at one point $\vec{x}$ in space. To account for the quantum mechanical nature of the electrons, we must integrate this energy times the average density of electrons at all points in space. This density is found by the operator $\hat{n}(\vec{x}) = \hat{\psi}^{\dagger}(\vec{x})\hat{\psi}(\vec{x})$, where $\hat{\psi}(\vec{x})$ is the spatial field operator for electrons defined in Section 13.1.3. We therefore write the total interaction energy operator as

$$V_{\text{e-phon}} = \int d^3x \, V_{\text{int}}(\vec{x})\psi^{\dagger}(\vec{x})\psi(\vec{x}). \tag{14.2.5}$$

From Section 13.1.3, the spatial field operators are written as

$$\hat{\psi}(\vec{x}) = \sum_{\vec{k}} \frac{e^{i\vec{k}\cdot\vec{x}}}{\sqrt{V}} b_{\vec{k}},$$

$$\hat{\psi}^{\dagger}(\vec{x}) = \sum_{\vec{k}} \frac{e^{-i\vec{k}\cdot\vec{x}}}{\sqrt{V}} b_{\vec{k}}^{\dagger}, \tag{14.2.6}$$

where we have written the fermion creation and destruction operators as $b_{\vec{k}}^{\dagger}$ and $b_{\vec{k}}$ to distinguish them from the boson operators $a_{\vec{k}}^{\dagger}$ and $a_{\vec{k}}$ for the phonons.

Each sum over a k-vector needs a separate label. Therefore, inserting (14.2.4) into (14.2.5), we have

$$\begin{aligned} V_{\text{e-phon}} &= \sum_{\vec{k}_1,\vec{k}_2} \int d^3x \, V_{\text{int}}(\vec{x}) \left(\frac{e^{-i\vec{k}_2\cdot\vec{x}}}{\sqrt{V}} b_{\vec{k}_2}^{\dagger} \right) \left(\frac{e^{i\vec{k}_1\cdot\vec{x}}}{\sqrt{V}} b_{\vec{k}_1} \right) \\ &= \sum_{\vec{k},\vec{k}_1,\vec{k}_2} Dk \sqrt{\frac{\hbar}{2\rho V \omega_k}} \left[i \, a_{\vec{k}} b_{\vec{k}_2}^{\dagger} b_{\vec{k}_1} \left(\frac{1}{V} \int d^3x \, e^{i(\vec{k}_1-\vec{k}_2+\vec{k})\cdot\vec{x}} \right) \right. \\ &\quad \left. - i \, a_{\vec{k}}^{\dagger} b_{\vec{k}_2}^{\dagger} b_{\vec{k}_1} \left(\frac{1}{V} \int d^3x \, e^{i(\vec{k}_1-\vec{k}_2-\vec{k})\cdot\vec{x}} \right) \right]. \end{aligned} \tag{14.2.7}$$

The integrals over $\vec{x}$ are equal to a Kronecker delta-function, equal to 1 when the k-vectors sum to zero, and 0 otherwise. We therefore have, finally,

$$V_{\text{e-phon}} = \sum_{\vec{k},\vec{k}_1} Dk \sqrt{\frac{\hbar}{2\rho V \omega_k}} \, i \left(a_{\vec{k}} b^{\dagger}_{\vec{k}_1+\vec{k}} b_{\vec{k}_1} - a^{\dagger}_{\vec{k}} b^{\dagger}_{\vec{k}_1-\vec{k}} b_{\vec{k}_1} \right). \tag{14.2.8}$$

The creation and destruction operators give us a simple way of looking at the interaction. The Hamiltonian can be viewed as the sum of all possible momentum-conserving scattering processes – in the first term in the parentheses, an electron starts with momentum $\hbar\vec{k}_1$ and absorbs a phonon with momentum $\hbar\vec{k}$, ending up with momentum $\hbar(\vec{k} + \vec{k}_1)$, and in the second term, the electron emits a phonon with momentum $\vec{k}$. Note that the electron–phonon interaction does not conserve the total number of phonons – phonons appear when they are emitted and disappear when they are absorbed by the electrons. Since the electron operators occur only in pairs $b^{\dagger}b$, the number of electrons is conserved.

Although this is a convenient picture, note that we do not have to think in terms of the particle picture. As we have seen, the a_k and b_k operators give measurements of the amplitude of a wave. Note also that momentum conservation has arisen here as a wave property (sometimes called *phase matching*), just as energy conservation did in Fermi's golden rule.

14.2.2 Electron–Photon Interactions

The same approach can be used for electron-photon interactions as used for electron-phonon interactions. Again, we need a term in the Hamiltonian which depends on both the photon and electron wave function. We can deduce this from the fundamental electromagnetic theory.

As discussed in many textbooks (e.g., Cohen-Tannoudji 1977), when there is an electromagnetic field, we must replace the p operator with $\vec{p} - q\vec{A}$, where $\vec{A}$ is the electromagnetic vector potential. We therefore rewrite the kinetic energy of an electron in the presence of a magnetic field by the rule

$$\frac{p^2}{2m} \rightarrow \frac{|\vec{p} - q\vec{A}|^2}{2m} = \frac{(p^2 - q\vec{p} \cdot \vec{A} - q\vec{A} \cdot \vec{p} + q^2 A^2)}{2m}. \tag{14.2.9}$$

Using the quantum mechanical definition $\vec{p} = -i\hbar\nabla$, we can write

$$\vec{p} \cdot \vec{A}\psi = -i\hbar(\nabla \cdot \vec{A})\psi - i\hbar\vec{A} \cdot (\nabla\psi). \tag{14.2.10}$$

In the Coulomb gauge, also known as the radiation gauge or transverse gauge, we set $\nabla \cdot \vec{A} = 0$, and therefore $\vec{p} \cdot \vec{A} = \vec{A} \cdot \vec{p}$. We thus have

$$H = \frac{p^2}{2m} + U - \frac{q}{m}\vec{A} \cdot \vec{p} + \frac{q^2}{2m}A^2. \tag{14.2.11}$$

The sum $p^2/2m + U$ gives the electron energy in the absence of the electromagnetic field. The last term is proportional to A^2, and therefore can be ignored for electromagnetic waves with low amplitude. This leaves us with a term for the interaction of charged particles with electromagnetic field,

$$V_{\text{int}} = -\frac{q}{m}\vec{A}\cdot\vec{p}. \tag{14.2.12}$$

The mass m is the vacuum electron mass, even if we are dealing with electrons in the solid.

The total Hamiltonian will be this energy weighted by the square of the amplitude of the electron wave function at each location, which gives the electron density. From Section 13.1.3, we can write the spatial field operator of the electrons in terms of a set of states n:

$$\hat{\psi}^\dagger(\vec{x}') = \sum_n \phi_n^*(\vec{x}')b_n^\dagger, \tag{14.2.13}$$

where $b_n^\dagger$ is the fermion creation operator for state n. The full Hamiltonian is then

$$V_{\text{e-phot}} = \int d^3x\,\hat{\psi}^\dagger(\vec{x})V_{\text{int}}(\vec{x})\hat{\psi}(\vec{x})$$

$$= \sum_{n,n'}\int d^3\,\phi_n^*(\vec{x})b_n^\dagger\,V_{\text{int}}(\vec{x})\phi_{n'}(\vec{x})b_{n'}. \tag{14.2.14}$$

Note that we put the interaction term between the two electron field operators that give the density, because the p-operator acts on the wave function of the electrons, not the square of their wave function.

Writing this out explicitly, we have

$$V_{\text{e-phot}} = -\sum_{n,n'}b_n^\dagger b_{n'}\frac{q}{m}\int d^3x\left(-i\hbar\phi_n^*(\vec{x})\vec{A}(\vec{x})\cdot\nabla\phi_{n'}(\vec{x})\right), \tag{14.2.15}$$

where we have used the quantum mechanical definition of the momentum acting on the electron states.

If the electromagnetic wavelength is long compared to the size of an electron state n, we can treat the electromagnetic field $\vec{A}$ as a constant in space and remove it from the integral. In this case, we have

$$V_{\text{e-phot}} = -\sum_{n,n'}b_n^\dagger b_{n'}\frac{q}{m}\vec{A}\cdot\langle n|\vec{p}|n'\rangle, \tag{14.2.16}$$

where

$$\langle n|\vec{p}|n'\rangle \equiv -i\hbar\int d^3x\,\phi_n^*(\vec{x})\nabla\phi_{n'}(\vec{x}) \tag{14.2.17}$$

is a *transition matrix element*, which is an intrinsic property of the states n and n'.

We can write this matrix element in terms of the quantum mechanical $\vec{x}$ operator by using the commutation relation $[x, H_0] = [x, p^2/2m + U(x)] = i\hbar p_x/m$, based on the commutation relation $[x, p_x] = i\hbar$. Therefore,

$$\langle n|[x, H_0]|n'\rangle = \langle n|(xH_0 - H_0 x)|n'\rangle = (E_{n'} - E_n)\langle n|x|n'\rangle$$

$$= \frac{i\hbar}{m}\langle n|p_x|n'\rangle, \tag{14.2.18}$$

and consequently

$$\langle n|\vec{p}|n'\rangle = im\omega_0\langle n|\vec{x}|n'\rangle, \tag{14.2.19}$$

where $\omega_0 = (E_n - E_{n'})/\hbar$. This relation is exact if $|n\rangle$ and $|n'\rangle$ are eigenstates of the Hamiltonian.

General electron–photon interaction. We now have a choice of how to treat the electromagnetic vector field $\vec{A}$. We could keep it as a classical field, which corresponds to the coherent-state limit of a boson field, or we could do as we did with phonons, and use the quantized version for $\vec{A}$.

In Section 14.3, we will adopt the coherent-state approach. Here, we follow the same approach as with phonons in Section 14.2. Using (12.4.11), we write

$$\vec{A}(\vec{x}) = \sum_{\vec{k}} \hat{\eta} \sqrt{\frac{\hbar}{2\varepsilon V\omega}} \left(a_{\vec{k}} e^{i\vec{k}\cdot\vec{x}} + a_{\vec{k}}^\dagger e^{-i\vec{k}\cdot\vec{x}} \right), \tag{14.2.20}$$

where $\hat{\eta}$ is a unit-length polarization vector of the wave, and we use the appropriate permittivity ε for the medium.

Since we have already made the assumption that the wavelength of $\vec{A}$ is long compared to the size of the electron states, which means $\vec{k}$ is small, we can approximate that the exponential factors are nearly equal to 1. This gives us

$$V_{\text{e-phot}} = -\sum_{n,n'} \sum_{\vec{k}} \left(b_n^\dagger b_{n'} - b_n^\dagger b_{n'} \right) \left(a_{\vec{k}} + a_{\vec{k}}^\dagger \right) \frac{q}{m} \hat{\eta} \cdot \langle n|\vec{p}|n'\rangle.$$

$$\tag{14.2.21}$$

There is no explicit momentum conservation in this case, because we have assumed that the electrons are in small, localized states, which allowed us to pull $\vec{A}$ out of the integral over space. If we did not make that approximation, we would have a spatial dependence of $\vec{A}$ that would give us a similar momentum conservation rule as with phonons.

The two operators for photons give either photon absorption or emission. Energy conservation, derived for Fermi's golden rule in Section 14.1.1, will determine which electron states can couple to which via emission or absorption.

14.2.3 Other Interactions

The same approach can be used to generate interaction terms for all kinds of fields. Without going through the derivations, we give here a few more results:[2]

For electrons in a solid interacting with localized defects, we have

$$H_{\text{def-e}} = \frac{\Delta U}{V} \left(\frac{4\pi a^3}{3} \right) \sum_{\vec{k}_1, \vec{k}_2} b_{\vec{k}_2}^\dagger b_{\vec{k}_1}, \tag{14.2.22}$$

where the electron states are plane-wave states with definite momenta $\hbar\vec{k}$. Note that in this case, as in the case of photons absorbed by localized states, momentum is not conserved – an electron can approach a defect at any angle and scatter out at any angle.

[2] For a derivation of these, see Snoke 2020, Chapter.

For electrons interacting with other electrons, we have

$$V_{\text{e-e}} = \frac{1}{2V} \sum_{\vec{k}_1, \vec{k}_2, \vec{k}_3} \frac{e^2/\varepsilon}{|\Delta\vec{k}|^2 + \kappa^2} \, b^\dagger_{\vec{k}_4} b^\dagger_{\vec{k}_3} b_{\vec{k}_2} b_{\vec{k}_1}, \qquad (14.2.23)$$

where $\Delta\vec{k} = \vec{k}_1 - \vec{k}_4$, and κ is a screening constant that depends on the properties of the medium. Here we see two destruction operators that eliminate incoming electrons in plane-wave states, and two creation operators that replace those electrons in two new states. In this case, momentum is strictly conserved, such that $\vec{k}_1 + \vec{k}_2 = \vec{k}_3 + \vec{k}_4$, so that there is no sum over $\vec{k}_4$.

In each of these, as in the previous two sections, we could adopt the particle picture and treat these interactions as processes in which incoming particles are destroyed and outgoing particles are created. But all of these interactions are primarily *energies* that depend on the *amplitudes of waves*, given by the a and b operators.

14.3 Optical Transitions

As discussed in Section 3.2, the wave equations of quantum mechanics give behavior that looks like quantum jumps, without the need to invoke discontinuities. Before we derive the equations used to generate the plots shown in that section, we first derive a general rule for interaction of electromagnetic waves and electrons.

14.3.1 Derivation of the Bloch Equations for a Two-Level System

Most of the math of quantum states interacting with electromagnetic fields comes originally from work in nuclear magnetic resonance in the 1950s. The same formalism applies to any two-level system, however, and, in particular, optical excitation of electrons in atoms. The same formalism is also used in the modern field of quantum information science, as discussed in Section 8.4 in which a *qubit* is defined as a system with two states.

We consider a transition between just two states, which we will label g and e for the ground and excited states, respectively. This is justified in many cases when the electromagnetic wave is tuned to have a frequency corresponding to the energy difference between two electron states.

Using our results from Section 14.2.2, we write the Hamiltonian including (14.2.16) for the electromagnetic interaction with the electrons as

$$H = \left(E_g b^\dagger_g b_g + E_e b^\dagger_e b_e \right) - iq\omega_0 A \left(b^\dagger_e b_g \langle e|x|g\rangle - b^\dagger_e b_g \langle g|x|e\rangle \right), \qquad (14.3.1)$$

where E_g and E_e are the energies of the states g and e.

We assume that the A-field is oscillating, that is,

$$A(t) = A_0 \sin \omega t = \frac{1}{2i} A_0 (e^{i\omega t} - e^{-i\omega t}), \qquad (14.3.2)$$

and use the relation $\vec{E} = \partial\vec{A}/\partial t = E_0 \cos\omega t$ to equate the amplitude $A_0 = E_0/\omega$. We then have

$$H = \left(E_g b_g^\dagger b_g + E_e b_e^\dagger b_e\right) - \frac{q}{m}\frac{E_0}{2i\omega}\left(e^{i\omega t} - e^{-i\omega t}\right) im\omega_0 \left(b_e^\dagger b_g \langle e|x|g\rangle - b_g^\dagger b_e \langle g|x|e\rangle\right).$$

(14.3.3)

In general, this will have four terms when multiplied out. We will drop two terms, however, which correspond to terms very far from resonance. This is known as the rotating-wave *approximation*, and will be discussed in greater detail at the end of Section 14.3.2. We also approximate $\omega \approx \omega_0$ for near-resonant excitation, and take $\langle e|x|g\rangle$ as real. We then have

$$H = \left(E_g b_g^\dagger b_g + E_e b_e^\dagger b_e\right) + \frac{q}{2}\langle e|x|g\rangle E_0 \left(e^{-i\omega t} b_e^\dagger b_g + e^{i\omega t} b_g^\dagger b_e\right).$$

(14.3.4)

We now define the *density matrix* for a general state $|\psi\rangle$, which can include superpositions of the states e and g, as

$$\hat{\rho} = \begin{pmatrix} \rho_{ee} & \rho_{ge} \\ \rho_{ge} & \rho_{gg} \end{pmatrix} \equiv \begin{pmatrix} \langle\psi|b_e^\dagger b_e|\psi\rangle & \langle\psi|b_g^\dagger b_e|\psi\rangle \\ \langle\psi|b_e^\dagger b_g|\psi\rangle & \langle\psi|b_g^\dagger b_g|\psi\rangle \end{pmatrix}.$$

(14.3.5)

The diagonal terms give the average number of electrons in each state at any point in time, and the off-diagonal terms can be viewed as measures of the *coherence* between the two states, since they can only be nonzero if $|\psi\rangle$ is in a superposition of both states.

We define the *Rabi frequency*

$$\omega_R = \frac{|qE_0\langle e|x|g\rangle|}{\hbar},$$

(14.3.6)

and use our definition of $\hbar\omega_0 = E_e - E_g$. The Hamiltonian then can be written in simple form as

$$H = \frac{\hbar\omega_0}{2}(\rho_{ee} - \rho_{gg}) - \frac{\hbar\omega_R}{2}(e^{-i\omega t}\rho_{eg} + e^{i\omega t}\rho_{ge}),$$

(14.3.7)

where we have subtracted off a constant $(E_e - E_g)/2$, and have used conservation of total electron number, $\rho_{ee} + \rho_{gg} = 1$.

The time evolution of any element of the density matrix is found by evolving the state $|\psi\rangle$ using the Schrödinger equation,

$$\frac{\partial\rho_{mn}(t)}{\partial t} = \frac{\partial}{\partial t}\langle\psi|b_m^\dagger b_n|\psi\rangle = \left(-\frac{1}{i\hbar}\langle\psi|H\right) b_m^\dagger b_n|\psi\rangle + \langle\psi|b_m^\dagger b_n\left(\frac{1}{i\hbar}H|\psi\rangle\right),$$

(14.3.8)

which gives

$$\frac{\partial\rho_{mn}(t)}{\partial t} = -\frac{1}{i\hbar}\langle\psi|[H, b_m^\dagger b_n]|\psi\rangle.$$

(14.3.9)

To solve the time dependence, it is convenient to define new terms, which can be viewed as the three components of a vector, known as a *Bloch vector*, in a three-dimensional space:

$$\begin{aligned} U_1 &= \rho_{eg} + \rho_{ge} \\ U_2 &= i(\rho_{eg} - \rho_{ge}) \\ U_3 &= \rho_{ee} - \rho_{gg}. \end{aligned}$$

(14.3.10)

The component U_3 gives the degree of *inversion*, that is, the fraction lifted from the ground to the excited state, and the other two components give the real and imaginary parts of the off-diagonal coherence. In terms of these, the time evolution equation (14.3.9) becomes

$$\frac{\partial U_1}{\partial t} = \omega_0 U_2 - \omega_R \sin \omega t \, U_3$$

$$\frac{\partial U_2}{\partial t} = -\omega_0 U_1 - \omega_R \cos \omega t \, U_3$$

$$\frac{\partial U_3}{\partial t} = \omega_R \sin \omega t \, U_1 + \omega_R \cos \omega t \, U_2. \tag{14.3.11}$$

We can switch to the *rotating frame* by choosing new coordinates U_1' and U_2', such that the old coordinates can be written as

$$U_1 = U_1' \cos \omega t + U_2' \sin \omega t$$

$$U_2 = -U_1' \sin \omega t + U_2' \cos \omega t.$$

$$U_3 = U_3'. \tag{14.3.12}$$

Substituting these into (14.3.11) and equating $\sin \omega t$ and $\cos \omega t$ terms on each side of the equations, we then have

$$\frac{\partial U_1'}{\partial t} = \tilde{\omega} U_2'$$

$$\frac{\partial U_2'}{\partial t} = -\tilde{\omega} U_1' - \omega_R U_3'$$

$$\frac{\partial U_3'}{\partial t} = \omega_R U_2', \tag{14.3.13}$$

where $\tilde{\omega} = \omega_0 - \omega$ is the *detuning* of the oscillating electromagnetic field from the resonance frequency.

14.3.2 The Bloch Vector Representation

The result of the derivation of the previous section is the set of equations in (14.3.13). These equations describe the motion of a vector of length $= 1$ moving around a sphere, as shown in Figure 14.2. The bottom of the sphere corresponds to the system being wholly in one of the two states, and the top of the sphere corresponds to the system being wholly in the other state. The positions between these two poles represent superpositions of the two states, given by

$$|\psi\rangle = \cos(\theta/2)|g\rangle + \sin(\theta/2)e^{i\phi}|e\rangle, \tag{14.3.14}$$

where θ and ϕ are the standard angles for spherical coordinates.

This is known as the *Bloch sphere*. The Bloch vector described by (14.3.13) will stay on this sphere and not change length, since the terms describe a pure rotation around the U_1 axis with angular frequency ω_R, which depends on the strength of the driving force, and another pure rotation around the U_3 axis with frequency $\tilde{\omega}$, which gives the mismatch of the driving frequency and the natural resonance frequency for oscillations between the two states. (We now drop the prime on the vector components from the previous section.)

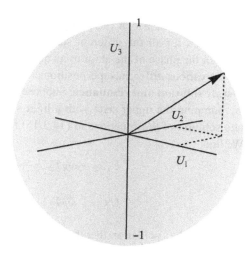

Figure 14.2 The superposition of two states represented as a vector of unit length on the Bloch sphere.

The Bloch equations can be modified to allow for *dissipation* and *decoherence*. We add terms with two timescales, T_1 and T_2, as follows:

$$\frac{\partial U_1}{\partial t} = -\frac{U_1}{T_2} + \tilde{\omega} U_2$$

$$\frac{\partial U_2}{\partial t} = -\frac{U_2}{T_2} - \tilde{\omega} U_1 - \omega_R U_3$$

$$\frac{\partial U_3}{\partial t} = -\frac{U_3 + 1}{T_1} + \omega_R U_2. \tag{14.3.15}$$

These are the equations solved to obtain Figures 3.3 and 3.4 in Section 3.2, for the case of $\tilde{\omega} = 0$ and $T_2 = T_1$. As seen in those figures, in the strong decoherence limit, T_2 gives the intrinsic timescale for quantum jumps.

The T_1 process, accounted for in the last equation, describes energy loss to the environment, which leads the upper, excited state to revert back down into the lower state. When $U_3 = -1$, the system is fully in the ground state, and no more energy loss occurs.

The T_2 process, accounted for by terms in the first two equations, describes decoherence, and leads to the Bloch vector becoming shorter over time, until it reaches zero length. There are two ways to visualize why this happens. The first is to imagine a large set of different atoms, which initially are all doing the same thing, but then small phase shifts lead to the Bloch vectors of the different atoms to all be pointing in different directions. The average of the many Bloch vectors in different directions then tends toward zero.

Energy dissipation also leads to decoherence, and therefore T_2 can never be longer than T_1. However, other nondissipative processes can lead to phase shifts, and so T_2 can be much shorter than T_1.

Another way to view the decoherence applies to a single atom. One can imagine that many interactions with the environment lead to a superposition of the same atom in many different states. Since the Bloch vector represents the quantum state of the atom, the average of this vectors for all the different states of the atom in its superposition gives the degree to which the phase of the atom in its quantum evolution has been scrambled, to be different in the various different superpositions.

Steady-state solution and resonance. Suppose that we control the value of U_3 directly, perhaps by pumping the upper state with a light source. Assuming U_3 is a constant, the steady-state solution of the equations in (14.3.15) is found by setting the time derivatives to zero. We find

$$U_1 = -\omega_R T_2 U_3 \frac{\tilde{\omega} T_2}{1 + \tilde{\omega}^2 T_2^2}$$

$$U_2 = -\omega_R T_2 U_3 \frac{1}{1 + \tilde{\omega}^2 T_2^2}. \tag{14.3.16}$$

The magnitude of the oscillation therefore has a resonance peak centered at $\tilde{\omega} = 0$, with a width determined by T_2. Note that this also tells us when we were justified in using the rotating-wave approximation in Section 14.3.1, that is, in dropping two of the exponential terms in (14.3.4). These two terms correspond to $\tilde{\omega} = 2\omega$. If this is much greater than the width $1/T_2$, then these terms will make very little contribution.

This effect makes the response of the atom to electromagnetic field very sensitive to the input frequency, which makes it possible to use the resonance to identify particular atoms. This is the basis of *magnetic resonance imaging* (often known just as MRI), used in hospitals.

14.4 Single-Photon Transitions and Fermi's Golden Rule

The analysis of Section 14.3 in terms of the Bloch equations may be dismissed by some as not directly relevant to the case of photon absorption, because it assumes that the electromagnetic field is a coherent wave, but a single photon is a Fock state, not a coherent state. The Bloch sphere analysis is actually relevant for any electromagnetic excitation, because it gives the intrinsic timescales. But we can also quantitatively analyze the case of an initial state with exactly one photon.

The single-photon analysis is nontrivial, because in general we must create a *wave packet* to determine the location of a photon; as discussed in Section 3.1, a photon has no natural "size." A wave packet is constructed as a superposition of many photon eigenstates. If the size of the wave packet in space is large, the time for a transition of an electron due to that photon will be given by the time duration of the wave packet hitting the atom.

We can study the opposite limit, of a very short wave packet, by taking the extreme case of a photon localized to one point in space, right where the atom is. In this case, the state of the electromagnetic field is given by the spatial creation operator introduced in Section 12.2.3,

$$|\psi\rangle = \psi^\dagger(r)|0\rangle = \frac{1}{\sqrt{V}} \sum_k e^{ikr} a_k^\dagger |0\rangle, \tag{14.4.1}$$

which creates a photon exactly at point r.

To calculate the timescale for the photon state (14.4.1) to turn into an electronic excitation, we can use a "reverse Fermi's golden rule." Fermi's golden rule was derived in Section 14.1, and is often used for an atom in a prepared excited state, which couples to a continuum of possible final states. In that analysis, the total rate of depletion of an initial state is given by

$$\frac{\partial}{\partial t}|\langle i|\psi_t\rangle|^2 = \frac{\partial}{\partial t}\left(1 - \sum_{f\neq i}|\langle f|\psi_t\rangle|^2\right) = -\frac{2\pi}{\hbar}\sum_{f\neq i}|\langle f|V_{\text{int}}|i\rangle|^2\delta(E_f - E_i). \quad (14.4.2)$$

As discussed in Section 14.1, the infinity in the δ-function will not cause problems, because when we sum over all possible final states $|f\rangle$, the integral will be finite.

Fermi's golden rule gives *irreversible* behavior, namely transitions from $|i\rangle$ to $|f\rangle$ and not back again. The rate calculated by this rule corresponds exactly to the rate $1/T_1$ for energy loss, used in Section 14.3. Getting irreversible behavior is a somewhat surprising result, because the underlying quantum mechanical equations are time-reversible. It occurs in this case because we assume an asymmetry: the initial, prepared state is just one atomic state, but this state couples to a vast number of possible final states $|f\rangle$, corresponding to all possible emitted photons. This allows us to treat the sum over final states as a continuum, which then allows us to resolve the δ-function to get a finite number.

Physically, what is happening in an irreversible transition is that the electron in the excited state couples to many other states, leading to oscillations between the initial state and all of these other states. Because there are so many oscillations, most of them cancel out by interference after some time, except for those that conserve energy between the initial and final states.

The electron–photon interaction term V_{int} is given, in general, by (14.2.12), but to account for photons, we must write the A-field in terms of the creation and destruction operators, which gives us formula (14.2.21) derived in Section 14.2,

$$V_{\text{e-phot}} = -\sum_{n,n'}\sum_{\vec{k}}\left(b_n^\dagger b_{n'} - b_n^\dagger b_{n'}\right)\left(a_{\vec{k}} + a_{\vec{k}}^\dagger\right)\frac{q}{m}\vec{\eta}\cdot\langle n|\vec{p}|n'\rangle. \quad (14.4.3)$$

The $a_{\vec{k}}^\dagger$ operator adds one photon, and the a_k operator subtracts one photon. For photon emission from an atom, we assume that the initial state has no photons and that the electron of the atom is in the excited state, and that the final state has one photon in some state k, and the electron in the ground state.

Reverse Fermi's golden rule. We are now ready to compute the rate for a transition the opposite way, for a photon localized on an atom. The initial state (14.4.1) corresponds to a continuum of k-states, and the final state is the single atom in its excited electron state. The same logic that gave Fermi's golden rule then gives us the rate of *increase* of the probability of being in the final state as

$$\frac{\partial}{\partial t}|\langle f|\psi_t\rangle|^2 = \frac{2\pi}{\hbar}\sum_{i\neq f}|\langle f|H_{\text{int}}|i\rangle|^2\delta(E_f - E_i), \quad (14.4.4)$$

where now the sum is over all the *initial* states in (14.4.1).

This formula gives the same timescale for transitions upward as given by Fermi's golden rule for transitions downward. The absorption process is therefore never instantaneous, even for a single photon. Any irreversible process has an intrinsic timescale, which we call the timescale for decoherence.

14.5 Nonlinear Optics and Nonunitarity

In Section 6.1, we discussed how the equations of quantum mechanics have strictly *unitary* evolution. This is a type of *linear* behavior, but not all nonlinear behavior is nonunitary. Many optical systems, and classical waves in general, have nonlinear behavior but still have unitary time evolution.

Following the procedure in Section 14.2, we can derive the nonlinear optical behavior starting from first principles of the electron–photon interaction. The total energy held by an electromagnetic field is (Jackson 1998)

$$H = \int dr^3 \varepsilon |E|^2, \tag{14.5.1}$$

where ε is the *permittivity*, which is a universal constant of nature for a vacuum, ε_0, and more generally has properties that depend on the details of the medium, including the possibility of properties that depend on the strength of the electric field. The standard way to write ε is to expand it as a Taylor series as a function of the electric field amplitude E,

$$\varepsilon = \varepsilon_0(1 + \chi) = \varepsilon_0(1 + \chi^{(1)} + 2\chi^{(2)}E + 4\chi^{(3)}E^2 + \cdots), \tag{14.5.2}$$

where χ is the *susceptibility*, which gives the effect of the optical medium (χ is nominally equal to zero in vacuum). The standard index of refraction n of a medium is given by the formula $n^2 = 1 + \chi^{(1)}$, and $\chi^{(2)}$ and $\chi^{(3)}$ are *nonlinear susceptibility* constants.

From Section 12.4, the electric field can be written in terms of photon operators as

$$E(r) = i \sum_k \sqrt{\frac{\hbar \omega_k}{2\varepsilon V}} \left(a_k e^{ikr} - a_k^\dagger e^{-ikr} \right). \tag{14.5.3}$$

The lowest-order term of (14.5.1) is then given by

$$H = -\int dr^3 \varepsilon_0(1 + \chi^{(1)}) \frac{\hbar \omega_k}{2\varepsilon V} \sum_{k,k'} \left(a_k e^{ikr} - a_k^\dagger e^{-ikr} \right) \left(a_{k'} e^{ik'r} - a_{k'}^\dagger e^{-ik'r} \right). \tag{14.5.4}$$

If $k = k'$, then two of the terms in the product will have exponents that cancel, leading to an integral over all space, which just gives the total volume V. The other two terms in the product will have oscillations that cancel out, to give a negligible total contribution. The Hamiltonian for the total energy, for this lowest order of the Taylor expansion, is then

$$H = \sum_k \frac{1}{2} \hbar \omega_k \left(a_k^\dagger a_k + a_k a_k^\dagger \right)$$

$$= \sum_k \hbar \omega_k \left(a_k^\dagger a_k + \frac{1}{2} \right), \tag{14.5.5}$$

using the commutation relations (12.2.15), which apply to both phonons and photons. This is, of course, the standard photon Hamiltonian, where $\hat{N}_k = a_k^\dagger a_k$ gives the number of photons in state k.

The next-higher-order term in expansion (14.5.2) gives an additional term to the Hamiltonian,

$$
V_{\text{int}} = -2i \int d^3r \, \varepsilon_0 \chi^{(2)} \left(\frac{\hbar \omega_k}{2\varepsilon V} \right)^{3/2}
$$

$$
\times \sum_{k,k',k''} \left(a_k e^{ikr} - a_k^\dagger e^{-ikr} \right) \left(a_{k'} e^{ik'r} - a_{k'}^\dagger e^{-ik'r} \right) \left(a_{k''} e^{ik''r} - a_{k''}^\dagger e^{-ik''r} \right). \quad (14.5.6)
$$

Without expanding this any further in detail, we can see that there will be terms that have products of three operators of the forms

$$
a_k^\dagger a_{k'}^\dagger a_{k''}, \quad a_k a_{k'}^\dagger a_{k''}, \quad \text{and so on.} \tag{14.5.7}
$$

These correspond to *three-photon* processes, also called *parametric* processes, in which one photon is removed and two are created in its place, or two photons are removed and one is created in its place. This is an example of a nonlinear optical process.

The many-photon system still obeys the Schrödinger equation,

$$
i\hbar \frac{\partial}{\partial t} |\psi\rangle = H|\psi\rangle, \tag{14.5.8}
$$

where the state of the system is a many-body state that can be written as a superposition of Fock states (introduced in Section 12.2),

$$
|\psi\rangle = \alpha | \ldots N_j, N_{j+1}, \ldots \rangle + \beta | \ldots N_j', N_{j+1}', \ldots \rangle + \cdots, \tag{14.5.9}
$$

where the α and β factors are complex numbers.

The nonlinearity comes in because any given bosonic Fock state can have any of the N_j larger than 1, and can have more than one state j have nonzero N_j. Such a state will allow two destruction operators $a_k a_{k'}$ to operate on it successively with a nonzero result. However, the time evolution of the system is still *unitary*, because it is always true that

$$
i\hbar \frac{\partial}{\partial t} (\alpha |\psi_1\rangle + \beta |\psi_2\rangle) = \alpha H |\psi_1\rangle + \beta H |\psi_2\rangle, \tag{14.5.10}
$$

which means that we could solve two separate equations,

$$
i\hbar \frac{\partial}{\partial t} |\psi_1\rangle = H|\psi_1\rangle,
$$

$$
i\hbar \frac{\partial}{\partial t} |\psi_2\rangle = H|\psi_2\rangle, \tag{14.5.11}
$$

and then just add the solutions together afterward, with the appropriate factors, to get the full solution.

A measurement with "collapse" is nonunitary, because if the system starts out in state $\alpha |\psi_1\rangle + \beta |\psi_2\rangle$, where $|\psi_1\rangle$ and $|\psi_2\rangle$ are two eigenstates of the operator corresponding to the measurement, then, after the measurement, the state will be in either state $|\psi_1\rangle$ or $|\psi_2\rangle$. Suppose that it collapses into state $|\psi_1\rangle$. This can't be written as the superposition of the two solutions for the initial states $|\psi_1\rangle$ and $|\psi_2\rangle$ separately.

References

C. Cohen-Tannoudji, B. Diu, and F. Laloë, *Quantum Mechanics*, (Wiley, 1977), Volume I, Appendix III.

H. Haug and A.-P. Jauho, *Quantum Kinetics in Transport and Optics of Semiconductors*, (Springer, 2004).

J. D. Jackson, *Classical Electrodynamics*, 3rd ed., (Wiley, 1998).

15

Feynman Diagrams

The diagrams used in quantum field theory are often mystifying to people, but actually are just shorthand ways of writing down the integrals used in transition rates and energy corrections. These energies arise in calculation of the *S-matrix* for the time evolution of a system.

The reason why energy is so important is that in the Schrödinger equation, the time evolution of a state is given by a phase factor that depends on the energy. In its most basic form, this reads

$$i\hbar\frac{\partial}{\partial t}|\psi\rangle = H|\psi\rangle, \tag{15.0.1}$$

where H is the Hamiltonian operator that gives the energy of the state $|\psi\rangle$. This implies, in general terms, that

$$|\psi\rangle = e^{-iHt/\hbar}|\psi_0\rangle. \tag{15.0.2}$$

A problem arises, however, when H acting on $|\psi\rangle$ does not give a simple number, but instead changes the state $|\psi\rangle$ into some other state. In general, to solve the problem, we can expand the exponential term as a series with greater and greater factors of H. We have already done this in a truncated way in Section 14.1. But if we want to account for many higher-order terms, we can quickly run into problems with bookkeeping. Richard Feynman and others developed a diagrammatic way of keeping track of all the terms.

In this chapter, we will study a simplified Feynman diagram approach commonly used in condensed matter physics (e.g., Fetter and Walecka 1971; Snoke 2020). It is not essential to study particle physics with quarks and so on to grasp the essential concepts of quantum field theory. For standard treatments of relativistic quantum field theory, see Mandl and Shaw 2010 and Peskin and Schroeder 1995.

15.1 The Expansion of the S-Matrix

We start with the general form for the time evolution of any given wave function,

$$|\psi(t)\rangle = e^{-(i/\hbar)\int V_{\text{int}}(t)dt}|\psi(0)\rangle$$
$$= \left(1 + (1/i\hbar)\int_0^t dt'\, V_{\text{int}}(t') + (1/i\hbar)^2 \int_0^t dt' \int_0^{t'} dt''\, V_{\text{int}}(t')V_{\text{int}}(t'') + \cdots\right)|\psi(0)\rangle, \tag{15.1.1}$$

which we first derived in Section 14.1. The exponential operator $e^{-(i/\hbar)\int_{t_0}^{t} V_{\text{int}}(t')dt'}$ is written as $S(t,0)$ and is called the *S-matrix*.

The S-matrix can be rewritten as

$$S(t,0) = 1 + \sum_{n=1}^{\infty}(1/i\hbar)^n \int_0^t dt_1 \int_0^{t_1} dt_2 \ldots \int_0^{t_{n-1}} dt_n \, V_{\text{int}}(t_1)V_{\text{int}}(t_2)\ldots V_{\text{int}}(t_n).$$

(15.1.2)

The upper limits of the time integrals are all different. We can rewrite the S-matrix with the same upper limit for all the time integrals as follows:

$$S(t,0) = 1 + \sum_{n=1}^{\infty}\frac{(1/i\hbar)^n}{n!} \int_0^t dt_1 \int_0^t dt_2 \ldots \int_0^t dt_n \, \mathsf{T}(V_{\text{int}}(t_1)V_{\text{int}}(t_2)\ldots V_{\text{int}}(t_n)),$$

(15.1.3)

in which $\mathsf{T}(V_{\text{int}}(t_1)V_{\text{int}}(t_2)\ldots V_{\text{int}}(t_n))$ is the *time-ordered product* of the operators $V_{\text{int}}(t_1)$, $V_{\text{int}}(t_2)$, and so on. In the time-ordered product, the operators are rearranged such that $t_1 > t_2 > \cdots > t_n$.

To see that (15.1.3) is equivalent to (15.1.2), consider just the second-order term,

$$\int_0^t dt_1 \int_0^t dt_2 \, \mathsf{T}(V_{\text{int}}(t_1)V_{\text{int}}(t_2)) = \int_0^t dt_1 \int_0^{t_1} dt_2 \, V_{\text{int}}(t_1)V_{\text{int}}(t_2)$$
$$+ \int_0^t dt_2 \int_0^{t_2} dt_1 \, V_{\text{int}}(t_2)V_{\text{int}}(t_1).$$

(15.1.4)

By the simple change of variables, $t_1 \to t_2, t_2 \to t_1$, the second integral is equivalent to the first. By normalizing the left-hand side by a factor of 1/2, we obtain the second-order term of (15.1.2). The same procedure works for all orders of the expansion.[1]

Wick's theorem. In Section 14.2, the interaction terms we wrote down were all written in terms of boson and fermion creation and destruction operators a_k, $a_k^\dagger$, b_k, $b_k^\dagger$, and so on. We would like to convert the time-ordered product of the interactions in the S-matrix into a *normal-ordered* product of the creation and destruction operators, that is, a product in which all creation operators are to the left of all destruction operators.

The key theorem that allows us to do this is *Wick's theorem*, which says, for operators $A, B, C, \ldots, X, Y, Z$,

$$\mathsf{T}(ABC\ldots YZ) = \mathsf{N}(ABCD\ldots YZ) + \mathsf{N}(\overset{\frown}{AB}CD\ldots YZ)$$

$$+\mathsf{N}(\overset{\frown}{AB}CD\ldots YZ) + \mathsf{N}(\overset{\frown}{AB}\overset{\frown}{CD}\ldots YZ)$$

$$+\mathsf{N}(\overset{\frown}{ABCD}\ldots YZ) + \cdots$$

(15.1.5)

where $\overset{\frown}{AB}$ is a complex number (a "*c*-number") known as the *contraction* of operators A and B, defined by

$$\overset{\frown}{AB} = \mathsf{T}(AB) - \mathsf{N}(AB),$$

(15.1.6)

---

[1] Note that (15.1.2) is equivalent to (15.1.3) only if there are an even number of fermion operators in the interaction V_{int}, for example, if, for every fermion created, another is destroyed, conserving the number of fermions.

and $N(ABC \ldots)$ is the *normal-ordered product* of any number of operators. The normal-ordered product is obtained by putting all the creation operators to the left of all the destruction operators. If some of the operators are fermionic, both the time-ordered product and the normal-ordered product include additionally a factor $(-1)^F$, where F is the number of interchanges of neighboring fermion operators needed to obtain the product from the original ordering.

In other words, Wick's theorem says that the time-ordered product is equal to the sum of all the possible normal-ordered products, for every possible contraction of two operators. On one hand, we have increased the complexity, since we now have a large sum of products instead of just one, but on the other hand, normal-ordered products are much easier to handle.

15.1.1 Justification of Wick's Theorem

We will not prove Wick's theorem, but we can see how it works with just two operators, and then extrapolate to more.

Wick's theorem for two operators amounts to just the statement that the contraction of our standard creation and destruction operators $a_{\vec{k}}^\dagger(t)$ and $a_{\vec{k}}(t)$ is a c-number. To show that, we must first determine the commutator of these two operators.

In the interaction picture, for any operator $A(t)$, we can differentiate with respect to time to obtain

$$i\hbar \frac{\partial}{\partial t} A(t) = e^{iH_0 t/\hbar}(AH_0 - H_0 A)e^{-iH_0 t/\hbar}$$
$$= [A(t), H_0]. \tag{15.1.7}$$

The unperturbed Hamiltonian H_0 by definition is diagonal in the number operators, that is,

$$H_0 = \sum_{\vec{k}} E_k \left(\hat{N}_{\vec{k}} + \frac{1}{2} \right) = \sum_{\vec{k}} \hbar\omega_k \left(a_{\vec{k}}^\dagger a_{\vec{k}} + \frac{1}{2} \right), \tag{15.1.8}$$

and therefore,

$$[a_{\vec{k}}, H_0] = E_k [a_{\vec{k}}, \hat{N}_{\vec{k}}] = \hbar\omega_k a_{\vec{k}} \tag{15.1.9}$$

$$[a_{\vec{k}}^\dagger, H_0] = E_k [a_{\vec{k}}^\dagger, \hat{N}_{\vec{k}}] = -\hbar\omega_k a_{\vec{k}}^\dagger, \tag{15.1.10}$$

using the commutation relations (12.1.7) and (12.1.8) from Section 12.1.1, which, surprisingly, are true for both bosons and fermions. Then

$$i\hbar \frac{\partial}{\partial t} a_{\vec{k}}(t) = e^{iH_0 t/\hbar}[a_{\vec{k}}, H_0]e^{-iH_0 t/\hbar}$$
$$= \hbar\omega_k a_{\vec{k}}(t). \tag{15.1.11}$$

Therefore, solving the differential equation with initial condition $a_{\vec{k}}(0) = a_{\vec{k}}$,

$$a_{\vec{k}}(t) = e^{-i\omega_k t} a_{\vec{k}}. \tag{15.1.12}$$

All of the equations (15.1.7)–(15.1.11) apply to both fermion and boson operators. Using these results, it is easy to show that

$$[a_{\vec{k}}(t_1), a_{\vec{k}}^{\dagger}(t_2)] = e^{-i\omega_k(t_1-t_2)} \quad \text{for bosons,}$$

$$\{b_{\vec{k}}(t_1), b_{\vec{k}}^{\dagger}(t_2)\} = e^{-i\omega_k(t_1-t_2)} \quad \text{for fermions,}$$

(15.1.13)

where the square brackets are the commutator and the curly brackets are the anticommutator, introduced in Section 13.1.1 for fermions.

Let us now compute the contraction for two boson operators. If $t_1 > t_2$, we have

$$\begin{aligned}
\mathsf{T}(a_{\vec{k}}(t_1)a_{\vec{k}}^{\dagger}(t_2)) - \mathsf{N}(a_{\vec{k}}(t_1)a_{\vec{k}}^{\dagger}(t_2)) &= a_{\vec{k}}(t_1)a_{\vec{k}}^{\dagger}(t_2) - a_{\vec{k}}^{\dagger}(t_2)a_{\vec{k}}(t_1) \\
&= [a_{\vec{k}}(t_1), a_{\vec{k}}^{\dagger}(t_2)] \\
&= e^{-i\omega_k(t_1-t_2)}.
\end{aligned}$$

(15.1.14)

If $t_1 < t_2$, then

$$\mathsf{T}(a_{\vec{k}}(t_1)a_{\vec{k}}^{\dagger}(t_2)) = a_{\vec{k}}^{\dagger}(t_2)a_{\vec{k}}(t_1) = \mathsf{N}(a_{\vec{k}}(t_1)a_{\vec{k}}^{\dagger}(t_2)),$$

(15.1.15)

and the contraction is equal to zero. Putting both possibilities $t_1 > t_2$ and $t_2 > t_1$ together, we have

$$\begin{aligned}
\mathsf{T}(a_{\vec{k}}(t_1)a_{\vec{k}}^{\dagger}(t_2)) - \mathsf{N}(a_{\vec{k}}(t_1)a_{\vec{k}}^{\dagger}(t_2)) &= a_{\vec{k}}(t_1)a_{\vec{k}}^{\dagger}(t_2) \\
&= a_{\vec{k}}^{\dagger}(t_2)a_{\vec{k}}(t_1) \\
&= e^{-i\omega_k(t_1-t_2)}\Theta(t_1 - t_2),
\end{aligned}$$

(15.1.16)

where $\Theta(t)$ is the Heaviside function, equal to 1 if $t > 0$, and equal to 0 if $t < 0$.

Equation (15.1.16) is Wick's theorem in the case of two boson operators. The contraction is a c-number regardless of the ordering of t_1 and t_2. The contraction is simply zero for the case of two creation operators or for two destruction operators, since they commute.

It should not be too hard to see that Wick's theorem will work for larger numbers of operators. The time-ordered product is converted to a normal-ordered product by switching the order of two operators, one pair at a time. Each time two operators are commuted, a commutator (which is a c-number) is generated, that is, a contraction.

15.1.2 Green's Functions

It is convenient to write the contraction of creation and destruction operators in terms of a *Green's function*, defined as follows:

$$G_{\vec{k}}(t_1 - t_2) \equiv -i \overline{a_k(t_1)a_{\vec{k}}^{\dagger}(t_2)}.$$

(15.1.17)

This definition implies

$$G_{\vec{k}}(t) = -ie^{-i\omega_k t}\Theta(t),$$

(15.1.18)

which has the Fourier transform

$$G(\vec{k}, \omega) = -i \int_{-\infty}^{\infty} dt \, e^{i\omega t} \, e^{-i\omega_k t} \Theta(t)$$

$$= \lim_{\varepsilon \to 0} -i \int_{0}^{\infty} dt \, e^{i(\omega - \omega_k)t} e^{-\varepsilon t}$$

$$= \frac{1}{\omega - \omega_k + i\varepsilon}. \tag{15.1.19}$$

The Green's function can also be expressed in terms of the vacuum expectation value. We write

$$G_{\vec{k}}(t_1 - t_2) \equiv -i \langle \text{vac} | \mathsf{T}(a_k(t_1) a_{\vec{k}}^{\dagger}(t_2)) | \text{vac} \rangle. \tag{15.1.20}$$

This automatically gives the right behavior for both cases $t_1 > t_2$ and $t_1 < t_2$ because a destruction operator in the rightmost position acting on vacuum always gives zero. The vacuum state here does not necessarily mean a true vacuum, but the ground state of whatever the system is.

The introduction of the small imaginary term in (15.1.19), which makes the integral tractable, deserves some careful attention. This will lead to imaginary terms that appear in many places. As seen in the first two lines of this calculation, it corresponds to decay of the wave amplitude over time.

Some authors (see, e.g., Cramer 2017 and Kastner 2021) have taken this to imply that there is real nonunitary behavior in quantum field theory, since decay of the wave function is an example of irreversible behavior. This is not actually implied by the existence of this term. Rather, we can use an argument of self-consistency. From calculations using this approach (see Section 15.3.2), one finds that every particle energy $E_k = \hbar\omega_k$ is always "renormalized" to have some imaginary part, giving us $E_k' = \hbar\omega_k' - i\hbar\gamma_k$. If we use this where ω_k appears in the exponential, we obtain

$$e^{-i(\omega_k - i\gamma_k)t} = e^{-i\omega_k t} e^{-\gamma_k t}, \tag{15.1.21}$$

which has exactly the same form as the decay term we have invoked. But the imaginary term $i\gamma_k$ does not come from nonunitary behavior, but from a calculation of the scattering rate out from the plane-wave state k into other states due to various interactions. Therefore, the use of the small $i\varepsilon$ term in (15.1.19) comes from accounting self-consistently for the fact that no state is truly eternal, but always has some scattering into other states over time, even in a perfectly unitary system.

We will see that the Green's functions have a central role in the perturbation theory; namely, they are the lines in the Feynman diagrams.

15.1.3 Example: S-Matrix for a Boson-Mediated Interaction

Wick's theorem gives us a recipe for determining the S-matrix, that is, the time evolution of a system, in any order of perturbation theory. Here, we work out an example explicitly.

Suppose that we have an initial state with two electrons in states $\vec{k}_1$ and $\vec{k}_2$, and a final state with two electrons in states $\vec{k}_3$ and $\vec{k}_4$. This transition can occur via an electron–phonon interaction, with an interaction term of the form (worked out in Section 14.2.1),

$$V_{\text{int}} = \sum_{\vec{k},\vec{k}_1} M_k\, i \left(a_{\vec{k}} b^{\dagger}_{\vec{k}_1+\vec{k}} b_{\vec{k}_1} - a^{\dagger}_{\vec{k}} b^{\dagger}_{\vec{k}_1-\vec{k}} b_{\vec{k}_1} \right), \tag{15.1.22}$$

where M_k is given by the constants in (14.2.8) for the electron–phonon interaction.

The S-matrix is given by

$$S(t,0) = 1 + \frac{1}{i\hbar} \int_0^t dt_1\, V_{\text{int}}(t_1) + \frac{1}{2}\frac{1}{(i\hbar)^2} \int_0^t dt_1 \int_0^t dt_2\, \mathsf{T}(V_{\text{int}}(t_1) V_{\text{int}}(t_2)) + \cdots, \tag{15.1.23}$$

where

$$V_{\text{int}}(t) = e^{iH_0 t/\hbar} V_{\text{int}} e^{-iH_0 t/\hbar} \tag{15.1.24}$$

$$= \sum_{\vec{k},\vec{k}_1} M_{\vec{k}}\, i \left(a_{\vec{k}}(t) b^{\dagger}_{\vec{k}_1+\vec{k}}(t) b_{\vec{k}_1}(t) - a^{\dagger}_{\vec{k}}(t) b^{\dagger}_{\vec{k}_1-\vec{k}}(t) b_{\vec{k}_1}(t) \right).$$

(It is easy to see that each of the operators in V_{int} can be converted into time-dependent operators in the interaction representation, because we can always insert pairs of exponentials $e^{-iH_0 t/\hbar} e^{iH_0 t/\hbar}$ between all the operators.)

Wick's theorem converts the time-ordered products to sums of normal-ordered products. We therefore just have to pick out each term in the S-matrix expansion that has a set of creation and destruction operators which turn the initial state into the final state.

To eliminate two electrons from the initial state, we must have a second-order term to get two electron destruction operators. An example of such a term is

$$S^{(2)}(t,0) = \frac{1}{(i\hbar)^2} \int_0^t dt_1 \int_0^t dt_2\, M_{\vec{k}_3-\vec{k}_2} M_{\vec{k}_1-\vec{k}_4}$$

$$\times \left(a_{\vec{k}_3-\vec{k}_2}(t_1) b^{\dagger}_{\vec{k}_3}(t_1) b_{\vec{k}_2}(t_1) a^{\dagger}_{\vec{k}_1-\vec{k}_4}(t_2) b^{\dagger}_{\vec{k}_4}(t_2) b_{\vec{k}_1}(t_2) \right), \tag{15.1.25}$$

and another is

$$S^{(2)}(t,0) = \frac{1}{(i\hbar)^2} \int_0^t dt_1 \int_0^t dt_2\, M_{\vec{k}_2-\vec{k}_3} M_{\vec{k}_4-\vec{k}_1}$$

$$\times \left(a^{\dagger}_{\vec{k}_2-\vec{k}_3}(t_1) b^{\dagger}_{\vec{k}_3}(t_1) b_{\vec{k}_2}(t_1) a_{\vec{k}_4-\vec{k}_1}(t_2) b^{\dagger}_{\vec{k}_4}(t_2) b_{\vec{k}_1}(t_2) \right). \tag{15.1.26}$$

All of the electron operators are needed to act on the initial state to turn it into the final state. But using Wick's theorem, we can perform a contraction of the two phonon operators in each case. Doing this forces $\vec{k}_1 - \vec{k}_4 = \vec{k}_3 - \vec{k}_2 \equiv \Delta\vec{k}$, that is, momentum conservation, $\vec{k}_1 + \vec{k}_2 = \vec{k}_3 + \vec{k}_4$. We obtain for the sum of (15.1.25) and (15.1.26),

$$S^{(2)}(t,0) = \frac{1}{(i\hbar)^2} \int_0^t dt_1 \int_0^t dt_2\, M^2_{\Delta\vec{k}}\, e^{i(\omega_3-\omega_2)t_1} e^{i(\omega_4-\omega_1)t_2}\, b^{\dagger}_{\vec{k}_4} b^{\dagger}_{\vec{k}_3} b_{\vec{k}_2} b_{\vec{k}_1}$$

$$\times \left[iG_{\Delta\vec{k}}(t_1 - t_2) + iG_{-\Delta\vec{k}}(t_2 - t_1) \right]. \tag{15.1.27}$$

We can now switch to the Fourier transforms of the Green's functions, using

$$G_{\vec{k}}(t) = \frac{1}{2\pi} \int_{-\infty}^{\infty} d\omega\, e^{-i\omega t} G(\vec{k}, \omega). \tag{15.1.28}$$

The pair of phonon Green's functions that appears in (15.1.27) commonly occurs in the computation of the S-matrix, because of the form of the interaction Hamiltonian (15.1.22). We can simplify this to

$$
\begin{aligned}
G_{\vec{k}}(t) + G_{-\vec{k}}(-t) &= \frac{1}{2\pi} \int_{-\infty}^{\infty} d\omega\, \left(e^{-i\omega t} G(\vec{k}, \omega) + e^{i\omega t} G(-\vec{k}, \omega) \right) \\
&= \frac{1}{2\pi} \int_{-\infty}^{\infty} d\omega\, e^{-i\omega t} \left(G(\vec{k}, \omega) + G(-\vec{k}, -\omega) \right) \\
&= \frac{1}{2\pi} \int_{-\infty}^{\infty} d\omega\, e^{-i\omega t} \tilde{G}(\vec{k}, \omega),
\end{aligned} \tag{15.1.29}
$$

where

$$
\begin{aligned}
\tilde{G}(\vec{k}, \omega) &= \frac{1}{\omega - \omega_k + i\varepsilon} + \frac{1}{-\omega - \omega_k + i\varepsilon} \\
&= \frac{2\omega_k - 2i\varepsilon}{\omega^2 - \omega_k^2 + 2i\omega_k\varepsilon + \varepsilon^2},
\end{aligned} \tag{15.1.30}
$$

where we have used $\omega_{-k} = \omega_k$, which follows from time-reversal symmetry. Since ε is an infinitesimal, we can rewrite this as

$$\tilde{G}(\vec{k}, \omega) = \frac{2\omega_k}{\omega^2 - \omega_k^2 + i\varepsilon}. \tag{15.1.31}$$

We can also define the Green's function in terms of energy,

$$\tilde{G}(\vec{k}, E) = \frac{2E_k}{E^2 - E_k^2 + i\varepsilon} = \frac{1}{\hbar} \tilde{G}(\vec{k}, \omega). \tag{15.1.32}$$

Going back to our example, (15.1.27) then becomes

$$
\begin{aligned}
S^{(2)}(t, 0) = \frac{1}{i\hbar^2} \int_0^t dt_1 \int_0^t dt_2\, M_{\Delta\vec{k}}^2\, e^{i(\omega_3 - \omega_2)t_1} e^{i(\omega_4 - \omega_1)t_2}\, b_{\vec{k}_4}^\dagger b_{\vec{k}_3}^\dagger b_{\vec{k}_2} b_{\vec{k}_1} \\
\times \frac{1}{2\pi} \int_{-\infty}^{\infty} d\omega\, e^{-i\omega(t_1 - t_2)} \tilde{G}(\Delta\vec{k}, \omega).
\end{aligned} \tag{15.1.33}
$$

If t is large enough, the integral over t_2 will converge to an energy-conserving δ-function, since then we can approximate the integral as

$$\int_{-\infty}^{\infty} dt_2\, e^{i(\omega_4 - \omega_1 + \omega)t_2} = 2\pi\, \delta(\omega_4 - \omega_1 + \omega). \tag{15.1.34}$$

This can then be used to eliminate the integration over ω, which is set to $\omega_1 - \omega_4$. We then have

$$S^{(2)}(t, 0) = \frac{1}{i\hbar} \int_0^t dt_1\, e^{i(\omega_4 + \omega_3 - \omega_2 - \omega_1)t_1} M_{\Delta\vec{k}}^2\, b_{\vec{k}_4}^\dagger b_{\vec{k}_3}^\dagger b_{\vec{k}_2} b_{\vec{k}_1}\, \tilde{G}(\Delta\vec{k}, \Delta E), \tag{15.1.35}$$

where $\Delta E = E_1 - E_4 = \hbar(\omega_1 - \omega_4)$.

The last integration over t_1 leads to an overall energy-conserving δ-function through the same procedure used to deduce Fermi's golden rule. The creation and destruction operators will vanish when we use them to couple the initial state to the final state. We write

$$\lim_{t\to\infty} |\langle f|S^{(2)}(t,t_0)|i\rangle|^2 = \frac{2\pi t}{\hbar}|\langle f|H^{(2)}|i\rangle|^2\delta(E_1 + E_2 - E_3 - E_4), \quad (15.1.36)$$

where

$$\langle f|H^{(2)}|i\rangle = M^2_{\Delta\vec{k}}\,\tilde{G}(\Delta\vec{k}, \Delta E) \quad (15.1.37)$$

has units of energy. As we did before for Fermi's golden rule, we then divide by the time interval to obtain the transition rate

$$\frac{\partial}{\partial t}|\langle f|\psi_t\rangle|^2 = \frac{\partial}{\partial t}\left[\lim_{t\to\infty} |\langle f|S^{(2)}(t,0)|i\rangle|^2\right]$$

$$= \frac{2\pi}{\hbar}|\langle f|H^{(2)}|i\rangle|^2\delta(E_1 + E_2 - E_3 - E_4). \quad (15.1.38)$$

The term (15.1.37) is a higher-order matrix element for the transitions between states $|i\rangle$ and $|f\rangle$. Although it is derived from the electron–phonon interaction, it plays a role very similar to the Coulomb interaction between two electrons given in Section 14.2.3. Looking at (15.1.35), we see that, if we strip off the time integral, the remaining part has a form just like the electron–electron interaction in (14.2.23), with units of energy times four fermion operators that take the initial state to the final state. It is also proportional to $1/V$, like the Coulomb interaction term, which we can see by putting in the details for the electron–phonon interaction. From (14.2.8), we have

$$M^2_{\Delta\vec{k}}\,\tilde{G}(\Delta\vec{k}, \Delta E) = \left(Dk\sqrt{\frac{\hbar}{2\rho V\omega_k}}\right)^2 \frac{2E_{\Delta\vec{k}}}{(\Delta E)^2 - (E_{\Delta\vec{k}})^2 + i\varepsilon}. \quad (15.1.39)$$

In the low-energy limit when $\Delta E = E_1 - E_4$ is very small, the real part of this is

$$M^2_{\Delta\vec{k}}\,\tilde{G}(\Delta\vec{k}, \Delta E) = -D^2k^2\frac{\hbar}{2\rho V\omega_k}\frac{2}{E_{\Delta\vec{k}}}$$

$$= -\frac{D^2k\hbar}{\rho Vv}\frac{1}{\hbar vk}$$

$$= -\frac{1}{V}\frac{D^2}{\rho v^2}, \quad (15.1.40)$$

where we have used $\omega_k = vk$, for the sound velocity v. Note that the minus sign implies an overall attraction between electrons due to this type of interaction. This is the phonon-assisted mechanism that can give Cooper pairing of electrons in superconductors.

15.2 Diagram Rules for Feynman Theory

In Section 15.1.3, we calculated some terms of the expansion of the S-matrix, which gave us a new, effective interaction energy between electrons. In general, there are many terms

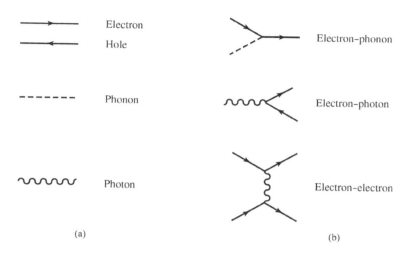

Figure 15.1

(a) Lines used for various quasiparticles in the diagrammatic approach. (b) Vertices for solid-state interactions. Electrons are represented by solid lines with arrows, phonons are represented by dashed lines, and photons are represented by wavy lines.

in the S-matrix expansion for interacting fields. Instead of doing each term from scratch, the bookkeeping for each term can be done easily using a method of diagrams. In what follows, we give a standard set of diagram rules. These Feynman diagram rules are not unique. In general, there are many different approaches to many-body theory in which different diagram rules are written down, depending on the system under study.

- Pin down the external "legs" for incoming and outgoing particles in the process of interest. Assign a momentum and energy for each.
- Draw all the topologically distinct diagrams that connect the external legs, using the vertices such as those shown in Figure 15.1. Assign a momentum and direction to each internal line, and an energy. The direction of electron lines, which is indicated by the arrow, must be continuous.

  In solids where there are many electrons in the ground state, we can also encounter *holes*, which correspond to electrons *removed* from below in the Fermi level (defined in Section 9.3). These act like positively charged particles, and are drawn as lines with arrows going backwards. They play the same role in solid-state physics that positrons (antielectrons) play in particle physics (see Section 13.2.1).
- Conserve both momentum and energy at each vertex. An outgoing hole at a vertex is counted as an incoming electron, that is, an electron moving backwards in time. (Elastic scattering from a defect is an exception; instead of conserving momentum, one multiplies by a phase factor for each vertex.)
- Sum over each momentum that is not determined by momentum conservation. If spin, polarization, or interband transitions are being taken into account, sum over all possible spin, polarization states, and different bands that are not determined by the external legs.

- Integrate over each energy not determined by energy conservation, and multiply by $i/2\pi$ for each integration over energy.
- For each internal line, write down the frequency-domain Green's function for the appropriate particle. These are also called *propagators* because they connect the vertices. For phonons or photons, we write

$$G(\vec{k}, E) = \frac{2E_{\vec{k}}}{E^2 - E_{\vec{k}}^2 + i\varepsilon}, \qquad (15.2.1)$$

to take into account the two terms that appear in the interaction Hamiltonian for absorption and emission, as for example in (15.1.27), leading to (15.1.31).

The basic form of the electron propagators is

$$G(\vec{k}, E) = \frac{1}{E - E_{\vec{k}} + i\varepsilon}, \qquad (15.2.2)$$

where $E_{\vec{k}}$ is the energy of the particle as calculated from the momentum $\vec{k}$. The hole propagator is

$$G(\vec{k}, E) = \frac{1}{E - E_{\vec{k}} - i\varepsilon}, \qquad (15.2.3)$$

with the opposite sign of the $i\varepsilon$ term. This makes a significant difference when the methods of analytical calculus are used to calculate the integrals.

- For each vertex multiply by the appropriate constant. These are given for some typical interactions in Table 15.1.
- For each crossing of fermion lines, multiply by -1. When working with external legs, do not cross fermion lines if there is a topologically equivalent diagram with uncrossed lines. Additionally, for each internal fermion loop, multiply by -1.

It is easy to check that the calculation done in Section 15.1.3 corresponds to the diagram shown in Figure 15.2, using these rules.

Table 15.1 Examples of interaction-vertex energy terms.	
Electron–phonon	$M_k^{\mathrm{Def}} = Dk\sqrt{\dfrac{\hbar}{2\rho V \omega_k}}$
Electron–photon	$M_k^{\mathrm{Dipole}} = \dfrac{e}{m}\langle p \rangle \sqrt{\dfrac{\hbar}{2\varepsilon V \omega_k}}$
Electron–electron	$M_{\Delta k}^{\mathrm{Coul}} = \dfrac{1}{V}\dfrac{e^2/\varepsilon}{|\Delta \vec{k}|^2 + \kappa^2}$

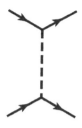

The Feynman diagram corresponding to the calculation done in Section 15.1.3, corresponding to an effective electron–electron interaction due to phonon exchange.

15.3 How to Interpret Feynman Diagrams

In Section 4.5, we looked at a Feynman diagram for a photon traveling in a vacuum. Figure 15.3 shows another possible Feynman diagram, in which the initial state and the final state are empty vacuum.

Physicists sometimes use loose language to describe such a process, talking of particles popping out of vacuum as though this was a random process that happened at various times. In that language, we could describe the process shown in Figure 15.3 as three particles appearing, namely a photon, an electron, and a positron, which then recombine. But the proper understanding of a diagram like this is that it gives a mathematical integral to calculate an *energy*, which arises due to interaction terms between fields. In the case of Figure 15.3, the state of interest is the vacuum, and like any eigenstate, it is by definition unvarying in time.

15.3.1 Example: Vacuum Energy

Using the simplified rules of Section 15.2, we can calculate some example diagrams. The first is the vacuum diagram in Figure 15.3, which is one of the terms generated from calculating the expectation value of the S-matrix in vacuum, $\langle \text{vac}|S(t,0)|\text{vac}\rangle$.

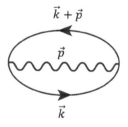

Feynman diagram for a process by which the vacuum state emits three particles, namely a photon (represented by the incoming squiggly line) and an electron–positron pair (represented by the solid lines with arrows in opposite directions), due to the coupling of the electromagnetic and matter fields, and then these three recombine.

Since the vacuum state is an eigenstate, we expect that the form of $S(t, 0)$ must be just a phase factor $e^{i(\Delta E)t/\hbar}$, where ΔE is an energy. This energy term corresponds exactly to the energy calculated by the diagram rules of Section 15.2, when the initial state and the final state are the vacuum. In general, there are infinitely many diagrams that contribute to this energy, but we can assume that the main contributions are from the diagrams with few vertices, such as the diagram in Figure 15.3.

To do this calculation, we use the propagator of the antielectron, or "hole" (defined in Section 15.2), indicated by the line with the arrow pointing backwards. Although the diagram method we use here is for holes in a solid, the same approach works for positrons in vacuum.

The integral corresponding to the diagram in Figure 15.3 is then, by the rules of Section 15.2,

$$\Delta E = \frac{1}{(2\pi)^2} \int_{-\infty}^{\infty} dE \int_{-\infty}^{\infty} dE' \sum_{\vec{k},\vec{p}} M^2 \left(\frac{1}{E - E_k + i\varepsilon} \right) \left(\frac{1}{E' - E_p - i\varepsilon} \right)$$

$$\times \left(\frac{1}{E + E' - E_{\vec{k}+\vec{p}} - i\varepsilon} \right). \tag{15.3.1}$$

The integrals over energy can be resolved by using the methods of analytical calculus. We will not discuss those methods here; instead, we present the final answer,

$$\Delta E = \sum_{\vec{k},\vec{p}} M^2 \left(\frac{1}{E_p + E_k - E_{\vec{k}+\vec{p}}} \right), \tag{15.3.2}$$

which is nonzero. This corresponds to a correction to the energy of a vacuum due to the nonlinear interactions of electrons and photons. Since it is just a constant, it doesn't matter what its value is, since it just shifts the definition of what we call zero energy.

It is convenient to use the shorthand language that the diagram in Figure 15.3 represents a superposition of an infinite number of different processes in which particles in plane-wave states spontaneously appear and then disappear. In the absence of "jumpy detectors" with decoherence processes, however, there is no reason to adopt that picture as anything more than a convenience, equating Green's functions with "particles." As discussed at the beginning of Section 15.3, there is no time dependence of the vacuum state other than an overall phase oscillation, which is undetectable experimentally; therefore, it is improper to think of particles appearing at various different times. It is far more natural to simply think of the Feynman diagrams as simply accounting for nonlinear processes that were initially ignored in writing down the energies for photons and electrons on their own. It is appropriate, however, to say that these diagrams indicate that the vacuum is "complicated," and not "nothing."

15.3.2 Example: Self-Energy

Figure 15.4 shows another possible diagram; in this case, the initial and final states are not the vacuum but instead are a state in which a plane wave has been created, namely $|\psi(0)\rangle = a_k^\dagger |0\rangle$.

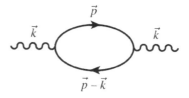

Self-energy Feynman diagram for a process by which a photon emits an electron–positron pair and then reabsorbs it.

The rules for lines in Feynman diagrams apply only to internal lines in this type of diagram; lines that extend off to one side or the other, sometimes called *external legs*, do not contribute propagator factors, because they correspond to creation and destruction operators that define the initial and final states. In the case shown in Figure 15.4, the initial and final states are the same; therefore, like the vacuum diagram, this diagram gives us an energy correction to an eigenstate, which is variously called a *self-energy* or *energy renormalization*.

In this case, there is just one unconstrained momentum and one unconstrained energy. Applying the diagram rules gives us

$$\Delta E = \frac{-i}{(2\pi)} \int_{-\infty}^{\infty} dE \sum_{\vec{k},\vec{p}} M^2 \left(\frac{1}{E - E_p + i\varepsilon} \right) \left(\frac{1}{E - E_k - E_{\vec{p}-\vec{k}} - i\varepsilon} \right), \quad (15.3.3)$$

which can again be resolved to a nonzero number. Using analytical calculus[2] gives

$$\Delta E = - \sum_{\vec{k},\vec{p}} M^2 \left(\frac{1}{E_p - E_k - E_{\vec{p}-\vec{k}} - i\varepsilon} \right). \quad (15.3.4)$$

The Dirac theorem of analytical calculus then says that this is equal to

$$\Delta E = \mathcal{P} \sum_{\vec{k},\vec{p}} M^2 \left(\frac{1}{E_k - E_p + E_{\vec{p}-\vec{k}}} \right) - i\pi \sum_{\vec{k},\vec{p}} M^2 \delta(E_k - E_p + E_{\vec{p}-\vec{k}}), \quad (15.3.5)$$

where the symbol $\mathcal{P}$ indicates the *principal value*, which removes any terms in which the denominator exactly equals zero.

The first term is real, and implies that a photon traveling in vacuum has a shifted energy due to interaction with the Dirac electron field. Again, one could talk of this in shorthand as a photon continuously virtually emitting and absorbing electrons and positrons, but there is no time variation or fluctuation of this.

The second term is imaginary, and is proportional to a decay rate; as discussed at the end of Section 15.1.2, it implies that the photon state $\vec{k}$ is not stable, that is, that it is not an eigenstate of the full system when interactions with electrons are taken into account.

---

[2] Specifically, taking a loop in the lower half of the complex plane for the energy E.

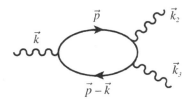

Figure 15.5 Feynman diagram for a process by which a photon in vacuum transforms into two photons.

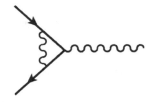

Figure 15.6 Feynman diagram for an electron interaction with a photon, with higher-order corrections.

15.3.3 Example: Nonlinear Optics in Vacuum

Figure 15.5 shows another possible diagram, in which the initial and final states are not the same; there is one photon in the initial state and two photons in the final state. The same rules apply as before: we do not write propagator factors for the external legs; instead, we just use them to determine the energies and wave vectors of the internal lines.

In this case, the energy found by the integral is not a self-energy; it plays the same role as the "matrix element" energy used in Fermi's golden rule (14.1.11), which is used in calculating the rate of transitions from the initial to the final state. The process shown in Figure 15.5 is very weak but not zero. Although we have used Feynman rules for a solid here, the same type of diagram exists in vacuum – photons in vacuum have nonlinear processes by which they can split and combine.

Note that the diagram in Figure 15.5 could be used as a vertex in another, larger diagram, since it connects three lines. Another example is shown in Figure 15.6, which shows an electron–photon vertex with higher-order corrections. Overall, there are three vertices in this diagram, but this vertex energy could be used in another diagram wherever a simple vertex appears. Thus, there can be many embedded Feynman diagrams within diagrams.

Diagrams like those shown in Figures 15.5 and 15.6 resemble the scattering patterns seen in cloud chambers (e.g., as seen in Figure 3.1). In cloud chambers, particular particle trajectories are selected out due to collapse of the wave function, which, as discussed in Chapter 5, is directly connected to both the spatial localization of the atoms with which energetic charged particles interact, and the decoherence that comes from the interaction of an atom with its environment.[3]

---

[3] Actually, these specific diagrams could not occur as cloud chamber patterns, because photons are absorbed by electronic transitions in atoms. Only charged particles that can shed energy to atoms continuously give such tracks.

References

J. Cramer and R. E. Kastner, "Quantifying absorption in the transactional interpretation," arXiv:1711.04501.

A. L. Fetter and J. D. Walecka, *Quantum Theory of Many-Particle Systems*, (McGraw-Hill, 1971).

R. E. Kastner, "The relativistic transactional interpretation and the quantum direct-action theory," arXiv:2101.00712.

F. Mandl and G. Shaw, *Quantum Field Theory*, 2nd ed., (Wiley, 2010).

M. E. Peskin and D. V. Schroeder, *An Introduction to Quantum Field Theory*, (CRC Press, 1995).

D. W. Snoke, *Solid State Physics: Essential Concepts*, 2nd ed., (Cambridge University Press, 2020).

Part IV

Mathematical Considerations of Philosophy of Quantum Mechanics

16

Mathematical Considerations of Quantum Interpretations

This chapter presents supplemental mathematical calculations relevant to the various "interpretations" of quantum mechanics that have been discussed in this book. In the first section, the local hidden-variables interpretation is discussed. This is the only interpretation that is widely viewed as disproven, although there are some philosophers who still hold out for "loopholes" that could allow it.

There are serious math-based problems with other interpretations, however. These do not have the character of absolutely disproving them, but of making them "ugly".

16.1 The Local Hidden-Variables Hypothesis

The local hidden-variables hypothesis posits that there are physical properties associated with individual particles that give different behavior for some particles compared to others even when the same experiment is done.

In many ways, this approach already was on the wrong footing because it treated particles as fundamental entities instead of just quantized excitations of the field amplitude. In the quantum field theory approach, the quanta must be indistinguishable because they are merely excitations of a single field.

The following sections give the standard methodology for showing that a hidden variable interpretation with distinguishable particles disagrees with experimental results.

16.1.1 Quantum Wave States of the EPR Experiment

Figure 16.1 reproduces the Einstein–Podalsky–Rosen (EPR) experiment described in Section 4.3. The two-particle state emitted from the source in this experiment can be written as

$$|\psi\rangle = \frac{1}{\sqrt{2}}(|H\rangle|H\rangle + |V\rangle|V\rangle), \tag{16.1.1}$$

where $|H\rangle$ is the horizontally polarized state and $|V\rangle$ is the vertically polarized state. The ordering of the two kets matters; the first ket represents the state of the particle that is

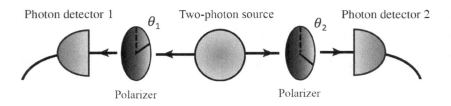

Figure 16.1 Layout of an EPR-type experiment, in which pairs of photons are sent in opposite directions by a two-photon source.

going to the left, and the second ket represents the particle that is going to the right. We can represent these in vector form as

$$|H\rangle = \begin{pmatrix} 1 \\ 0 \end{pmatrix}, \quad |V\rangle = \begin{pmatrix} 0 \\ 1 \end{pmatrix}. \tag{16.1.2}$$

The action of a polarizer at angle θ relative to the horizontal acting on these states is

$$\hat{P} = \begin{pmatrix} \cos^2\theta & \sin\theta\cos\theta \\ \cos\theta\sin\theta & \sin^2\theta \end{pmatrix}. \tag{16.1.3}$$

Acting on a single photon, this operator gives the action of *Malus' law* for polarizers, namely, the probability of a photon passing through the polarizer is equal to $\cos^2(\theta - \theta_1)$, where θ_1 is the angle of the photon's polarization relative to the horizontal. We can see this, for example, by starting with a photon in the state $|H\rangle$ and finding the final state,

$$|\psi'\rangle = \hat{P}|H\rangle = \cos\theta(\cos\theta|H\rangle + \sin\theta|V\rangle); \tag{16.1.4}$$

the Born rule gives the probability of a photon passing through as

$$\langle\psi'|\psi'\rangle = \cos^2\theta(\cos^2\theta + \sin^2\theta) = \cos^2\theta. \tag{16.1.5}$$

Here, we have used the orthogonality of the two polarization states, namely $\langle V|H\rangle = 0$.

If we have the two-particle state (16.1.1) and place just one polarizer at angle θ in the path of the photon moving to the right, we obtain the state

$$|\psi'\rangle = \hat{P}_1\frac{1}{\sqrt{2}}(|H\rangle|H\rangle + |V\rangle|V\rangle) \tag{16.1.6}$$

$$= \frac{1}{\sqrt{2}}\left(|H\rangle|(\cos^2\theta|H\rangle + \cos\theta\sin\theta|V\rangle) + |V\rangle(\sin\theta\cos\theta|H\rangle + \sin^2\theta|V\rangle)\right),$$

once again using $\langle V|H\rangle = 0$ for each photon. The magnitude of this is

$$\langle\psi'|\psi'\rangle = \frac{1}{2}(\cos^4\theta + 2\cos^2\theta\sin^2\theta + \sin^4\theta)$$

$$= \frac{1}{2}(\cos^2\theta + \sin^2\theta)^2 = \frac{1}{2}; \tag{16.1.7}$$

that is, by the Born rule, we see a "coincidence" (i.e., photons being detected at both detectors) half the time, no matter what the angle of the polarizer is. One of the photons is always detected, since there is no polarizer acting on it, and the other passes through its polarizer half the time, since it is equally likely to be at all polarization angles.

If now we act on both photons with two separate polarizers set at θ_1 and θ_2, respectively, we obtain

$$|\psi'\rangle = \hat{P}_2\hat{P}_1 \frac{1}{\sqrt{2}}(|H\rangle|H\rangle + |V\rangle|V\rangle) \tag{16.1.8}$$

$$= \frac{1}{\sqrt{2}}(\cos^2\theta_2|H\rangle + \cos\theta_2\sin\theta_2|V\rangle)(\cos^2\theta_1|H\rangle + \cos\theta_1\sin\theta_1|V\rangle)$$

$$+(\sin\theta_2\cos\theta_2|H\rangle + \sin^2\theta_2|V\rangle)(\sin\theta_1\cos\theta_1|H\rangle + \sin^2\theta_1|V\rangle).$$

After resolving the many trigonometric functions, and using the orthogonality of the vertical and horizontal states, this has the simple result

$$\langle\psi'|\psi'\rangle = \frac{1}{2}\cos^2(\theta_1 - \theta_2). \tag{16.1.9}$$

One can interpret this as saying that the first photon has a probability of 1/2 of passing through its polarizer, since its polarization is equally likely to be all angles, and then the second photon has a probability given by Malus' law of passing through its polarizer, assuming that it had exactly the same polarization as the first photon. This implies that, if the two polarizers have angles perpendicular to each other, with $\theta_1 - \theta_2 = \pi/2$ (which we called *crossed*), then there will be zero coincidences. Experiments like this have been done (e.g., Aspect 1982) and are in agreement with this prediction.

On the other hand, suppose that the photons were always sent out from the source with some definite polarization angle θ from the very start, and θ varied randomly. The probability for a coincidence with crossed polarizers in this case can be found by setting the polarizers to $\theta_1 = 0$ and $\theta_2 = \pi/2$, and taking the average over θ for all states with the same angle θ for both photons. We write the two-particle state in this case as

$$|\psi_D\rangle = (\cos\theta|H\rangle + \sin\theta|V\rangle)(\cos\theta|H\rangle + \sin\theta|V\rangle), \tag{16.1.10}$$

and the polarizer operators are

$$\hat{P}_1 = \begin{pmatrix} 1 & 0 \\ 0 & 0 \end{pmatrix}, \qquad \hat{P}_2 = \begin{pmatrix} 0 & 0 \\ 0 & 1 \end{pmatrix}. \tag{16.1.11}$$

This results in

$$|\psi_D'\rangle = \hat{P}_2\hat{P}_1|\psi_D\rangle = (\cos\theta|H\rangle)(\sin\theta|V\rangle), \tag{16.1.12}$$

and

$$\langle\psi_D'|\psi_D'\rangle = \cos^2\theta\sin^2\theta. \tag{16.1.13}$$

The average value of this is

$$\frac{1}{2\pi}\int d\theta \, \cos^2\theta\sin^2\theta = \frac{1}{8}, \tag{16.1.14}$$

which, obviously, is not zero.

The state (16.1.11) is called a *factorized* state, because it can be written as a product of two terms that each depends only on the properties of one particle. The state (16.1.1), on the other hand, is called an *entangled* state, because it cannot be factorized in this way.

The difference in the coincidence predictions means that the entangled state (16.1.1) gives experimental correlation results that cannot be described as a result of two particles acting independently, which would be represented by a factorized state.

16.1.2 Bell's Inequality

We have already seen in Section 16.1 that the results of quantum mechanics do not agree with the expectation for particles acting independently. That result was based on the assumption that the particles were emitted from the source with random polarization angles, with equal probability for all angles. One could imagine, however, a "rigged" situation in which the source did not emit particles at random polarizations. For example, if the source emitted only particles aligned along $\theta = 0$ or $\theta = \pi/2$, then the experimental result of zero coincidences would be achieved for crossed polarizations also aligned along $\theta = 0$ and $\theta = \pi/2$.

One would not normally expect that the source would "know" what direction the polarizers were aligned along; so this possibility seems unlikely. But can we rule out the possibility that the source does not emit equally at all polarization angles, and we have accidentally chosen in our experiment the angle that gives us the quantum mechanical result for entangled states?

Bell's (1964) inequality, which can actually be formulated in several different ways, puts a boundary on the number of coincidences that can occur for a set of measurements with "definite" properties, that is, properties that do not depend on nonlocal interactions with things far away.

The inequality is based on a general observation about sets. Suppose that set A is equal to some series $\{1, 0, 0, 1, 1, \ldots\}$, in which each member of the set is either 0 or 1. Set B has the same length, and also is either 0s or 1s. If we compare the two sets one element at a time, in order, we can define the number of *anticoincidences* N_{AB} as the total number of elements that are not the same at any given location in the set.

Suppose that we now compare both A and B to a third set of the same length, C. The numbers of anticoincidences in each of these cases are N_{AC} and N_{BC}, respectively. We can then know that

$$N_{AB} \leq N_{AC} + N_{BC}, \tag{16.1.15}$$

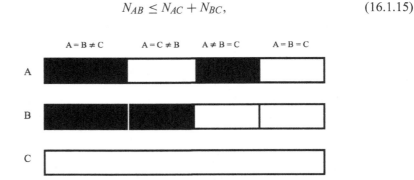

Figure 16.2 The sets used in the proof of the Bell inequality.

because there cannot be more differences between two sets than the sum of the differences each has with a third. As illustrated in Figure 16.2, there are four possible cases in regard to the elements of sets, which we can imagine rearranged into groups together. The total number of differences between A and B lie in the middle two sets. The sum of the middle two sets gives N_{AB}. The maximum number in these two sets occurs when the other two sets have no elements. In this case, N_{AB} is exactly equal to the sum of N_{AC} and N_{BC}.

So far, this is just a general statement about any sets of distinguishable elements. To apply this to the EPR experiment, we identify the elements as detector-counting events. We now imagine four different experiments.

In Experiment 1, the polarizer on the left leg is set to $\theta = 0$ and the polarizer on the right leg is set to some other angle, θ'. This gives us a series of clicks at the detectors corresponding to each pair sent out from the source. We equate a click of a detected photon with the number 1, and no click with the number 0.[1] This gives us two sets of numbers ordered in time, which we can call L_1 and R_1. We can define the coincidences and anticoincidences of these two sets.

We now imagine a second experiment, in which we use the *same* set of photons from the source but have the polarizers set both at θ'. In this case, we can assume that the data set for the right leg, R_2, is equal to the data set R_1, since the polarizer was in the same position. This is an example of *contrafactual definiteness*, that is, the assumption that we can be sure about the results of an experiment that we did not, in fact, do. This premise is precisely what is rejected by positivist interpretations of quantum mechanics. But it is an entirely reasonable assumption, if we assume local causation. The only thing different between Experiment 1 and Experiment 2 is what happened to the polarizer on the left leg, and if that is far away, there is no reason to expect that it will change the behavior of the polarizer and the detector on the right leg.

We now imagine a third experiment in which we keep the polarizer on the *left* leg the same, and change the angle of the polarizer on the right to a new angle, θ'', but assume that we use the same set of photons from the source. Once again, we assume by contrafactual definiteness that $L_3 = L_2$, and get a new set R_3 for the clicks of the detector on the right.

Finally, in our fourth experiment, we keep the polarizer on the right at θ'', and return the polarizer on the left to $\theta = 0$. Table 16.1 summarizes the four experiments.

Now let us put in some actual numbers. We choose $\theta' = 18°$ and $\theta'' = 36°$. From the predictions of quantum mechanics given by (16.1.9) in Section 16.1.1, the coincidence rate of L_1 and R_1 is $C_1 = \cos^2 18° = 0.9$. We assume that we have done an experiment, and have gotten this coincidence rate as a result.

---

[1] This assumes that we have detectors with 100% efficiency, so that we can know that no photon arrived. Since that is not the case in real experiments, real experiments are modified to replace the polarizers discussed here with polarizing beamsplitters, which have the action that instead of destroying a rejected photon, they send the rejected photon along a different path to another detector. In this case, we can know that a photon definitely was sent from the source when we get a click at one or the other detector, for the accepted or rejected photon. This is sometimes called "heralding" of a photon. We can put this apparatus on both legs of the EPR experiment, for the photons going to the right and to the left, and keep in our data set only those events in which a photon of either polarization was detected both on the right and the left. Thus, any events in which we record a click only on one leg and not the other, due to inefficiency of the detectors, are eliminated from the data sets, and the experiment becomes the equivalent of the one discussed here.

Table 16.1	The sets used in the proof of the Bell inequality.	
	Right polarizer	Left polarizer
1	θ'	0
2	θ'	θ'
3	θ''	θ'
4	θ''	0

By the same quantum mechanical formula (16.1.9), we predict that the result of the second experiment is $C_2 = 1$, that is, 100% coincidence since the polarizers are at the same setting. We now have three data sets, for 0, θ', and θ'', assuming that sets recorded with the same polarizer setting on both left and right are identical. In the notation of the inequality (16.1.15), we will say that set A is given by $L_1 = L_4$, set B is given by $R_3 = R_4$, and set C is given by $R_1 = L_2 = R_2 = L_3$.

The anticoincidence rate between sets A and C is equal to $N_{AC} = 1 - C_1 = 0.1$. The coincidence rate between sets B and C is the same value, because $36° - 18° = 18°$, so that we have the anticoincidence rate $N_{BC} = 1 - C_3 = 0.1$. By the inequality (16.1.15), we then have

$$N_{AB} \le N_{AC} + N_{BC} = 0.2. \tag{16.1.16}$$

But if we actually do Experiment 4, we find that C_4 is equal to $\cos^2 36° = 0.65$, and therefore $N_{AB} = 1 - 0.65 = 0.35$, as predicted by formula (16.1.9).

What went wrong? One approach is to say that our assumption of contrafactual definiteness is wrong – that we cannot talk about the results of experiments that we didn't actually do, as argued by positivists. But consider what dropping this assumption means. It means that the apparatus somehow takes notice of what we are measuring, and gives us different results based on what we are looking for.

The other approach is to say that we cannot equate sets R_1 and R_2, for example, because of *nonlocality*. In other words, the setting of the polarizer on the right affects what happens on the left, even though it is far away.

Experiments like this force us to "pick our poison" – to drop one of three things that physicists hold dear. Either the assumption of nonlocality is wrong, or quantum mechanics is wrong, or our belief that reality does not change in response to our measurements, just to mess with us, is wrong. Most physicists pick the first option, and accept that there is something intrinsically nonlocal about quantum mechanics.

16.2 The Many-Worlds Hypothesis

The many-worlds approach to quantum mechanics is in some ways the simplest, in that it takes the existing, unitary equations of quantum mechanics and holds them to be absolute.

As discussed in Section 5.1, however, it has no way to generate the Born rule within a purely unitary theory, although this is an empirical fact in our experience. The following two sections discuss other issues of the many-worlds approach.

16.2.1 The Spectral Weight Problem

As discussed in Sections 12.2.2 and 13.1.1, the quantum state of a many-particle system can be written in terms of *Fock states*, which give an exact number of particles in each allowed single-particle state. We write these as

$$|\psi\rangle = |N_1, N_2, \ldots\rangle, \tag{16.2.1}$$

where the numbers N_1, N_2, and so on, give the numbers of particles in state 1, state 2, and so on, all the way to a very large number of possible single-particle states.

When an interaction occurs, this is equivalent to acting on such a state with destruction operators and creation operators, which take the original state to a superposition of other states. For example, scattering of electrons from impurities can be represented by an interaction term

$$\hat{V} = \sum_{k_1, k_2} V_{k_1, k_2} a_{k_2}^\dagger a_{k_1}, \tag{16.2.2}$$

which corresponds the sum of all possible ways of removing an electron in state k_1 and placing it in state k_2; V_{k_1, k_2} is a factor that depends on the details of the microscopic processes (see Section 14.2). After a short amount of time with this interaction, the Schrödinger equation (11.4.1) implies that a system that started in state (16.2.1) would then be in the superposition

$$|\psi'\rangle = \alpha|N_1, N_2, \ldots\rangle + \beta|N_1 - 1, N_2 + 1, \ldots\rangle + \gamma|N_1 - 1, N_2, N_3 + 1, \ldots\rangle + \ldots, \tag{16.2.3}$$

where the sum runs over every possible other Fock state that the interaction (16.2.2) connects state $|\psi\rangle$ to, and α, β, γ, and so on are complex numbers that give the amplitude ("weight factor") of each possibility. Each of these new states in the superposition will in turn be connected to other Fock states by later interactions.

In a typical incoherent gas of atoms, after a very short time, the interactions will have caused every Fock state of the system to have some nonzero weight, since sums like that in (16.2.2) connect each state to a vast number of other states. The exact values of the weight factors α, β, γ, and so on may fluctuate in time, but the average number of particles in any state will stay the same, in equilibrium (Snoke 2012).[2]

For a disordered, homogeneous system like a gas, the average number of particles per state is the only relevant information. We can write this in terms of the number operator as

$$\langle \hat{N}_k \rangle = \langle a_k^\dagger a_k \rangle = \sum_n |\alpha_n|^2 \langle n|a_k^\dagger a_k|n\rangle, \tag{16.2.4}$$

---

[2] In recent years, there has been significant study of the "eigenstate thermalization hypothesis," which posits that, in equilibrium, the prefactors α, β, and γ do *not* fluctuate, in which case the equilibrium state is an eigenstate of the total Hamiltonian (see, e.g., Nandkishore 2015).

where the sum over n is over all possible Fock states, and α_n are the complex numbers represented as α, β, γ, and so on, for each Fock state as we have discussed for each Fock state as in (16.2.3). The sum in this case depends only on the magnitude of the α_n factors. This makes it fairly robust against "phase noise." In other words, if we write a given factor as $\alpha = |\alpha|e^{i\theta}$, where $|\alpha|$ is the amplitude of the complex number and θ is the phase, fluctuations in θ will not contribute to the particle numbers, because the phase factors cancel out when computing $|\alpha|^2 = \alpha^*\alpha$. The phase information could be entirely lost, and we would still have the same information about the numbers of particles in the states.

To put it another way, in a system like this, splits of the wave function into superpositions are "recycled" back into the whole system, because every state scatters into every other state, with different phase factors, but the phase information is irrelevant.

By contrast, in the case of *highly organized matter*, or *information-rich matter*, such as living creatures (e.g., cats), photomultipliers, or electrical circuits, not only the average numbers of particles matter, but also *correlations* of states. For example, suppose that a collection of atoms is at a specific set of locations, r_1, r_2, and r_3. We can create the three-particle state with definite positions for each particle by applying the spatial creation operator (introduced in Section 13.1) for each particle successively to the vacuum state $|vac\rangle$, as follows:

$$|\psi''\rangle = \hat{\psi}^\dagger(r_3)\hat{\psi}^\dagger(r_2)\hat{\psi}^\dagger(r_1)|vac\rangle \tag{16.2.5}$$
$$= \frac{1}{V^{3/2}} \sum_{k_1,k_2,k_3} e^{-ik_1r_1}e^{-ik_2r_2}e^{-ik_3r_3}a_{k_3}^\dagger a_{k_2}^\dagger a_{k_1}^\dagger|vac\rangle,$$

where we have generalized to a three-dimensional system, and $V = L^3$ is the volume of the system. This corresponds to a sum over a large number of Fock states like (16.2.3), with coefficients of the form

$$\alpha'' = \frac{1}{V^{3/2}}e^{-ik_1r_1}e^{-ik_2r_2}e^{-ik_3r_3}. \tag{16.2.6}$$

After a quantum measurement event in the many-worlds scenario, the state of the system will have the form

$$A\big[\alpha|N_1, N_2, \ldots\rangle + \beta|N_1 - 1, N_2 + 1, \ldots\rangle + \gamma|N_1 - 1, N_2, N_3 + 1, \ldots\rangle + \ldots\big]$$
$$+ B\big[\alpha'|N_1, N_2, \ldots\rangle + \beta'|N_1 - 1, N_2 + 1, \ldots\rangle + \gamma'|N_1 - 1, N_2, N_3 + 1\ldots\rangle + \ldots\big]$$
$$+ \ldots, \tag{16.2.7}$$

where A, B, and so on are overall amplitudes multiplying all the states that make up the separate "worlds," and α, β, and so on, are the amplitudes of the states within each world.

For one set to be a complete "world," with locations r_1, r_2, and r_3 different from another "world," each of the subsets with amplitude A, B, and so on must keep the phase information that allows us to recover r_1, r_2, and r_3. Thus, if terms like $A\alpha$ and $B\alpha'$ are of order $2^{-1,000,000}$, as would be the case after using a photon counter to collect 1,000,000 counts (as is quite common in experiments), the phase information must be accurate at this level. Thus, while this is the case if Schrödinger's equation is strictly true, this analysis shows that Schrödinger's equation must be *fantastically*, *perfectly* true for the different, information-rich worlds to be preserved.

16.2.2 The Many-Worlds Hypothesis and Nonlocality

As discussed in Sections 4.1 and 4.2, quantum mechanics involves nonlocal correlations; that is, events separated in space can be correlated even though no signal can have traveled from one to the other at the speed of light or slower. Some have assumed (e.g., DeWitt 1973) that the many-worlds approach removes the nonlocality problem. In this section, we will see, through a simple thought experiment, that the many-worlds scenario still involves issues of nonlocality.

To see this, consider the EPR experiment discussed in Section 16.1.1, with the layout shown in Figure 16.1. We again suppose that the two-photon state (16.1.1) is emitted from the source,

$$|\psi\rangle = \frac{1}{\sqrt{2}}(|H\rangle|H\rangle + |V\rangle|V\rangle), \tag{16.2.8}$$

where $|H\rangle$ is the horizontally polarized state and $|V\rangle$ is the vertically polarized state.

We now write the initial state of the system with explicit time and position dependence as

$$\frac{1}{\sqrt{2}}(|H(r_1, t_1)\rangle|H(r_2, t_1)\rangle + |V(r_1, t_1)\rangle|V(r_2, t_1)\rangle)|E_1\rangle|E_2\rangle, \tag{16.2.9}$$

where r_1 represents the position of a moving wave packet on the left, and r_2 represents the position of a moving wave packet on the right. $|E_1\rangle$ and $|E_2\rangle$ are the many-body states of the detectors and environment prior to any interaction with the photons.

We assume that the photon wave packet on the left encounters its polarizer/detector assembly first. If this polarizer is set to pass horizontally polarized photons, then after its encounter, the state of the system, in the many-worlds approach, is

$$\frac{1}{\sqrt{2}}(|D(r_1 = R_1, t_2)\rangle|H(r_2, t_2)\rangle|E_2\rangle + |N(r_1 = R_1, t_2)\rangle|V(r_2, t_2)\rangle|E_2\rangle), \tag{16.2.10}$$

where D indicates a many-body wave function of the many particles in the detector and its environment that make up the detection event of a horizontally polarized photon, and N indicates no detection event, and only heat dissipated in the polarizer. (The slight separation of the polarizers and detectors will be treated as negligible compared to the distance between the detectors on opposite sides, so that these all are assigned the position R_1 on the right and R_2 on the left.)

At a later time, the wave packet on the right encounters a horizontal polarizer and a detector, at which point the system wave function is

$$\frac{1}{\sqrt{2}}(|D(R_1, t_3)\rangle|D(R_2, t_3)\rangle + |N(R_1, t_3)\rangle|N(R_2, t_3)\rangle). \tag{16.2.11}$$

If we pick the world in which detection of a photon occurs on the left, the state is

$$|D(R_1, t_3)\rangle|D(R_2, t_3)\rangle, \tag{16.2.12}$$

while, if no photon detection occurs on the left, there will be no photon detection on the right. In other words, there is deterministic behavior of a photon on the right, within this world.

Suppose now that, at the last moment before the photon hits the polarizer on the left, we suddenly changed the polarizer position to 45°. Then at time t_2, the state would be

$$\frac{1}{\sqrt{2}}\left(\frac{1}{\sqrt{2}}(|D(R_1,t_2)\rangle + |N(R_1,t_2)\rangle)|H(r_2,t_2)\rangle|E_2\rangle\right.$$
$$\left. +\frac{1}{\sqrt{2}}(|D(R_1,t_2)\rangle - |N(R_1,t_2)\rangle)|V(r_2,t_2)\rangle|E_2\rangle\right). \tag{16.2.13}$$

Then when the other wave packet encounters the horizontal polarizer and detector on the right at time t_3, the state would be

$$\frac{1}{\sqrt{2}}\left(\frac{1}{\sqrt{2}}(|D(R_1,t_3)\rangle + |N(R_1,t_3)\rangle)|D(R_2,t_3)\rangle + \frac{1}{\sqrt{2}}(|D(R_1,t_3)\rangle - |N(R_1,t_3)\rangle)|N(R_2,t_3)\rangle\right)$$
$$=\frac{1}{2}\left(|D(R_1,t_3)\rangle|D(R_2,t_3)\rangle + |N(R_1,t_3)\rangle|D(R_2,t_3)\rangle\right.$$
$$\left. +|D(R_1,t_3)\rangle|N(R_2,t_3)\rangle - |N(R_1,t_3)\rangle|N(R_2,t_3)\rangle\right). \tag{16.2.14}$$

If we once again pick the world in which detection of a photon occurred on the left, the state is then

$$|D(R_1,t_3)\rangle\frac{1}{\sqrt{2}}\left(|D(R_2,t_3)\rangle + |N(R_2,t_3)\rangle\right),$$

which is a 50% superposition; if no photon is detected on the left, the corresponding state on the right is also a 50% superposition. In other words, if I am on the left, I can know that, in *my* world, I can control whether the other detector is in a superposition or not, by whether I rotate my polarizer. Since in the many-worlds framework, the wave function of the system is a real thing, this is a true nonlocal change within my world due to the action of rotating the polarizer at R_1. This is guaranteed no matter how little time elapses between t_2 and t_3, that is, even if the detection events are spacelike separated. I cannot use this to send signals faster than the speed of light, because the outcome at the detector on the right is still random, but I have controlled how the branching of the worlds occurs.

In general, the relative weight of the two superposition states, from zero weight for one option up to equal weight for both options, can be controlled continuously by the angle between the two polarizers at the time the photons pass through them, which can be set after the emission of the two photons from the source but before any signal can have communicated what that setting is. A person experiencing one of these superpositions will have no way to know the relative weight of the superposition, or if there is a new superposition at all. But in the many-worlds approach, there is still a "fact of the matter" that the exact splitting into different worlds depends on measurements far away.

Frank Tipler (2014) has presented an argument that the many-worlds hypothesis does not require nonlocality. In that work, he assumed that a measurement apparatus can act to always give the definite state of particle, that is,

$$\hat{U}|\psi\rangle|M(0)\rangle = |\psi\rangle|M(\psi)\rangle, \tag{16.2.15}$$

where $\hat{U}$ is a unitary evolution operator giving the interaction with the measurement system, $|\psi\rangle$ is the state of the particle of interest, and $|M(0)\rangle$ and $|M(\psi)\rangle$ are the quantum states of the measurement apparatus before and after the measurement. Crucially, the detector state $|M(\psi)\rangle$ is uniquely identified with the state $|\psi\rangle$ that the particle had before the measurement.

In general, this is only possible if the measurement apparatus is set to detect exactly the state $|\psi\rangle$. For example, in the case of a photon hitting a polarizer and a detector considered, if the photon is polarized at $0°$ and the polarizer is set at $0°$, then the process in (16.2.15) will hold true. However, if the photon is polarized at $45°$, then for the setting of the polarizer at $0°$, it will *not* be true that the detector goes into a state of having definitely detected a photon with polarization at $45°$. Instead, it will "project" the system into either detection of a photon with polarization at $0°$ or detection of a photon with polarization at $90°$. In traditional quantum mechanics, one or the other of these states will occur with a probability given by the Born rule; in the many-worlds approach, the detector goes into a superposition of both possibilities. But this superposition is not the equivalent of having a single definite measurement of a photon with polarization at $45°$; a person living inside one of these two superposed worlds will see only either $0°$ or $90°$.

This can be seen as an example of environmentally induced selection, or *einselection*, discussed in Section 6.3. The decoherence of the detector allows it to only be one or the other of detecting the polarization states $0°$ or $90°$; in the language of Dirac notation, the detection apparatus forces a preferred set of "basis states," unlike the propagation of the photon through free space.

The nonlocality of quantum mechanics comes fundamentally from the fact that entangled states of spacelike separated wavepackets can be created. This is intrinsic to the mathematical structure and not removable by any of the interpretations of quantum mechanics that agree with experimental results.

16.2.3 Many-Worlds and Spontaneous Symmetry Breaking

In Section 7.4, we discussed the possibility of viewing the early universe as a superposition of spatially asymmetric states, which add up to a totality that is spatially symmetric. In this case, one might posit a many-worlds scenario in which the overall symmetry is never broken, but each parallel universe has broken symmetry.

If the early universe was thermalized (i.e., if it had a well-defined temperature) and spatially homogeneous, this is, in general, not possible. We can see this by writing the state of the many-particle system as

$$|\psi\rangle = \sum_{\{N_k\}} \alpha_{\{N_k\}} |\ldots N_{k-1}, N_k, N_{k+1}, \ldots\rangle, \qquad (16.2.16)$$

where the sum is over all possible Fock states with all possible different integer values of all the occupation numbers N_k. The equilibrium temperature assumption gives the magnitude of each factor $|\alpha_{\{N_k\}}|$ according to the distributions derived in Section 10.1, but there is a free phase factor $e^{i\theta(\{N_k\})}$ for each.

A *single*-particle wave function that is homogeneous can, in general, be written as a superposition of two or more other single-particle wave functions. But in the context of many-body field theory, using a single wave function is equivalent to writing a coherent state, defined in Section 12.5. A superposition of coherent states with different wavelengths would be written as

$$|\psi\rangle = \sum_k \sum_{N_k} \alpha_{N_k} |\ldots 0, N_k, 0, \ldots\rangle, \tag{16.2.17}$$

and the state (16.2.16) clearly cannot be written as such a sum. The thermal state (16.2.16) contains many terms with "cross-talk," that is, occupation of multiple single-particle states. In the language of entanglement (see Section 10.3), it is *massively entangled*.

A thermalized many-body also state cannot be written as a sum of spatially localized states. To see this, consider the case of a volume of size $2L$ broken into two half-volumes, each with size L in one direction. This is a simple example of introducing spatial asymmetry.

According to the prescription of Section 13.1.3, for both fermions and bosons, one can define creation operators for localized states using

$$a_n^\dagger = \int d^3r \, \phi_n(\vec{r}) \hat{\psi}^\dagger(\vec{r}), \tag{16.2.18}$$

where ϕ_n is the localized wave function. For particles restricted to a half-space, the available plane-wave states will be

$$\phi_n(x) = \frac{1}{\sqrt{V}} \sin \pi nx/L, \tag{16.2.19}$$

for integer values of n, while the states in the full-space will have $2L$ instead of L. The Fourier weight for the k-states of the half-space in terms of the full-space states will then be proportional to

$$\int_0^L dx \, \phi_n(x) e^{-ikx} = \frac{i}{\sqrt{V}} \int_0^L dx \sin(\pi nx/L) \sin(kx). \tag{16.2.20}$$

In the full volume of size $2L$, there will be k-states with magnitude $k = \pi n'/2L$. For $n' = 1$, this will give zero Fourier weight for all n of the half-space. In other words, the states of the half-space are spaced twice as far apart in k-space, and therefore there is no way to construct all the states of the full-space from them.

Yet, in the full-space, a thermally equilibrated distribution will have occupation of all its k-states, including the ones that cannot be built as superpositions of states in the half-space. Both of the two half-spaces will have zero occupation of these states of the full-space.

16.3 Bohmian Hydrodynamics

The Bohmian approach to quantum mechanics has experienced somewhat of a resurgence in recent years, as the problems with the Copenhagen approach have caused many people to question it. In the Bohmian approach, both the waves of the quantum equations and indivisible particles exist at the same time, with equal ontological status. As shown in the following sections, these particles cannot be equated with the quanta of field theory, which means that *two* types of particles must exist. Other strange implications of the theory lead most physicists to conclude that the Bohmian approach is "ugly," aesthetically. It is also incomplete, as discussed in the following, although some adherents will point out that other approaches such as spontaneous collapse theory are also incomplete. But in the case of Bohmian pilot waves, the method seems to depend on a "mathematical trick" that only works for the Schrödinger equation for particles with mass.

16.3.1 Derivation of the Bohmian Flow Equations

The Bohmian pilot wave approach starts with the Schrödinger equation,

$$i\hbar \frac{\partial}{\partial t} \psi = -\frac{\hbar^2}{2m} \nabla^2 \psi + U\psi, \tag{16.3.1}$$

and writes the wave function in the form $\psi = Re^{iS/\hbar}$, where R and S are two scalar, real functions. The derivatives are then resolved explicitly as

$$
i\hbar \left(\frac{\partial R}{\partial t} e^{iS/\hbar} + \frac{i}{\hbar} Re^{iS/\hbar} \frac{\partial S}{\partial t} \right) = -\frac{\hbar^2}{2m} \nabla \cdot \left((\nabla R)e^{iS/\hbar} + R\frac{i}{\hbar} e^{iS/\hbar}(\nabla S) \right) + URe^{iS/\hbar}
$$

$$
= -\frac{\hbar^2}{2m} \left((\nabla^2 R)e^{iS/\hbar} + \frac{2i}{\hbar}(\nabla R) \cdot (\nabla S)e^{iS/\hbar} \right.
$$

$$
\left. - R\frac{1}{\hbar^2} e^{iS/\hbar}|\nabla S|^2 + R\frac{i}{\hbar} e^{iS/\hbar}(\nabla^2 S) \right) + URe^{iS/\hbar}. \tag{16.3.2}
$$

We next multiply by $e^{-iS/\hbar}$ to eliminate the phase factors, and equate the real and imaginary parts on both sides to make two equations,

$$
-R\frac{\partial S}{\partial t} = -\frac{\hbar^2}{2m} \nabla^2 R + \frac{1}{2m} R|\nabla S|^2 + UR,
$$

$$
\frac{\partial R}{\partial t} = -\frac{1}{m}(\nabla R) \cdot (\nabla S) - \frac{1}{2m} R\nabla^2 S. \tag{16.3.3}
$$

We define the scalar density $\rho = R^2$, and the velocity $\vec{v} = \nabla S/m$. The *continuity equation* for any fluid described by a velocity and density is

$$\frac{\partial \rho}{\partial t} = -\nabla \cdot (n\vec{v}). \tag{16.3.4}$$

Integrating both sides of this equation over some closed volume V and using Stoke's formula shows that the continuity equation is equivalent to saying that the rate of change of the total mass inside any closed surface is given by the net current through that surface. In terms of S and R, this can be expanded as

$$\frac{\partial R^2}{\partial t} = -\nabla \cdot \left(R^2 \frac{1}{m} \nabla S \right)$$

$$\Rightarrow 2R\frac{\partial R}{\partial t} = -\frac{2}{m} R(\nabla R) \cdot \nabla S - \frac{1}{m} R^2 \nabla^2 S. \tag{16.3.5}$$

Comparing this to the second equation of (16.3.3), we see that it is equivalent to the continuity equation, if we multiply both sides by R. This gives us a picture of flow of the probability density through space like a fluid, with a speed given by the gradient of the phase. Because the continuity equation is obeyed, this means that the total probability density is conserved, and no particles disappear.

We can divide the first equation of (16.3.3) by R, to rewrite it as

$$\frac{\partial S}{\partial t} = -\frac{\hbar^2}{2m}\frac{\nabla^2 R}{R} + \frac{1}{2m}|\nabla S|^2 + U. \qquad (16.3.6)$$

The advance of the phase in time is proportional to the sum of three energy terms: the potential energy U, the kinetic energy $K = \frac{1}{2}mv^2$, and the "quantum potential" proportional to $\nabla^2 R/R$, which can also be written as

$$Q = -\frac{\hbar^2}{2m}\frac{(R\nabla^2 R)}{R^2}, \qquad (16.3.7)$$

which can be viewed as the kinetic energy due to density variation.

This equation can be viewed as giving a force, since the rate of change of v is the acceleration. We write

$$a = \frac{\partial v}{\partial t} = -\nabla(Q + K + U), \qquad (16.3.8)$$

which gives the acceleration proportional to the net gradient in the total energy, including the quantum potential.

If the second derivative is negative, which is the case at a density minimum, and the density $\rho = R^2$ is nearly zero, this makes a huge quantum potential energy barrier, which acts like a large potential energy barrier to repel any flow into that region.

In the standard Bohmian interpretation, the velocity v gives the "particle" velocity, which, as discussed in Section 5.2, cannot be equated to the particles derived in quantum field theory. But the velocity v could also simply be interpreted as a flow of "probability density," which is conserved in a sort of hydrodynamic flow. This connects to the concern of Dirac, discussed in Sections 9.1 and 13.2, that fields associated with particles with mass must have total probability density that is conserved at all times.

As noted in Section 5.2, the above formalism presented here will not work for a particle with zero mass, such as a photon, or a particle with mass at relativistic velocities. In both cases, the energy is proportional to the gradient of the wave function, rather than the square of the gradient; that is, they are linearly proportional to the momentum.

16.3.2 Comparison of Quantum Field Theory and Bohmian Particles in a Standing Wave

As discussed in Section 9.1, it is possible to make a standing wave in a region of finite size; these correspond to the "first-quantized" resonances discussed in Sections 2.1 and 2.2. A standing wave can be written as the sum of two traveling waves, which for the Schrödinger equation is

$$\psi_0 e^{i(kx-\omega t)} + \psi_0 e^{i(-kx-\omega t)} = 2\psi_0 \cos kx e^{-i\omega t}. \qquad (16.3.9)$$

In quantum field theory notation, this is written as

$$|\psi\rangle = (a_k^\dagger + a_{-k}^\dagger)|0\rangle; \tag{16.3.10}$$

that is, as the superposition of an electron in the states k and $-k$, each of which corresponds to a particle traveling in a straight line with speed $v = \pm\hbar k/m$. If a velocity measurement was done, there would be only two possible outcomes, with value $\pm v$.

In the Bohmian equations, the velocity of a "particle" is proportional to ∇S. From (16.3.7), the quantum potential for a standing wave is given by

$$Q = -\frac{\hbar^2}{2m}\frac{(\cos kx \nabla^2 \cos kx)}{\cos^2 kx} = \frac{\hbar^2 k^2}{2m}, \tag{16.3.11}$$

which implies that there is no acceleration of the particles due to the quantum potential, and the velocity is given by $\nabla S = 0$, since there is no x-dependence of the phase factor in (16.3.9). The "particles" sit in the regions of space between the antinodes and do not move.

Since the two pictures give completely different descriptions of the behavior, the "particles" of Bohmian theory cannot be equated with the particles of quantum field theory, used for a great number of experiments and calculations such as cloud chamber tracks. Therefore, the Bohmian approach, if it is to describe the results of field theory, must keep track of *two* sets of particles, which are unrelated to each other. As noted at the end of Section 3.3, the canonical experiments that convinced people of the existence of particles, namely the photoelectric effect and the Planck radiation spectrum, arise from the quanta of field theory; they do *not* arise from the Bohmian particles.

16.4 The Transactional Interpretation

The transactional approach to quantum mechanics posits that *advanced waves*, that is, waves traveling backwards in time, exist in reality, and that the interaction of the advanced waves and forward-going waves (called *retarded* waves) leads to an instability that causes irreversible particle jumps.

In recent years, John Cramer (2017) and Ruth Kastner (2021) have argued that the structure of quantum field theory intrinsically includes irreversible, nonunitary behavior, and that it includes advanced waves. Here we look at both of these claims.

16.4.1 Are There Advanced Waves in Quantum Field Theory?

As discussed in Section 15.1.2, the Green's function for electron or photon propagation is written as

$$G_{\vec{k}}(t) \equiv -i\langle\text{vac}|a_k(t)a_{\vec{k}}^\dagger(0)|\text{vac}\rangle \Theta(t)$$
$$= -ie^{-i\omega_k t}\Theta(t), \tag{16.4.1}$$

where $|\text{vac}\rangle$ is the vacuum state. This can be understood physically as the overlap amplitude for two processes: one that starts with definite creation of an excitation in state $\vec{k}$ at time 0 and allows the system to evolve to a later time t, and another process in which the vacuum evolves on its own until time t, and at that time an excitation is created in state $\vec{k}$. In probability language, it is the probability amplitude for a particle remaining in state $\vec{k}$ after a time t has elapsed.

The Green's function for *holes* (defined in Section 15.2) is defined as

$$G_{\vec{k}}(t) \equiv i\langle \text{vac}|a_k^\dagger(t)a_k(0)|\text{vac}\rangle \Theta(-t)$$
$$= ie^{i\omega_k t}\Theta(-t). \tag{16.4.2}$$

This superficially looks like a backward-in-time traveling wave, because it asks the probability of first *removing* a particle, and then at a later time creating it. However, this makes sense as a forward-going process in the context of holes, because holes are absences of fermions below the Fermi level. In the standard theory, the energy states of a system are filled up (by Pauli exclusion) with fermions up to some cutoff energy level E_F, known as the Fermi level. A hole *creation* operator therefore corresponds to the *removal* of an electron in state $\vec{k}$, that is, an electron destruction operator for a state $\vec{k}$ below the Fermi level. In the same way, a hole *destruction* operator corresponds to an electron *creation* operator for a state below the Fermi level. The Green's function (16.4.2) therefore does not correspond to a wave actually traveling backwards in time; it corresponds to hole creation and destruction operators in the same order as in the case of the electron Green's function. In the case of electrons in the vacuum of free space, the same applies to positrons with negative energy.

In the case of bosons, Section 15.1.3 showed that typical interactions also give two types of Green's function. However, in this case, there is no Fermi level; so there is no switch to a destruction operator as an effective creation operator. Also, the boson operators do not pick up a $-$ sign when they are commuted. The complementary Green's function is then

$$G_{\vec{k}}(t) \equiv -i\langle \text{vac}|(a_{\vec{k}}^\dagger(0))a_k(t)|\text{vac}\rangle \Theta(-t), \tag{16.4.3}$$

which is the same as (16.4.1) but with t switched to $-t$.

As discussed in Section 15.1.3, the boson Green's function (16.4.1) can be switched to the frequency domain by the Fourier transform

$$G(\vec{k}, \omega) = -i \int_{-\infty}^{\infty} dt \, e^{i\omega t} \, e^{-i\omega_k t}\Theta(t)$$
$$= \lim_{\varepsilon \to 0} -i \int_{0}^{\infty} dt \, e^{i(\omega - \omega_k)t}e^{-\varepsilon t}$$
$$= \frac{1}{\omega - \omega_k + i\varepsilon}, \tag{16.4.4}$$

and the complementary term corresponding has the transform

$$G(-\vec{k}, \omega) = -i \int_{-\infty}^{\infty} dt \, e^{i\omega t} \, e^{i\omega_k t}\Theta(-t)$$
$$= \lim_{\varepsilon \to 0} -i \int_{-\infty}^{0} dt \, e^{i(\omega + \omega_k)t}e^{\varepsilon t}$$
$$= \frac{1}{-\omega - \omega_k + i\varepsilon}. \tag{16.4.5}$$

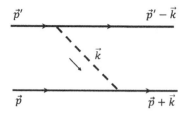 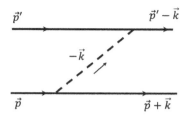

Figure 16.3 Two processes for virtual phonon exchange between two electrons.

This term accounts for the fact that a particle *emitting* a boson with momentum $\vec{k}$ and energy $\hbar\omega$ has the same effect as *absorbing* the same type of boson with momentum $-\vec{k}$ and energy $-\hbar\omega$. Figure 16.3 shows these two processes separately, which are accounted together as a single, effective interaction between the two electrons in the diagram shown in Figure 15.2.

What do we mean by a photon with negative frequency? One interpretation is to treat this as an advanced wave traveling backwards in time. But if we remember the reason why we have two terms, it is because, as discussed in Section 12.5, phonon (and photon) waves are *real-valued*, and to have a Hermitian operator corresponding to a real-valued amplitude, we need the sum of $a_k^\dagger + a_k$.

The first Green's function corresponds to the traveling wave $e^{i(\vec{k}\cdot x - \omega t)}$, while the second, complementary wave corresponds to $e^{i(-\vec{k}\cdot x + \omega t)} = e^{-i(\vec{k}\cdot x - \omega t)}$, which is just the complex conjugate of the first wave. Both of these propagate in the same direction. The sum of the two is $\cos(\vec{k}\cdot x - \omega t)$. In other words, the field theory simply ensures that the interaction of the electrons responds to the real part of the phonon wave. The use of negative frequency is common in optics to account for the complex conjugate part that gives the real part of traveling waves.

We thus see that for both fermions and bosons, the Green's functions that are often written as backwards-in-time-traveling waves are not really tachyons! They are simply bookkeeping conveniences in the theory.

We have done this calculation for phonons and electrons in a solid, for simplicity, but the same argument applies to the case of photons in vacuum, worked out by Paul Davies (1971).

16.4.2 Is There Nonunitarity in Quantum Field Theory?

As discussed in Section 6.1, a unitary system cannot give nonunitary behavior; the mathematical approximations of the S-matrix expansion and Fermi's golden rule do not change this. The Carver Mead model for coupled superconductors shown in Figure 5.4 does not involve dissipation and irreversibility of this type – the jumpy dynamics are time-reversible. As discussed in Section 14.1, Fermi's golden rule appears to give irreversible behavior, but this is an approximation based on the assumption of strong decoherence due to interaction with an external environment.

Does the inclusion of the $i\varepsilon$ term in the Green's functions of Section 16.4 mean that there is irreversible, nonunitary behavior intrinsic to quantum field theory? Nominally, this term corresponds to decay proportional to $e^{-\varepsilon t}$, which is nonunitary. But as discussed in Sections 15.1.2 and 15.3.2, this imaginary term can be seen as arising from a small imaginary self-energy of the states of interest, which in turn corresponds to dissipation due to decoherence derived within the fully unitary quantum field theory, calculated using exactly the same method as Fermi's golden rule. The introduction of the $i\varepsilon$ term can be seen as arising from the need for self-consistency when higher-order terms in the field theory are taken into account, and not as an ad hoc introduction of something nonunitary.

As noted in Davies 1971, and as discussed in Section 18.1, in an infinite system, unitary evolution gives irreversible behavior which looks like nonunitary behavior, because energy can flow outward forever without returning. In the discussion of Davies, this corresponds to outgoing photons that are never absorbed. This does not mean that there is a general nonunitarity of standard quantum mechanics, but rather that part of the system (at $t = \infty$) has been placed "off the books," in the same way that an "environment" is often placed off the books in decoherence theory, as discussed in Part V.

References

A. Aspect, P. Grangier, and G. Roger, "Experimental realization of Einstein–Podolsky–Rosen–Bohm gedankenexperiment: A new violation of Bell's inequalities," *Physical Review Letters* **49**, 91 (1982).

J. S. Bell, "On the Einstein Podalsky Rosen paradox," *Physics* (N.Y.) **1**, 195 (1964).

J. Cramer and R. E. Kastner, "Quantifying absorption in the transactional interpretation," arXiv:1711.04501.

P. C. W. Davies, "Extension of Wheeler–Feynman quantum theory to the relativistic domain. I. Scattering processes," *Journal of Physics A* **4**, 836 (1971).

B. S. DeWitt and N. Graham, *The Many-Worlds Interpretation of Quantum Mechanics*, (Princeton University Press, 1973), p. 149.

R. E. Kastner, "The relativistic transactional interpretation and the quantum direct-action theory," arXiv:2101.00712.

R. Nandkishore and D. Huse, "Many-body localization and thermalization in quantum statistical mechanics," *Annual Review of Condensed Matter Physics* **6**, 15 (2015).

D. W. Snoke, G.-Q. Liu, and S. M. Girvin, "The basis of the second law of thermodynamics in quantum field theory," *Annals of Physics* **327**, 1825 (2012).

D. W. Snoke, *Solid State Physics: Essential Concepts*, 2nd ed., (Cambridge University Press, 2020).

F. J. Tipler, "Quantum nonlocality does not exist," *Proceedings of the National Academy of Sciences* (USA) **111**, 11281 (2014).

17

Entanglement in a Classical System

In 1935, Erwin Schrödinger wrote,

> When two systems, of which we know the states by their respective representatives, enter into temporary physical interaction due to known forces between them, and when after a time of mutual influence the systems separate again, then they can no longer be described in the same way as before, viz. by endowing each of them with a representative of its own. I would not call that one but rather the characteristic trait of quantum mechanics, the one that enforces its entire departure from classical lines of thought. By the interaction the two representatives (or ψ-functions) have become entangled. (Schrödinger 1935)

Schrödinger's statement is correct when taken in the full context of this quote. But the property of entanglement is often stated in a mathematical formalism, and the conclusion is sometimes drawn that this mathematical formalism cannot apply to any classical system. As we will see in this chapter, that is incorrect. A classical system can be constructed which has exactly the same mathematical structure and exactly the same probability predictions as a quantum entangled system. There are key differences, however.

Several theoretical and experimental works (Spreeuw 1998; Shen 2021; Zhan 2021) have shown that optical states can have the essential mathematical property of being non-factorizable sums of products of functions. The wave functions of the electromagnetic states in those cases differed quite a bit from those of quantum mechanics. In this chapter, we will see that a classical system can be made which has exactly the same mathematical description as an entangled quantum system. This work originally appeared in Snoke 2021.

17.1 An Optical System with Second Quantization

The canonical case of entanglement in quantum mechanics is given by a superposition of two states that cannot be factored into a product, for example, the state

$$\Psi(x,y) = \frac{1}{\sqrt{2}}[\psi_0(x)\psi_1(y) + i\psi_1(x)\psi_0(y)], \qquad (17.1.1)$$

where $\psi_i(x)$ and $\psi_i(y)$ are the wave functions for two states labeled 0 and 1. In bra–ket notation, the relation (17.1.1) is written as

$$|\Psi\rangle = \frac{1}{\sqrt{2}}(|0\rangle|1\rangle + i|1\rangle|0\rangle), \qquad (17.1.2)$$

where $|0\rangle$ and $|1\rangle$ represent the two states available to two different subsystems.

The entangled state (17.1.2) is physically realized, for example, in the case of a beamsplitter which has one photon impinging on it. In this case, $|0\rangle$ corresponds to one output of the beamsplitter having no photon, and $|1\rangle$ corresponds to the output having one photon. The product state gives the total state of both outputs of the beamsplitter. This state is the result of the standard 50-50 beamsplitter matrix operator (Mandel and Wolf 1995)

$$M = \frac{1}{\sqrt{2}}\begin{pmatrix} 1 & i \\ i & 1 \end{pmatrix} \qquad (17.1.3)$$

acting on the input state $|1\rangle|0\rangle$, which is written in vector form as $(1,0)$, and corresponds to one photon entering the beamsplitter from one direction.

Particle operators as operators on continuous functions. To see how to simulate the state (17.1.2) classically, we must begin by recalling how particle operators and states are defined. Particles such as photons are defined as the eigenstates of the Hamiltonian

$$H = \sum_k \hbar\omega_k \left(\hat{N}_k + \frac{1}{2}\right) = \sum_k \hbar\omega_k \left(a_k^\dagger a_k + \frac{1}{2}\right), \qquad (17.1.4)$$

where $a_k^\dagger$ and a_k are the creation and destruction operators for the wave mode k, and ω_k is the frequency of the mode k. As discussed in Sections 12.1 and 12.2, the creation and destruction operators obey the commutation relation

$$[a_k, a_k^\dagger] = 1; \qquad (17.1.5)$$

this relation follows from the underlying wave equation for the harmonic oscillator,

$$H\psi = i\hbar\frac{\partial\psi}{\partial t} = \left[\frac{p_k^2}{2m} + \frac{1}{2}\gamma x_k^2\right]\psi = \left[-\frac{\hbar^2}{2m}\frac{\partial^2}{\partial x_k^2} + \frac{1}{2}\gamma x_k^2\right]\psi, \qquad (17.1.6)$$

where ψ is a wave function. Here we have used an effective mass m and spring constant γ, which are appropriate for phonons in a system of coupled atoms, but photons in a vacuum have exactly the same mathematical structure (Section 12.4), if we substitute

$$m/a^3 \to \varepsilon_0, \quad a/\gamma \to \mu_0, \quad x_k \to A_k, \qquad (17.1.7)$$

where a is the size of the local oscillator with mass m, ε_0 and μ_0 are the permittivity and permeability of free space used in Maxwell's equations, and A_k is the vector potential of electromagnetism. Instead of x_k for the spatial displacement of the oscillator k, we have the strength of the electromagnetic field A_k. The wave function ψ gives the probability of a given value of x_k or A_k.

In this algebra, the destruction operator is defined as

$$a_k = \frac{1}{\sqrt{2}}\left(\sqrt{\frac{m\omega_k}{\hbar}}x_k + \frac{i}{\sqrt{m\hbar\omega_k}}p_k\right) = \frac{1}{\sqrt{2}}\left(\sqrt{\frac{m\omega_k}{\hbar}}x_k + \sqrt{\frac{\hbar}{m\omega_k}}\frac{\partial}{\partial x_k}\right)$$

$$= \frac{1}{\sqrt{2}}\left(\tilde{x}_k + \frac{\partial}{\partial\tilde{x}_k}\right), \qquad (17.1.8)$$

where $\tilde{x}_k = (\sqrt{m\omega_k/\hbar})x_k$. The creation operator is then

$$a_k^\dagger = \frac{1}{\sqrt{2}}\left(\tilde{x}_k - \frac{\partial}{\partial \tilde{x}_k}\right), \tag{17.1.9}$$

and the number operator is

$$\hat{N}_k \equiv a_k^\dagger a_k = \frac{1}{2}\left(\tilde{x}^2 - \frac{\partial^2}{\partial \tilde{x}^2} - 1\right). \tag{17.1.10}$$

With these definitions, it is easy to show that the eigenstates $|0\rangle$ and $|1\rangle$ correspond to

$$\psi_0(x_k) = \langle x_k|0\rangle = \frac{1}{\pi^{1/4}}e^{-\tilde{x}_k^2/2} \tag{17.1.11}$$

$$\psi_1(x_k) = \langle x_k|1\rangle = \frac{\sqrt{2}}{\pi^{1/4}}\tilde{x}_k e^{-\tilde{x}_k^2/2}, \tag{17.1.12}$$

with $\omega_k = \sqrt{\gamma/m}$, and $a_k^\dagger$ and a_k have the standard actions $a_k^\dagger|0\rangle = |1\rangle$ and $a_k|1\rangle = |0\rangle$.

Thus, the ground state of the photon mode k, corresponding to no photon, is a wave function ψ_0 which is a Gaussian, and the first excited state, corresponding to one photon, is the wave function ψ_1 which is a Gaussian multiplied by $\sqrt{2}x_k$. The wave functions here are not the same as the electromagnetic field functions of the mode k in real space. The electromagnetic field of mode k is given by $A(z,t) = A_k e^{i(kz-\omega_k t)}$; the wave function $\psi(A_k)$, which maps to $\psi(x_k)$ here, gives the probability of finding a particular amplitude A_k. If no measurement is made of A_k, however, then ψ is a continuous function which satisfies the wave equation (17.1.6).

Cavity resonators with effective mass and spring constant. The question is then whether there is a classical system that obeys the wave equation (17.1.6). The answer is yes; we can construct a system with this wave equation using a classical optical resonator.

We imagine a classical resonator made of two parallel mirrors separated by a distance L. The classical Maxwell wave equation which applies in this system is

$$\nabla^2 E = \frac{1}{c^2}\frac{\partial^2 E}{\partial t^2}, \tag{17.1.13}$$

where we ignore the polarization of the electric field; in all of the following we assume that the electric field is always polarized in one direction. We write the solution of this wave equation subject to the cavity boundary conditions as

$$E = \psi \cos(k_\perp z)e^{-i\omega t}, \tag{17.1.14}$$

where $k_\perp = N\pi/L$; only integer values of N are allowed, because the perpendicular component $k_\perp$ is quantized by the boundary condition that the electric field must vanish at the surface of the mirrors. The amplitude ψ may vary in time and in space along the plane of the cavity. We write this envelope amplitude suggestively as ψ because we will see that it plays the same role as the harmonic oscillator wave function ψ.

Keeping only leading terms in frequency (known as the slowly varying envelope approximation), we have for the time derivative of E,

$$\frac{\partial^2 E}{\partial t^2} \simeq \left(-\omega^2\psi - 2i\omega\frac{\partial \psi}{\partial t}\right)\cos(k_\perp z)e^{-i\omega t}, \tag{17.1.15}$$

The Maxwell wave equation (17.1.13) then becomes

$$(-k_\perp^2 \psi + \nabla_\parallel^2 \psi) = \frac{1}{c^2}\left(-\omega^2 \psi - 2i\omega \frac{\partial \psi}{\partial t}\right). \tag{17.1.16}$$

We allow that $k_\perp$ may vary slowly along the plane of the cavity, due to varying cavity thickness L. In particular, if we arrange to have a maximum of the thickness L at position $x = 0$, with parabolic variation of the thickness away from $x = 0$, we can write

$$k_\perp^2 = \frac{N^2\pi^2}{L^2(x)} = N^2\pi^2 \frac{1}{(L_0 - bx^2)^2} \simeq \frac{N^2\pi^2}{L_0^2}(1 + 2bx^2/L_0) \equiv \frac{\omega_0^2}{c^2}(1 + 2bx^2/L_0), \tag{17.1.17}$$

where b is a constant that gives the variation of $L(x)$ in the plane, and $\omega_0 = N\pi/L_0$. Picking $\omega \simeq \omega_0$, the Maxwell wave equation (17.1.16) becomes

$$\nabla_\parallel^2 \psi - \frac{2\omega_0^2 b}{c^2 L_0}x^2 \psi = \frac{1}{c^2}\left(-2i\omega_0 \frac{\partial \psi}{\partial t}\right). \tag{17.1.18}$$

Rearranging, we have

$$-\frac{c^2}{2\omega_0}\nabla_\parallel^2 \psi + \frac{b\omega_0}{L_0}x^2 \psi = i\frac{\partial \psi}{\partial t}. \tag{17.1.19}$$

This is equivalent to (17.1.6) if we assign $m = \hbar\omega_0/c^2$ and $\gamma = 2\hbar\omega_0(b/L_0)$. The solutions of this equation are already well known, namely the solutions of the quantum harmonic oscillator already discussed, with evenly spaced frequencies.

We have made two assumptions to arrive at this result, namely that the cavity thickness is thin enough that ω_0 is well above the rate of change of the envelope function ψ, and the gradient of the cavity thickness is small enough that the cavity can be treated as locally planar. Both of these limits are easily achieved in experiments, and such experiments have been done in at least two cases. One possibility is to vary the index of refraction in a parabolic fashion, giving the equivalent behavior by changing the effective velocity c instead of L in (17.1.13)–(17.1.19). This was invoked in a proposal (Gordon 2002; Zhang 2012) for modelocking of a very small cavity laser using the evenly spaced frequencies for the lateral modes in the plane of the cavity instead of the standard modelocking method of using the evenly spaced longitudinal modes. This limit has also been used in the recent "photon condensate" experiments (Klaers 2010); the variation of the cavity thickness gave a harmonic potential in the plane which could be used to trap the photons in the ground state at the center of the cavity, which is a Gaussian mode.

This type of resonator is therefore standard optics, not exotic, and can easily be fabricated for experimental studies using either varying cavity thickness or index of refraction variation. If the optical modes are coupled to electronic transitions, this leads to a nonlinear term which makes (17.1.19) become a standard Gross–Pitaevskii equation, also known as a nonlinear Schrödinger equation (see Snoke 2020, Section 11.13.1).

17.2 Entangled States of a Resonator

The fact that the resonator discussed in Section 17.1 has two spatial dimensions in the plane allows us to create entangled states exactly equivalent to (17.1.2). Since linear waves obey the principle of superposition, we can make superpositions of macroscopic electromagnetic waves just as we do with quantum mechanical wave functions. The state equivalent to (17.1.2) is

$$\psi(x,y,t) = \frac{1}{\pi^{1/2}}e^{-x^2/2}ye^{-y^2/2}e^{i\tilde{\omega}t} + \frac{i}{\pi^{1/2}}xe^{-x^2/2}e^{-y^2/2}e^{i\tilde{\omega}t}, \tag{17.2.1}$$

where $\tilde{\omega} = \sqrt{\gamma/m} = \sqrt{2(b/L_0)}c$. This frequency can be quite low compared to the frequency ω_0 at which the electromagnetic field oscillates, if the curvature of the mirrors is low.

The state (17.2.1) is a physically possible classical electromagnetic state, since each of the two terms is allowed in a two-dimensional system, and a superposition of the two is therefore also possible. This wave function is plotted in Figure 17.1 for various times.

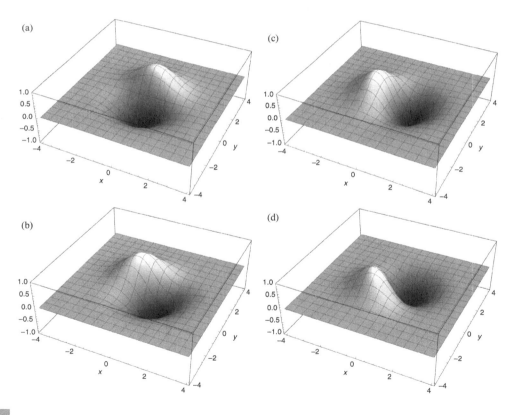

Figure 17.1   The real part of the entangled classical wave (17.2.1) at four times corresponding to phase of 0, $\pi/4$, $\pi/2$, and $3\pi/2$ radians during the period of oscillation $T = 2\pi/\tilde{\omega}$. The distribution rotates at constant frequency $\tilde{\omega}$ in the two-dimensional plane. From Snoke 2021.

Note that the wave function ψ plotted here, which corresponds to the electromagnetic wave amplitude in our classical analog, maps to the wave function ψ in the single-photon states (17.1.12), while the position x or y here corresponds to the electromagnetic wave amplitude in the mapping (17.1.7). The two spatial dimensions map to the electromagnetic wave amplitude along the two output legs of the beamsplitter discussed in the introduction.

With this state, it is manifest that the expectation value for having both axes in a $|1\rangle$ state is

$$\langle \Psi | a_x^\dagger a_x a_y^\dagger a_y | \Psi \rangle = \frac{1}{4} \int dx \int dy \, \psi^* \left(x^2 - \frac{\partial^2}{\partial x^2} - 1 \right) \left(y^2 - \frac{\partial^2}{\partial y^2} - 1 \right) \psi = 0. \quad (17.2.2)$$

17.3 Bell Inequality for the Classical Resonator

The entangled nature of the system allows violation of a Bell inequality, for example, the CHSH inequality (Clauser 1974)

$$\langle \mathcal{O}_a^{(x)} \mathcal{O}_a^{(y)} \rangle + \langle \mathcal{O}_a^{(x)} \mathcal{O}_b^{(y)} \rangle + \langle \mathcal{O}_b^{(x)} \mathcal{O}_a^{(y)} \rangle - \langle \mathcal{O}_b^{(x)} \mathcal{O}_b^{(y)} \rangle \leq 2, \quad (17.3.1)$$

where we pick

$$\mathcal{O}_a^{(x)} = S_z^{(x)} = \begin{pmatrix} 1 & 0 \\ 0 & -1 \end{pmatrix}, \qquad \mathcal{O}_b^{(x)} = S_x^{(x)} = \begin{pmatrix} 0 & 1 \\ 1 & 0 \end{pmatrix}, \quad (17.3.2)$$

which are spin-Pauli matrices acting on the $|1\rangle$ and $|0\rangle$ states of the x-axis. (Here the x and z subscripts have nothing to do with the x- and y-axes of the cavity, which are indicated by the superscripts.) For the y-axis, we use

$$\mathcal{O}_a^{(y)} = -\frac{1}{\sqrt{2}} \left(S_z^{(y)} + S_x^{(y)} \right) = \frac{1}{\sqrt{2}} \begin{pmatrix} -1 & -1 \\ -1 & 1 \end{pmatrix},$$

$$\mathcal{O}_b^{(y)} = \frac{1}{\sqrt{2}} \left(S_z^{(y)} - S_x^{(y)} \right) = \frac{1}{\sqrt{2}} \begin{pmatrix} 1 & -1 \\ -1 & -1 \end{pmatrix}, \quad (17.3.3)$$

which act on the $|1\rangle$ and $|0\rangle$ states of the y-axis. In terms of the continuous functions (17.1.12), the S_x operator is equivalent to

$$S_x^{(x)} = a_x^\dagger (1 - a_x^\dagger a_x) + a_x$$

$$= \frac{1}{2\sqrt{2}} \left(x - \frac{\partial}{\partial x} \right) \left(3 - x^2 + \frac{\partial^2}{\partial x^2} \right) + \frac{1}{\sqrt{2}} \left(x + \frac{\partial}{\partial x} \right)$$

$$= \frac{1}{2\sqrt{2}} \left(7x - x^3 + (x^2 - 1) \frac{\partial}{\partial x} + x \frac{\partial^2}{\partial x^2} - \frac{\partial^3}{\partial x^3} \right), \quad (17.3.4)$$

while the S_z operator is equivalent to

$$S_z^{(x)} = 2a_x^\dagger a_x - 1$$

$$= x^2 - \frac{\partial^2}{\partial x^2} - 2. \quad (17.3.5)$$

To measure the state of the system to see if it violates the Bell inequality, we can measure the electric field amplitude $\psi(x,y)$ everywhere in the cavity, and then perform these operations on it analytically, and integrate over the plane,

$$\langle S_z^{(x)} S_z^{(y)} \rangle = \int dx dy \; \psi^*(x,y) \left(x^2 - \frac{\partial^2}{\partial x^2} - 2 \right) \left(y^2 - \frac{\partial^2}{\partial y^2} - 2 \right) \psi(x,y). \quad (17.3.6)$$

The electric field can be measured by a set of small linear detectors adjacent to the cavity, for example, polarized antennas connected to tank circuits resonant at the cavity frequency ω_0. This is hard to do in the optical frequency range, but it is easy to implement linear detection for electromagnetic fields in the microwave range. Since we have assumed in all of the analyses of this chapter so far that there is only one polarization of interest in the cavity, all the antennas will point in the same direction (though the orthogonal polarization could also be used to give four degrees of freedom, namely the two spatial coordinates and the two polarizations).

For the choice of operators (17.3.2) and (17.3.3), this type of measurement gives a violation of the CHSH inequality, with the left side of (17.3.1) equal to $2\sqrt{2}$, as expected since we have mapped the system one-to-one to the quantum entangled system. It is not actually necessary to measure the electric field amplitude everywhere in the plane. A violation of the Bell inequality can be obtained for a reasonable sampling of the electric field at different sites in the plane, giving a good approximation of integrals of the form (17.3.6).

We could, if we wanted, rig the detection system to simulate the discrete detection properties of particle detectors, so that the electric field amplitude $\psi(x,y)$ actually does act as a probability amplitude. There are many ways to create a classical detector that produces random clicks or counts with probability proportional to the intensity of input light, in accordance with the Born rule. One way is simply to have a classical digital computer measure the input light intensity using an analog-to-digital converter, and then use this information to generate function calls to a random number generator in a way that follows the Born rule. However, it might be objected that digital computers typically generate pseudo-random numbers, not truly random numbers. Another method would be to use an electrical circuit with classical chaos, which is truly unpredictable,[1] and have a photocell control the period of the driving oscillation of the chaotic circuit via a voltage-controlled oscillator.

Figure 17.2 shows wiring patterns for a few linear antennas in the plane which could be used to be sensitive only to the $|0\rangle$ or $|1\rangle$ states for either the x- or y-axis. The detector for any given state could be designed to trigger with probability proportional to the square of the current, and then immediately to jam other detectors. This would give data with the same probability distributions as a standard particle counting experiment for an entangled state.

Furthermore, we could rig the system to collapse the wave function after a measurement. In any physical cavity, there is a loss of the energy of the cavity through the cavity mirrors leaking radiation (this is what allows the antenna detection of the electric field). To keep the wave function normalized, energy must be pumped into the system, as in any optical

---

[1] See, for example, Su 1989; for further discussion, see Snoke 2015, Section 4.6.

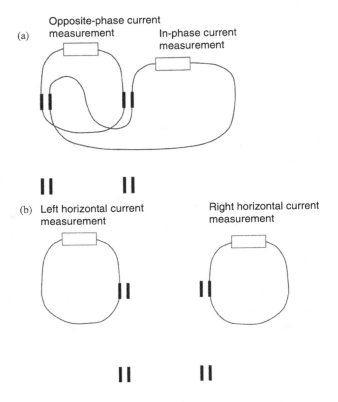

Figure 17.2 (a) The detection of the $|0\rangle$ (symmetric) and $|1\rangle$ (antisymmetric) states using linear antennas distributed in the $x - y$ plane, for a y-polarized electromagnetic wave. The wiring shown measures the $|0\rangle$ and $|1\rangle$ states in the x-direction. A similar wiring can be done in the y-direction. (b) Wiring for detecting two orthogonal superpositions $(|0\rangle \pm |1\rangle)/\sqrt{2}$ in the x-direction. From Snoke 2021.

cavity. One possibility with this system is to use the antenna detector array to pump energy into the system. In this case, positive feedback could be used to force the system into one or the other of the $|0\rangle$ or $|1\rangle$ states.

Comparison to quantum systems. The existence of this analog for quantum systems can help us to identify what is truly quantum and what is simply a consequence of the wave nature of quantum systems, in common with all wave systems. As we have seen here, a classical system can have mathematical probability predictions which are identical to those of an entangled quantum system. The mathematical structure of nonfactorizability of (17.1.2) is therefore not uniquely quantum, nor are the violations of Bell inequalities which follow for entangled states. Quantum systems can have many more possible degrees of entanglement, however. In quantum mechanics, each degree of freedom corresponds to a new dimension, that is, a new orthogonal Hilbert space, with no upper limit to the number of dimensions. In classical mechanics, the number of entangled degrees of freedom is limited by the number of spatial dimensions, in a three-dimensional universe. The system described here allows up to four entangled degrees of freedom, accounting for two electric field polarizations and two spatial dimensions, but there is a definite upper bound of the number of entangled degrees of freedom in classical systems.

The CHSH inequality and other Bell's inequalities are derived for classical "objects" with finite countability and continuous histories (as in the derivation in Section 16.1.2). Such inequalities are not universal statements for all classical systems; rather they are applicable to a subset of classical systems consisting of distinguishable objects. In the context of classical waves, violation of a Bell inequality is not surprising. When the Bell inequalities are mapped to quantum systems, however, it is assumed that quantum systems also count "objects" which we call particles. But if we keep in mind only the continuous quantum wave functions, the violation of the Bell inequalities in quantum systems is no more surprising than in classical wave systems.

References

J. F. Clauser and M. A. Horne, "Experimental consequences of objective local theories," *Physical Review D* **10**, 526 (1974).

R. Gordon, A. P. Heberle, and J. R. A. Cleaver, "Transverse mode-locking in microcavity lasers," *Applied Physics Letters* **81**, 4523 (2002).

J. Klaers, J. Schmitt, F. Vewinger, and M. Weitz, "Bose–Einstein condensation of photons in an optical microcavity," *Nature* **468**, 545 (2010).

L. Mandel and E. Wolf, *Optical Coherence and Quantum Optics*, (Cambridge University Press, 1995), Section 12.12.

E. Schrödinger, "Discussion of probability relations between separated systems," *Mathematical Proceedings of the Cambridge Philosophical Society* **31**, 555 (1935).

Y. Shen, I. Nape, X. Lang et al., "Creation and control of high-dimensional multi-partite classically entangled light," *Light: Science and Applications* **10**, 50 (2021).

D. W. Snoke, *Electronics*, (Pearson, 2015).

D. W. Snoke, *Solid State Physics: Essential Concepts*, 2nd ed., (Cambridge University Press, 2020).

D. W. Snoke, "A macroscopic classical system with entanglement," *International Journal of Quantum Foundations* **7**, 34 (2021).

R. J. C. Spreeuw, "A classical analogy of entanglement," *Foundations of Physics* **28**, 361 (1998).

R. W. Rollins, and E. R. Hunt, "Simulation and characterization of strange attractors in driven diode resonator systems," *Physical Review A* **40**, 2968 (1989).

Q. Zhan, "Entanglement goes classically high-dimensional," *Light: Science and Applications* **10**, 81 (2021).

B. Zhang, D. W. Snoke, and A. P. Heberle, "Towards the transverse mode-locking of oxide-confined VCSELs," *Optics Communications* **285**, 4117 (2012).

Part V

Decoherence, Spontaneous Coherence, and Spontaneous Collapse

18

Irreversibility in Unitary Quantum Field Theory

A major question of quantum mechanics is how one can get irreversible behavior in a physical system that is fundamentally time-reversible in its governing equations. As proved in Section 18.1, any finite, unitary system will always have periodic behavior, over some timescale. Yet much of our experience gives behavior that seems completely irreversible, such as an egg splattering on a pavement.

The general approach to this in quantum mechanics is to treat carefully the physical effect of *decoherence*, also known as *dephasing*. Sections 18.2 and 18.4 show that irreversible behavior occurs in quantum mechanics when the phase coherence of many-particle states decays away to zero due to interactions in a large system. There are two ways to think about what really happens in this case. One is to say that most systems we look at are sufficiently large that the Poincaré time for periodic recurrence is so long that we can effectively set it to infinity. Another is to say that in a truly infinite system, the Poincaré time is truly infinite, so that there is no recurrence, and we live in a truly infinitely large universe. Any real finite system is coupled to outer space (e.g., by heat going into infrared radiation), and so will have truly irreversible behavior.

All of these statements, and the math of this chapter, are completely standard, unitary quantum theory. One can also speculate that there is nonunitary behavior in the physical world, which is not yet accounted for by the equations we use for quantum mechanics. The "collapse" in a measurement process may be like this. It is also introduced as a mathematical convenience in the quantum trajectories method, discussed in Chapter 19. Collapse is off the books of the standard equations of quantum mechanics, but can be introduced into those equations in a way that looks much like random noise (see, e.g., Jacobs 2006).

As discussed in Chapter 4, "measurements" are ill-defined, and some interpretations of quantum mechanics make an explicit commitment to the need for human knowledge to give a measurement. It can also be speculated that there are spontaneous events that have the same effect, without the need for a human observer to define a measurement. A quantitative proposal for this scenario is given in Chapter 20.

18.1 The Poincaré Recurrence Theorem

This section follows the proof of Boccieri and Loinger (1957). We suppose that the many-body wave function of a system evolves in time according to

$$|\psi(t)\rangle = \sum_{n=0}^{\infty} c_n \exp(-iE_n t/\hbar)|\phi_n\rangle, \tag{18.1.1}$$

where the E_n are the energy eigenvalues and the $|\phi_n\rangle$ are the eigenstates. If the system starts evolving at time $t = 0$ and evolves until time T, the magnitude of the difference of the wave function at the two times is given by

$$||\psi(T)\rangle - |\psi(0)\rangle|^2 = 2 \sum_{n=0}^{\infty} |c_n|^2 [1 - \cos(E_n T/\hbar)].$$ (18.1.2)

Consider now the same sum but up to a finite value of N. We first show that any such finite sum can be made arbitrarily close to zero by picking the correct time T. We start with just two frequencies, $\omega_1 = E_1/\hbar$ and $\omega_2 = E_2/\hbar$. If the ratio of these frequencies is a rational number,

$$\frac{\omega_1}{\omega_2} = \frac{N_1}{N_2},$$ (18.1.3)

then it will always be true that the time $T = N_1 T_1 = N_2 T_2 \equiv T_{12}$ will be an integer number of periods for each frequency, where the periods are given by $T_1 = 2\pi/\omega_1$ and $T_2 = 2\pi/\omega_2$. (Note that the times T_1 and T_2 defined here are not the T_1 and T_2 times defined elsewhere in this book for decoherence processes – here, they are simply two different time intervals.) Both of the waves will return to their starting value after an integer number of periods, which means that $||\psi(T)\rangle - |\psi(0)\rangle|^2$ will be zero. If the ratio of the two wave frequencies is not a rational number, but instead is irrational, we can always find a rational number that is arbitrarily close to the irrational number, allowing arbitrarily small difference between the waves.

If we add in a third frequency, we can still always choose a time T for which all the waves return to their original value, because the first two frequencies define a new period T_{12} for recurrence, and we can then follow the same procedure for the recurrence time of the combination of T_{12} and T_3. We can do this for any finite number of frequencies.

We now write the original sum as two parts,

$$||\psi(T)\rangle - |\psi(0)\rangle|^2 = 2 \sum_{n=0}^{N} |c_n|^2 [1 - \cos(E_n T/\hbar)]$$

$$+ 2 \sum_{n=N+1}^{\infty} |c_n|^2 [1 - \cos(E_n T/\hbar)].$$ (18.1.4)

The first term on the right can be made to vanish, as we have seen. The second term is bounded, since all the terms in the sum are nonnegative:

$$2 \sum_{n=N+1}^{\infty} |c_n|^2 [1 - \cos(E_n T/\hbar)] \leq 2 \sum_{n=N+1}^{\infty} |c_n|^2.$$ (18.1.5)

As $N \to \infty$, this term must approach zero, because we assume that the initial state is normalized, that is,

$$\langle \psi_i | \psi_i \rangle = \sum_{n=0}^{\infty} |c_n|^2 = 1.$$ (18.1.6)

Since all the contributions to the sum are nonnegative, reducing the number of terms in the sum must make it get smaller. Therefore, the sum of the terms of (18.1.4) becomes arbitrarily close to zero for large N and correct choice of T.

This result depends, first, on the *unitary* nature of the system, because every quantum state evolves deterministically with its oscillation frequency ω_n, independent of what the other states are doing. It also depends on the system being *finite*, not infinite, because if the system is infinite, then the energy difference between states, $E_n - E_{n'}$, goes to zero. In that case, the sum over states is a continuum, with an infinite number of states within any energy interval ΔE_n, and the first term of (18.1.4) cannot be made to vanish in a finite time. In other words, the more frequency terms that are added, the longer the recursion time, so that in an infinite system, the recursion time T goes to infinity. This is fundamentally no different from the result that, in an infinite lake, a circular wave from a pebble will propagate outward and never return, having nothing to reflect from.

18.2 The Quantum Boltzmann Equation

In this section, we derive the quantum Boltzmann equation, which gives apparently irreversible behavior derived from a fully unitary quantum theory, and in Section 19.3, we will prove that the main assumption made in this derivation, of strong dephasing, is valid for most systems. There are some systems that do not have strong dephasing, and in fact have the opposite, increase of phase coherence, or "enphasing." These will be discussed in Chapter 21.

18.2.1 Derivation of the Quantum Boltzmann Equation for a Many-Particle System

This section follows closely a derivation published in Snoke 2012. We assume that the state of the system can be written as a superposition (sum) of Fock states, each with a definite number of particles in each single-particle state. We assume that the Hamiltonian of the system is given by

$$H = H_0 + \hat{V}, \tag{18.2.1}$$

where

$$H_0 = \sum_k E_k \hat{N}_k \tag{18.2.2}$$

gives the energies of the unperturbed eigenstates of the system, and $\hat{V}$ gives the interactions between eigenstates. In general we assume that the interaction term can be written as a sum of terms of the form

$$U a_1^\dagger a_2^\dagger \ldots a_{N-1} a_N, \tag{18.2.3}$$

where U is some complex number and the $a_i^\dagger$ and a_i operators are various creation and destruction operators acting on the unperturbed eigenstates. These can be seen as removing particles in the initial state of the system and creating particles in new eigenstates.

If the initial state of the system is $|\psi_i\rangle$, and the state of the system at some later time t is $|\psi_t\rangle$, the change in the average number of particles in state $\vec{k}$ is given by

$$
\begin{aligned}
d\langle \hat{N}_k \rangle &= \langle \psi_t | \hat{N}_k | \psi_t \rangle - \langle \psi_i | \hat{N}_k | \psi_i \rangle \\
&= \langle \psi(t) | e^{iH_0 t/\hbar} \hat{N}_k e^{-iH_0 t/\hbar} | \psi(t) \rangle - \langle \psi_i | \hat{N}_k | \psi_i \rangle \\
&= \langle \psi_i | e^{(i/\hbar)\int \hat{V}(t)dt} \hat{N}_k e^{-(i/\hbar)\int \hat{V}(t)dt} | \psi_i \rangle - \langle \psi_i | \hat{N}_k | \psi_i \rangle \\
&= \langle \psi_i | e^{(i/\hbar)\int \hat{V}(t)dt} [\hat{N}_k, e^{-(i/\hbar)\int \hat{V}(t)dt}] | \psi_i \rangle,
\end{aligned}
\tag{18.2.4}
$$

where we use the interaction representation, introduced in Section 11.4, with $|\psi(t)\rangle = e^{iH_0 t/\hbar} |\psi_t\rangle$ and $\hat{V}(t) = e^{iH_0 t/\hbar} \hat{V} e^{-iH_0 t/\hbar}$, and $e^{(i/\hbar)\int \hat{V}(t)dt}$ is given by the expansion (14.1.5).

The operator $\hat{N}_k$ commutes with H_0, by definition. If it commutes with $\hat{V}$, there is no change in the occupation numbers over time. Substituting in the expansion (14.1.5) for the exponentials, we have

$$
\begin{aligned}
d\langle \hat{N}_k \rangle = \langle \psi_i | & \left(1 - (1/i\hbar) \int_0^t \hat{V}(t')dt' + \cdots \right) \\
& \times \left((1/i\hbar) \int_0^t dt' [\hat{N}_k, \hat{V}(t')] + (1/i\hbar)^2 \int_0^t dt' \int_0^{t'} dt'' [\hat{N}_k, \hat{V}(t')\hat{V}(t'')] + \cdots \right) | \psi_i \rangle.
\end{aligned}
\tag{18.2.5}
$$

The lowest-order term of this expansion is

$$
\begin{aligned}
d\langle \hat{N}_k \rangle &= \frac{1}{i\hbar} \int_0^t dt' \langle \psi_i | [\hat{N}_k, \hat{V}(t')] | \psi_i \rangle \\
&= \frac{t}{i\hbar} \langle \psi_i | [\hat{N}_k, \hat{V}] | \psi_i \rangle.
\end{aligned}
\tag{18.2.6}
$$

If $|\psi_i\rangle$ is a pure Fock state, then the operator $\hat{N}_k$ acting on the initial state either to the left or to the right gives the same number, and therefore this term will vanish. If the state is not a pure Fock state, then the interaction must contain terms that change the number of particles in a given state. For some bosons such as photons and phonons, the number of particles does not need to be conserved, and so the interaction can have terms of the form $[A]a_k$ or $a_k^\dagger[A]$, where $[A]$ is some operator that can contain various other creation and destruction operators; for fermions, there must be interaction terms of the form $a_{k'}^\dagger[A]a_k$, to conserve the number of particles. In either case, the interaction will create nonzero values of "off-diagonal" correlation functions with unbalanced operators, which can be called phase-coherence factors. These will be discussed in Section 19.3; the underlying assumption of the quantum Boltzmann equation is that these are negligible. We can therefore ignore the first-order term.

We therefore move on to the second-order terms of the expansion (18.2.5). The first of these is

$$
\begin{aligned}
d\langle \hat{N}_k \rangle &= \frac{1}{\hbar^2} \int_0^t dt' \int_0^t dt'' \langle \psi_i | \hat{V}(t') [\hat{N}_k, \hat{V}(t'')] | \psi_i \rangle \\
&= \frac{1}{\hbar^2} \int_0^t dt' \int_0^t dt'' \langle \psi_i | e^{iH_0 t'/\hbar} \hat{V} e^{-iH_0(t'-t'')/\hbar} [\hat{N}_k, \hat{V}] e^{-iH_0 t''/\hbar} | \psi_i \rangle.
\end{aligned}
\tag{18.2.7}
$$

We write the quantum state as a superposition of Fock states:

$$|\psi\rangle = \sum_n \alpha_n |n\rangle, \qquad (18.2.8)$$

where the states $|n\rangle$ are Fock states and α_n is the phase factor for each state. Using this for state $|\psi_i\rangle$, and inserting a sum over the projection operators for all Fock states, $\sum_m |m\rangle\langle m| = 1$, we obtain

$$\langle \hat{N}_k \rangle = \sum_{m,n,n'} \frac{1}{\hbar^2} \int_0^t dt' \int_0^t dt'' \, \alpha_{n'}^* \alpha_n e^{i(E_{n'}-E_m)t'/\hbar} e^{-i(E_n-E_m)t''/\hbar}$$

$$\times \langle n'|\hat{V}|m\rangle\langle m|[\hat{N}_k, \hat{V}]|n\rangle, \qquad (18.2.9)$$

where we have used the fact that the unperturbed Hamiltonian H_0 acting on a pure Fock state n gives a well-defined energy E_n. The term $\langle n'|\hat{V}|m\rangle\langle m|[\hat{N}_k, \hat{V}]|n\rangle$ defined for pure Fock states is a complex number that depends on the number of particles in each state k; in general, this will give final-states factors $(1 + N)$ for bosons and $(1 - N)$ for fermions, as in Fermi's golden rule, discussed in Section 14.1. Examples of specific interactions are given in Section 14.2.

If $n' \neq n$, then we will once again have off-diagonal terms that we assume are negligible, by the arguments in Section 19.3. If $n' = n$, then the time-dependent factors can be resolved, just as we did for Fermi's golden rule, as

$$\left(\int_0^t dt' \, e^{i\omega t'} \right) \left(\int_0^t dt'' \, e^{-i\omega t''} \right)$$

$$= \left| \frac{e^{i\omega t} - 1}{\omega} \right|^2 = \frac{\sin^2(\omega t/2)}{\omega^2}, \qquad (18.2.10)$$

which, as discussed in Section 14.1 and under the same assumptions, becomes

$$\lim_{t\to\infty} \frac{\sin^2(\omega t/2)}{\omega^2} = \delta(\omega)2\pi t. \qquad (18.2.11)$$

If we set t to some small interval dt, we then have the general quantum Boltzmann equation:

$$\frac{d\langle \hat{N}_k \rangle}{dt} = \frac{2\pi}{\hbar} \sum_{m,n} |\alpha_n|^2 \langle n|\hat{V}|m\rangle\langle m|[\hat{N}_k, \hat{V}]|n\rangle \delta(E_n - E_m). \qquad (18.2.12)$$

Last, we consider the other second-order term in (18.2.5). This is

$$-\frac{1}{\hbar^2} \int_0^t dt' \int_0^{t'} dt'' \langle \psi_i|[\hat{N}_k, \hat{V}(t')\hat{V}(t'')]|\psi_i\rangle$$

$$= -\frac{1}{\hbar^2} \int_0^t dt' \int_0^{t'} dt'' \sum_{n,n'} \alpha_{n'}^* \alpha_n \left(\langle n'|\hat{N}_k \hat{V}(t')\hat{V}(t'')|n\rangle - \langle n'|\hat{V}(t')\hat{V}(t'')\hat{N}_k|n\rangle \right)$$

$$-\frac{1}{\hbar^2} \int_0^t dt' \int_0^{t'} dt'' \sum_n \alpha_{n''}^* \alpha_n (N_k'' - N_k)\langle n''|\hat{V}(t')\hat{V}(t'')|n\rangle. \qquad (18.2.13)$$

If we consider terms in which the state $|n''\rangle$ equals $|n\rangle$, then this term vanishes, since N_k'', which is the value of N_k in the state $|n''\rangle$, is then equal to N_k. If we consider other terms, then we must deal with the same type of off-diagonal terms that have appeared earlier in this section, which we will argue in Section 19.3 are negligible.

18.2.2 Quantum Boltzmann Equation for an Interacting Gas

We now pick the specific example of a gas with two-body interactions of the form

$$\hat{V} = \frac{1}{2V} \sum_{k_1,k_2,k_3} U_{k_1,k_2,k_3,k_4} a_{k_4}^\dagger a_{k_3}^\dagger a_{k_2} a_{k_1}, \tag{18.2.14}$$

where the summation is not over $\vec{k}_4$ because it is implicitly assumed that momentum is conserved, so that $\vec{k}_4 = \vec{k}_1 + \vec{k}_2 - \vec{k}_3$. We assume the interaction energy U is symmetric on exchange of $\vec{k}_1$ with $\vec{k}_3$ or $\vec{k}_2$ with $\vec{k}_4$. This interaction has the same form as the electron–electron interaction introduced in Section 14.2.3, but can apply to short-range interactions of bosons as well.

Using relations (12.1.7) and (12.1.8), which are true for both bosons and fermions, we compute

$$[\hat{N}_k, V_{\text{int}}] = \frac{1}{2V} \sum_{k_1,k_2} \left(U_{k_1,k_2,k',k}\, a_k^\dagger a_{k'}^\dagger a_{k_2} a_{k_1} + U_{k_1,k_2,k,k'}\, a_{k'}^\dagger a_k^\dagger a_{k_2} a_{k_1} \right.$$
$$\left. - U_{k',k,k_2,k_1}\, a_{k_1}^\dagger a_{k_2}^\dagger a_k a_{k'} - U_{k,k',k_2,k_1}\, a_{k_1}^\dagger a_{k_2}^\dagger a_{k'} a_k \right)$$
$$= \frac{1}{2V} \sum_{k_1,k_2} (U_D \pm U_E)(a_k^\dagger a_{k'}^\dagger a_{k_2} a_{k_1} - a_{k_1}^\dagger a_{k_2}^\dagger a_k a_{k'}), \tag{18.2.15}$$

where $\vec{k}' = \vec{k}_1 + \vec{k}_2 - \vec{k}$, and $U_D = U_{k_1,k_2,k',k}$ is the direct interaction term and $U_E = U_{k_1,k_2,k,k'}$ is the "exchange" term, and the $+$ sign is for bosons and the $-$ sign is for fermions.[1] If U is a constant, then the scattering rate is enhanced by a factor of 4 for bosons and is forbidden for fermions.

Since the Fock states are orthonormal, the matrix element $\langle \psi_m|[\hat{N}_k, V_{\text{int}}]|\psi_n\rangle$ for a given term in the sum (18.2.15) determines the state $|\psi_m\rangle$, so that we no longer sum over all final states $|\psi_m\rangle$ in (18.2.12). The summation over k-states in definition (18.2.14) of $\hat{V}$ is eliminated in the term $\langle \psi_n|\hat{V}|\psi_m\rangle$ (18.2.12) because only four terms that couple $|\psi_n\rangle$ to $|\psi_m\rangle$ survive. Since a destruction operator a_k acting to the right on a state with N_k particles gives a factor $\sqrt{N_k}$, and a creation operator $a_k^\dagger$ gives a factor $\sqrt{1 \pm N_k}$ (where the $+$ sign is for bosons and the $-$ sign is for fermions), we have

$$\frac{d\langle N_k\rangle}{dt} = \frac{2\pi}{\hbar} \frac{1}{2V^2} \sum_{k_1,k_2} (U_D \pm U_E)^2 \delta(E_{k_1} + E_{k_2} - E_k - E_{k'}) \tag{18.2.16}$$
$$\times \langle N_{k_1} N_{k_2}(1 \pm N_k)(1 \pm N_{k'}) - N_k N_{k'}(1 \pm N_{k_1})(1 \pm N_{k_2})\rangle,$$

---

[1] The assumption $U_{k_1,k_2,k',k} = U_{k,k',k_2,k_1}$ is a result of time-reversal symmetry.

where the average $\langle \ldots \rangle$ is the weighted average over the Fock states $|\psi_n\rangle$ which are included in the initial state $|\psi_i\rangle$. We have dropped the hat on N_k, because the number operator acting on Fock states just gives us integer numbers. Equation (18.2.12) has the same condition of validity as Fermi's golden rule, namely the timescale for depletion of any given state must be long compared to $\hbar/\Delta E_f$, where ΔE_f is the range of final states which can be considered smooth.

We can make a very powerful approximation by treating the average of the product which occurs in (18.2.12) as a product of averages:

$$\langle N_{k_1} N_{k_2}(1 \pm N_k)(1 \pm N_{k'}) - N_k N_{k'}(1 \pm N_{k_1})(1 \pm N_{k_2})\rangle$$
$$\simeq \langle N_{k_1}\rangle\langle N_{k_2}\rangle(1 \pm \langle N_k\rangle)(1 \pm \langle N_{k'}\rangle)) - \langle N_k\rangle\langle N_{k'}\rangle(1 \pm \langle N_{k_1}\rangle)(1 \pm \langle N_{k_2}\rangle)). \quad (18.2.17)$$

This relies on the assumption that there are no special correlations between the occupation numbers of different states; this in turn follows from the assumption that the correlation functions $\langle a_{k_1}^\dagger a_{k_2}^\dagger a_{k_3} a_{k_4}\rangle$ are negligible (Snoke 2012). We therefore can write the equation entirely in terms of the distribution function $\langle N_k\rangle$ for the k-states:

$$\frac{d\langle N_k\rangle}{dt} = \frac{2\pi}{\hbar}\frac{1}{2V^2}\sum_{\vec{k}_1,\vec{k}_2}(U_D \pm U_E)^2\left(\langle N_{k_1}\rangle\langle N_{k_2}\rangle(1 \pm \langle N_k\rangle)(1 \pm \langle N_{k'}\rangle)\right.$$
$$\left. - \langle N_k\rangle\langle N_{k'}\rangle(1 \pm \langle N_{k_1}\rangle)(1 \pm \langle N_{k_2}\rangle)\right)\delta(E_{k_1} + E_{k_2} - E_k - E_{k'}). \quad (18.2.18)$$

This is the quantum Boltzmann equation for two-body scattering, which gives the total rate of change of the probability distribution. We see that it has the same final states factors $(1 \pm N_k)$ as we found in computing the transition rates for single particles using Fermi's golden rule in Section 14.1.

18.3 Experimental Verification of the Quantum Boltzmann Equation

The nonequilibrium time evolution predicted by the quantum Boltzmann equation (18.2.12) has been verified experimentally. Section 18.2 showed that, for the case of two-particle scattering, it depends only on the numbers N_k for each state $\vec{k}$. The same can be shown for all kinds of other processes, such as phonon emission. This makes it relatively easy to do numerical calculations for the time evolution of many-particle systems. For an isotropic system, that is, one in which properties of the system do not depend on the direction in space, we can write the *distribution function* $f(E_k) = \langle N_k\rangle$ which depends only on the magnitude of k. We can then computing the rate of change $df(E)/dt$ for all E at any point in time, and update $f(E)$ according to

$$f(E) \to f(E) + \frac{df(E)}{dt}dt, \quad (18.3.1)$$

where dt is some small time interval, chosen such that the change of $f(E)$ is small during any one update.

The quantum Boltzmann equation has been used extensively in experimental physics (for a review, see Snoke 2011). Figure 18.1 shows an example of the evolution of a population

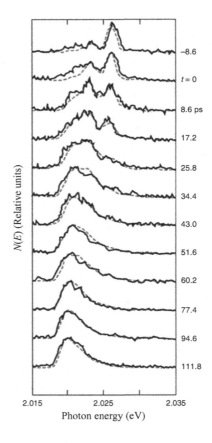

N(E) (Relative units)

−8.6

t = 0

8.6 ps

17.2

25.8

34.4

43.0

51.6

60.2

77.4

94.6

111.8

2.015   2.025   2.035

Photon energy (eV)

Figure 18.1 Solid lines: Energy distribution of excitons in the semiconductor Cu_2O measured at various times following a laser pulse with temporal width 2 ps and maximum intensity at time $t = 0$. Dashed lines: Solution of the quantum Boltzmann equation for the time evolution of the population using a deformation-potential theory for exciton–phonon scattering. The theory gives a Maxwell–Boltzmann distribution at all late times. From Snoke 1991.

of particles (excitons) inside a semiconductor. Modern ultrafast optics experiments can resolve the energy distribution of these particles on timescales short compared to the time to equilibrate. Therefore, we can observe the approach to equilibrium and compare it to the evolution given by the quantum Boltzmann equation. As seen in this figure, there is a very good fit of the experiment to the theoretical prediction of the quantum Boltzmann equation. This theoretical prediction was found by setting the initial state to a peaked distribution determined by the laser pulse in the experiment, and then updating the distribution in time according to the procedure (18.3.1). Long after the laser pulse, the particles equilibrate to a Maxwell–Boltzmann distribution, which we already have seen in Section 10.1, is the predicted equilibrium distribution for a many-particle system at low density.

The numerical solution in this case leads to irreversibility – at all later times, the solution, like the experimental data, remains in a Maxwell–Boltzmann distribution. This behavior arises *deterministically*, which is to say, no random numbers were used in the calculation;

the time evolution was found by evaluating an integral at each point in time. Given any initial distribution $f(E)$ for all E, we can find the distribution at all later times t this method. This provides an important connection with the field of thermodynamics. Historically, at the time of Maxwell and Boltzmann there was considerable debate over the "H-theorem," which says that a system far from equilibrium must approach equilibrium. Boltzmann justified this assumption using a statistical argument that was never fully accepted (contributing eventually to Boltzmann's suicide). The quantum Boltzmann equation gives a deterministic evolution of the quantum wave function, however. There is no need to invoke the Born rule for particle statistics at all.

18.4 Proof of the Quantum H-Theorem

The quantum Boltzmann equation implies Boltzmann's "H-theorem," which is the basis of the second law of thermodynamics, namely, that entropy never decreases in a closed system. We can define a semiclassical entropy (also known as *diagonal entropy* (Polkovnikov 2011) in terms of the distribution function of the particles. For classical particles, this is

$$S = -k_B \sum_k \langle N_k \rangle \ln \langle N_k \rangle. \tag{18.4.1}$$

For fermions and bosons, this is modified to (Band 1955)

$$S = -k_B \sum_k \Big(\langle N_k \rangle \ln \langle N_k \rangle \mp (1 \pm \langle N_k \rangle) \ln(1 \pm \langle N_k \rangle) \Big), \tag{18.4.2}$$

where the upper sign is for bosons and the lower sign is for fermions. Assuming conservation of the total number of particles, the time derivative of this is

$$\frac{\partial S}{\partial t} = -k_B \sum_k \frac{\partial \langle N_k \rangle}{\partial t} \ln \left(\frac{\langle N_k \rangle}{1 \pm \langle N_k \rangle} \right). \tag{18.4.3}$$

Let us now compute the time derivative of $\langle N_k \rangle$ using the quantum Boltzmann equation, for a specific choice of the interaction term. From the results of Fermi's golden rule in Section 14.1, which also apply to the quantum Boltzmann equation, we know that there will be final-states factors of the form $(1 + N)$ for bosons and $(1 - N)$ for fermions, for any interaction. For the case of collisions of two particles used in Section 18.2, this gives us terms proportional to

$$\langle N_{k_1} \rangle \langle N_{k_2} \rangle (1 \pm \langle N_k \rangle)(1 \pm \langle N_{k'} \rangle)$$

for collisions of two particles in states k_1 and k_2 ending up in states k and k', and proportional to

$$\langle N_k \rangle \langle N_{k'} \rangle (1 \pm \langle N_{k_1} \rangle)(1 \pm \langle N_{k_2} \rangle)$$

for the reverse process. This implies that the rate of change of the entropy is

$$
\frac{\partial S}{\partial t} = -k_B \sum_{k,k_1,k_2} C \ln\left(\frac{\langle N_k \rangle}{1 \pm \langle N_k \rangle}\right) \left[\langle N_{k_1} \rangle \langle N_{k_2} \rangle (1 \pm \langle N_k \rangle)(1 \pm \langle N_{k'} \rangle) \right.
$$
$$
\left. - \langle N_k \rangle \langle N_{k'} \rangle (1 \pm \langle N_{k_1} \rangle)(1 \pm \langle N_{k_2} \rangle) \right], \tag{18.4.4}
$$

where C is a positive number that contains the system-specific details of the collision process.

For any choice of the four states k_1, k_2, k, and k', the total of all terms in the sum involving the same four states is

$$
\left(\ln\left(\frac{\langle N_k \rangle}{1 \pm \langle N_k \rangle}\right) + \ln\left(\frac{\langle N_{k'} \rangle}{1 \pm \langle N_{k'} \rangle}\right) - \ln\left(\frac{\langle N_{k_1} \rangle}{1 \pm \langle N_{k_1} \rangle}\right) - \ln\left(\frac{\langle N_{k_2} \rangle}{1 \pm \langle N_{k_2} \rangle}\right) \right)
$$
$$
\times \left[\langle N_{k_1} \rangle \langle N_{k_2} \rangle (1 \pm \langle N_k \rangle)(1 \pm \langle N_{k'} \rangle) - \langle N_k \rangle \langle N_{k'} \rangle (1 \pm \langle N_{k_1} \rangle)(1 \pm \langle N_{k_2} \rangle) \right]
$$
$$
= \ln\left(\frac{\langle N_k \rangle \langle N_{k'} \rangle (1 \pm \langle N_{k_1} \rangle)(1 \pm \langle N_{k_2} \rangle)}{\langle N_{k_1} \rangle \langle N_{k_2} \rangle (1 \pm \langle N_k \rangle)(1 \pm \langle N_{k'} \rangle)}\right)
$$
$$
\times \left[\langle N_{k_1} \rangle \langle N_{k_2} \rangle (1 \pm \langle N_k \rangle)(1 \pm \langle N_{k'} \rangle) - \langle N_k \rangle \langle N_{k'} \rangle (1 \pm \langle N_{k_1} \rangle)(1 \pm \langle N_{k_2} \rangle) \right]. \tag{18.4.5}
$$

If the in-scattering term in the square brackets is larger than the out-scattering term, then the denominator of the logarithm is larger than the numerator, making the logarithm negative. Conversely, if the in-scattering is less than the out-scattering term, the term in the square brackets is negative. Since the whole sum in (18.4.4) consists of terms like this, the total sum is less than or equal to zero, and therefore $\partial S/\partial t > 0$. This is the standard form of the H-theorem.[2]

We know that quantum mechanics is time-reversible on the microscopic scale. Yet we have derived an equation with irreversibility. How did the irreversibility enter in?

In deriving (18.2.12), we argued at several points that we could drop terms with "off-diagonal" correlation functions, and keep only the information about the average number of particles in each state, $\langle N_k \rangle$. The quantum Boltzmann equation solution method amounts to setting all these correlation terms strictly to zero after each time step dt. This amounts to an erasure of information, which ultimately leads to the irreversibility in the system.

This loss of reversibility in the quantum Boltzmann equation is illustrated in Figure 18.2. The system starts in state 1, in which we assume that all of the correlation functions with phase information are strictly zero. The true evolution of the full quantum mechanical solution gives state 2, which has slightly nonzero values for these correlation functions. If this state 2 were run backwards in time, the true evolution would take this state back to state 1. However, in the iterated quantum Boltzmann equation approach, we approximate state 2 by state 2', which has all off-diagonal correlation functions strictly equal to zero. State 2' then evolves to state 3', which is a good approximation of state 3, the real state reached by state 2. Thus, if we continue forward in time, we have a series of successive approximations that are very good approximations of the evolution of the full quantum mechanical solution. If we evolve backward in time, however, our dropping of the correlation information

---

[2] We have used a specific choice for the interaction term here, with two incoming particles in a collision giving the same two particles in different outgoing states, but it is easy to show that any scattering process will lead to the same result (see Snoke 2013 for one example).

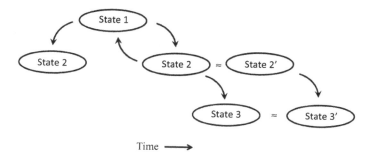

Time ⟶

Figure 18.2 Approximation steps in the iterative method of evolving the many-body wave state using the quantum Boltzmann equation.

means that the backward evolution will not be a good approximation of the real backwards evolution. The tiny values of the correlation functions, which are dropped in the iterated Boltzmann approach, carry information that is crucial for recovering the true time-reversed behavior.

References

W. Band, *An Introduction to Quantum Statistics*, (D. van Nostrand, 1955), pp. 154, 162.

P. Bocchieri and A. Loinger, "Quantum recurrence theorem," *Physical Review* **107**, 337 (1957).

K. Jacobs and D. A. Steck, "A straightforward introduction to continuous quantum measurement," *Contemporary Physics* **47**, 279 (2006).

A. Polkovnikov, "Microscopic diagonal entropy and its connection to basic thermodynamic relations," *Annals of Physics* **326**, 486 (2011).

D. W. Snoke, D. Braun, and M. Cardona, "Carrier thermalization in Cu_2O: Phonon emission by excitons," *Physical Review B* **44**, 2991 (1991).

D. W. Snoke, "The quantum Boltzmann equation in semiconductor physics," *Annalen der Physik* **523**, 87 (2011).

D. W. Snoke, G.-Q. Liu, and S. M. Girvin, "The basis of the second law of thermodynamics in quantum field theory," *Annals of Physics* **327**, 1825 (2012).

D. W. Snoke and S. M. Girvin, "Dynamics of phase coherence onset in Bose condensates of photons by incoherent phonon emission," *Journal of Low Temperature Physics* **171**, 1 (2013).

D. W. Snoke, *Solid State Physics: Essential Concepts*, 2nd ed., (Cambridge University Press, 2020).

As we saw in Chapter 18, decoherence is crucially related to irreversibility. We do not have to just assume that decoherence occurs, however; we can compute it and show how it arises in a unitary quantum field theory. To discuss coherence in quantum systems, we make use of the *density matrix*, introduced in Section 14.3.1.

19.1 Density Matrix Formalism

The density operator in quantum mechanics is another way of representing the same information that is held in the quantum wave function. It is defined as

$$\hat{\rho} = |\psi\rangle\langle\psi|. \tag{19.1.1}$$

Like any operator defined in Section 11.1, it can be expressed as a matrix acting on a complete set of states, with elements $\langle\psi_n|\rho|\psi_m\rangle$. The elements of the density matrix can therefore be written as

$$\rho_{mn} = \langle\psi_n|\hat{\rho}|\psi_m\rangle = \langle\psi_n|\psi\rangle\langle\psi|\psi_m\rangle,$$
$$= c_m^* c_n, \tag{19.1.2}$$

where $|\psi_m\rangle$ and $|\psi_n\rangle$ are members of a complete set of states for the system, and c_m and c_n are the probability amplitudes for occupation of those states. Since the diagonal elements of the density matrix give the fractional weight of the individual states, the trace of the density matrix must add up to 100%, that is,

$$\text{Tr}\,\tilde{\rho} = \sum_n \rho_{nn} = \sum_n |c_n|^2 = 1. \tag{19.1.3}$$

The time evolution of the density matrix can be determined using the Schrödinger equation $(i\hbar)\partial/\partial t|\psi\rangle = H|\psi\rangle$:

$$\frac{\partial\hat{\rho}}{\partial t} = \left(\frac{\partial}{\partial t}|\psi\rangle\right)\langle\psi| + |\psi\rangle\left(\frac{\partial}{\partial t}\langle\psi|\right)$$
$$= \left(\frac{1}{i\hbar}H|\psi\rangle\right)\langle\psi| + |\psi\rangle\left(-\frac{1}{i\hbar}\langle\psi|H\right), \tag{19.1.4}$$

which is equivalent to

$$\frac{\partial\hat{\rho}}{\partial t} = -\frac{i}{\hbar}[H, \hat{\rho}]. \tag{19.1.5}$$

This is known as the *Liouville* equation.

The density matrix formalism gives us a natural way to account for coherent and incoherent populations. We distinguish between a system in a *pure* quantum state, which can be written as a linear superposition of other quantum states, and one that is in a *mixed* state, which consists of a random ensemble of pure states. In a pure state, there is uncertainty in the outcome of a measurement if the state is a linear superposition of other states, and this comes from the intrinsic uncertainty of quantum mechanics. A pure state can be called *coherent*, since by a change of basis we could describe the system simply in terms of the amplitude and phase of a single quantum state. In a mixed state, there is randomness due to standard statistical uncertainty, such as thermal fluctuations or simple ignorance about the initial state of the system.

In general, we suppose that there is some number of pure states that can be occupied with probability P_i, subject to the normalization

$$\sum_i P_i = 1. \tag{19.1.6}$$

The density operator for the mixed state in this case will be the sum of the density operators for all the separate pure states, weighted by the probability of being in each state:

$$\hat{\rho} = \sum_i P_i \hat{\rho}^{(i)} = \sum_i P_i |\psi_i\rangle \langle \psi_i|. \tag{19.1.7}$$

The probability of finding the system in state $|\psi_n\rangle$ is therefore

$$\sum_i P_i |\langle \psi_n|\psi_i\rangle|^2 = \sum_i P_i \langle \psi_n|\psi_i\rangle \langle \psi_i|\psi_n\rangle$$
$$= \langle \psi_n|\hat{\rho}|\psi_n\rangle, \tag{19.1.8}$$

that is, the diagonal elements of the density operator, just as in the pure case. Since the density operator (19.1.7) is a linear superposition of pure-state operators, it still obeys the Liouville equation.

Suppose now that the system is in a pure state that is a linear superposition of states $|\psi_m\rangle$, $|\psi_n\rangle$, and so on, with amplitudes c_n, c_m, respectively. The diagonal elements are $\rho_{nn} = |c_n|^2$, while the off-diagonal elements are $\rho_{nm} = c_m^* c_n$.

Compare this to a statistical ensemble of states, each of which is a pure state $|\psi_n\rangle$, and the probability for each is $P_n = |c_n|^2$, where the probabilities are chosen to be exactly the same as in the pure linear superposition. If we look only at the diagonal elements of the density matrix, we cannot distinguish between these two cases. The off-diagonal elements in the statistical mixture, however, are

$$\rho_{mn} = \sum_i P_i \langle \psi_n|\psi_i\rangle \langle \psi_i|\psi_m\rangle$$
$$= \sum_i |c_i|^2 \delta_{ni}\delta_{im}$$
$$= |c_n|^2 \delta_{mn}. \tag{19.1.9}$$

The off-diagonal elements of the density matrix are therefore good measures of the "pureness" of the system, which can also be called its *coherence*.

19.2 Correlation Functions in Quantum Field Theory

We can also generate a density matrix formalism for a many-body system in terms of creation and destruction operators. Suppose that we define a pure single-particle state as the superposition

$$|\psi\rangle = c_1 a_1^\dagger |0\rangle + c_2 a_2^\dagger |0\rangle + \cdots = \sum_i c_i a_i^\dagger |0\rangle, \tag{19.2.1}$$

where $|0\rangle$ is the vacuum (ground) state. Then if we define the density matrix element

$$\rho_{mn} = \langle \psi | a_m^\dagger a_n | \psi \rangle, \tag{19.2.2}$$

we obtain

$$\rho_{mn} = \left(\langle 0 | \sum_i c_i^* a_i \right) a_m^\dagger a_n \left(\sum_j c_j a_j^\dagger |0\rangle \right) = c_m^* c_n, \tag{19.2.3}$$

just the same as for definition (19.1.2). This allows us to write the operator

$$\hat{\rho}_{mn} = a_m^\dagger a_n \tag{19.2.4}$$

for the individual matrix elements.

One advantage of using this notation is that we can now define correlation functions for a many-particle state with more than just two states. For example, if we write

$$|\psi\rangle = \sum_n \alpha_n |n\rangle, \tag{19.2.5}$$

where the $|n\rangle$ kets are all the possible pure Fock states, and the α_n are complex numbers which the relative weight and phase of each, we can select out two states k and k', and define the generalized density matrix element

$$\rho_{k,k'} = \sum_{n,n'} \alpha_{n'}^* \alpha_n \langle n' | a_k^\dagger a_{k'} | n \rangle. \tag{19.2.6}$$

If $k = k'$, this gives the average number in state k,

$$\langle \hat{N}_k \rangle = \sum_n |\alpha_n|^2 \langle n | a_k^\dagger a_k | n \rangle. \tag{19.2.7}$$

Therefore, the average occupation numbers of the states are also called the "diagonal" elements of the system. If $k \neq k'$, $\rho_{k,k'}$ can only be nonzero if the weights of more than one state $|n'\rangle$ and $|n\rangle$ are nonzero, so that when the operators act on $|n\rangle$ to change it into a different Fock state, that new state exists already in the total superposition of states. If it does not, the term will vanish, because we assume that the eigenstates are orthonormal, that is, $\langle n' | n \rangle = \delta_{n,n'}$. This density matrix term with $k \neq k'$ is called an "off-diagonal" element, and is a measure of the coherence in the system.

Two states culled out for this off-diagonal element can be any two out of a great many; we can treat these two as the system under study and the rest of the states as the "environment." We do not need to stop at correlations between two states, however. In a many-particle quantum state, there is an infinite number of possible correlations. For example, we could write down a fourth-order correlation,

$$\rho_{k,k',k'',k'''} = \sum_{n,n'} \alpha_{n'}^* \alpha_n \langle n' | a_k^\dagger a_{k'}^\dagger a_{k''} a_{k'''} | n \rangle. \tag{19.2.8}$$

As shown in Snoke 2012, this correlation function becomes nonzero whenever there is a two-body collision term, which takes two incoming particles and turns them into two outgoing particles in different states. There is no limit to how many operators we put into a correlation function. However, if there are more destruction operators in the correlation function than the number of particles in the system it acts on, the correlation function will be zero, because a destruction operator acting on a state with $N_k = 0$ always returns the value zero.

For these higher-order terms, we can also distinguish between "diagonal" elements, which have all of the states the same (which, e.g., would make the term (19.2.8) return the average value of $\langle N_k^2 \rangle$), and "off-diagonal" terms, which measure coherence between multiple different states.

All of these correlation functions do not provide any additional information to what is already in the full quantum wave function. They are simply ways of extracting various subsets of information from that full wave function.

19.3 Time-Evolution Equations for Correlation Functions of a Many-Particle System

Following a similar approach to that of Section 18.2, we write for a given off-diagonal term $(k \neq k')$,

$$d\langle \hat{\rho}_{k,k'} \rangle = \langle \psi_t | \hat{\rho}_{k,k'} | \psi_t \rangle - \langle \psi_i | \hat{\rho}_{k,k'} | \psi_i \rangle$$
$$= \langle \psi(t) | \hat{\rho}_{k,k'}(t) | \psi(t) \rangle - \langle \psi_i | \hat{\rho}_{k,k'} | \psi_i \rangle. \tag{19.3.1}$$

As we did for (18.2.5), we expand the time-dependent exponential operators as

$$d\langle \hat{\rho}_{k,k'} \rangle = \langle \psi_i | \left(1 - \frac{1}{i\hbar} \int_0^t \hat{V}(t') dt' + \frac{1}{(i\hbar)^2} \int_0^t dt' \int_0^{t'} dt'' \hat{V}(t') \hat{V}(t'') + \cdots \right) \hat{\rho}_{k,k'}(t)$$
$$\times \left(1 + \frac{1}{i\hbar} \int_0^t \hat{V}(t') dt' + \frac{1}{(i\hbar)^2} \int_0^t dt' \int_0^{t'} dt'' \hat{V}(t') \hat{V}(t'') + \cdots \right) | \psi_i \rangle$$
$$- \langle \psi_i | \hat{\rho}_{k,k'} | \psi_i \rangle. \tag{19.3.2}$$

The lowest-order term is

$$\langle \psi_i | (\hat{\rho}_{k,k'}(t) - \hat{\rho}_{k,k'}) | \psi_i \rangle; \tag{19.3.3}$$

the first-order terms are

$$\frac{1}{i\hbar} \int_0^t dt' \langle \psi_i | [\hat{\rho}_{k,k'}(t), \hat{V}(t')] | \psi_i \rangle, \tag{19.3.4}$$

and the second-order terms are

$$
\frac{1}{\hbar^2} \int_0^t dt' \int_0^t dt'' \langle \psi_i | \hat{V}(t') \hat{\rho}_{k,k'}(t) \hat{V}(t'') | \psi_i \rangle
$$

$$
- \frac{1}{\hbar^2} \int_0^t dt' \int_0^{t'} dt'' \langle \psi_i | \hat{V}(t') \hat{V}(t'') \hat{\rho}_{k,k'}(t) | \psi_i \rangle
$$

$$
- \frac{1}{\hbar^2} \int_0^t dt' \int_0^{t'} dt'' \langle \psi_i | \hat{\rho}_{k,k'}(t) \hat{V}(t') \hat{V}(t'') | \psi_i \rangle. \tag{19.3.5}
$$

We cannot assume that $\hat{\rho}_{k,k'}$ commutes with H_0, as we did for N_k in the derivation of the quantum Boltzmann equation, so we must keep all of the $e^{iH_0 t/\hbar}$ terms in the definitions of the time-dependent operators for the moment. In the following sections, we compute each order of the expansion (19.3.2) separately.

Zero-order off-diagonal evolution. As before, we write $|\psi_i\rangle$ as a sum of Fock states, so that the zero-order term becomes

$$
d\langle \hat{\rho}_{k,k'} \rangle = \sum_{n,n'} \alpha_{n'}^* \alpha_n \left(\langle n' | (e^{iH_0 t/\hbar} \hat{\rho}_{k,k'} e^{-iH_0 t/\hbar} | n \rangle - \langle n' | \hat{\rho}_{k,k'}) | n \rangle \right). \tag{19.3.6}
$$

Applying the time-varying exponential factors to the Fock states gives us

$$
d\langle \hat{\rho}_{k,k'} \rangle = \sum_{n,n'} \alpha_{n'}^* \alpha_n (e^{i(E_k - E_{k'})t/\hbar} - 1) \langle n' | \hat{\rho}_{k,k'} | n \rangle
$$

$$
\simeq \frac{it}{\hbar} (E_k - E_{k'}) \sum_{n,n'} \alpha_{n'}^* \alpha_n \langle n' | \hat{\rho}_{k,k'} | n \rangle. \tag{19.3.7}
$$

Applying the action of H_0, this is then, for small $t = dt$,

$$
\frac{d}{dt} \langle \hat{\rho}_{k,k'} \rangle = \frac{i}{\hbar} \langle \psi_i | [H_0, \hat{\rho}_{k,k'}] | \psi_i \rangle. \tag{19.3.8}
$$

This has the opposite sign of the Liouville equation (19.1.5), because it is an equation for the elements of the density matrix, not the evolution of the whole density operator.

First-order off-diagonal evolution. If $n' = n$, this becomes

$$
d\langle \hat{\rho}_{k,k'} \rangle = \sum_n |\alpha_n|^2 \frac{1}{i\hbar} \int_0^t dt' \langle n | (e^{iH_0 t/\hbar} \hat{\rho}_{k,k'} e^{-iH_0(t-t')/\hbar} \hat{V} e^{-iH_0 t'/\hbar}
$$

$$
- e^{iH_0 t'/\hbar} \hat{V} e^{-iH_0(t'-t)/\hbar} \hat{\rho}_{k,k'} e^{-iH_0 t/\hbar}) | n \rangle
$$

$$
= \sum_n |\alpha_n|^2 \frac{1}{i\hbar} e^{i(E_k - E_{k'})t/\hbar} \int_0^t dt' (\langle n_- | e^{iH_0 t'/\hbar} \hat{V} e^{-iH_0 t'/\hbar} | n \rangle
$$

$$
- \langle n | e^{iH_0 t'/\hbar} \hat{V} e^{-iH_0 t'/\hbar} | n_+ \rangle)
$$

$$
= \sum_n |\alpha_n|^2 \frac{1}{i\hbar} e^{i(E_k - E_{k'})t/\hbar} \int_0^t dt' e^{-i(E_k - E_{k'})t'/\hbar} (\langle n_- | \hat{V} | n \rangle - \langle n | \hat{V} | n_+ \rangle), \tag{19.3.9}
$$

where $|n_+\rangle$ and $|n_-\rangle$ are the Fock state $|n\rangle$ with one particle removed from state k' and added to state k, and vice versa; in the last line, we have used the fact that the interaction term $\hat{V}$ must undo whatever changes the density matrix element has made. The time integral can be resolved as

$$e^{i\omega t} \int_0^t dt'\, e^{-i\omega t'} = e^{i\omega t} \frac{e^{-i\omega t} - 1}{-i\omega} = \frac{1 - e^{i\omega t}}{-i\omega} \simeq t. \qquad (19.3.10)$$

The first-order term therefore becomes, for a small time interval $t = dt$,

$$\frac{d}{dt}\langle \hat{\rho}_{k,k'} \rangle = \frac{i}{\hbar} \sum_n |\alpha_n|^2 \langle n|[\hat{V}, \hat{\rho}_{k,k'}]|n\rangle. \qquad (19.3.11)$$

If $n' \neq n$, then this term will involve higher-order correlation functions with more than two creation and destruction operators. We make the assumption that we can consider correlation terms in a hierarchy, with higher-order correlation functions treated as small compared to lower-order correlation functions. This then allows us to write

$$\frac{d}{dt}\langle \hat{\rho}_{k,k'} \rangle = \frac{i}{\hbar} \langle \psi_i|[\hat{V}, \hat{\rho}_{k,k'}]|\psi_i\rangle, \qquad (19.3.12)$$

similar to (19.3.8).

Second-order off-diagonal evolution. We now proceed to the second-order terms (19.3.5). We can write the interaction generally as

$$\hat{V} = \sum_{i,j,q} a_j^\dagger [A_q] a_i, \qquad (19.3.13)$$

where $[A_q]$ is some set of operators that give an interaction with the environment, and $a_j^\dagger$ and a_i are creation and destruction operators that act on the states of interest. Terms with $i \neq j$ correspond to T_1 processes, since they deplete the initial state, while terms with $i = j$ corresponds to pure T_2 processes. Since $\hat{V}$ is Hermitian (real-valued), there must be balanced terms, with one term in the sum having exactly creation and destruction operators with the opposite action as those in another term. To resolve the terms (19.3.5), we will select only those cases in which some term in $\hat{V}(t')$ undoes the action of the term from $\hat{V}(t'')$. If that were not true, we would once again have higher-order correlation functions, which we assume are small.

If $\hat{V}(t')$ undoes whatever changes have been done to $|\psi_i\rangle$ by $\hat{V}(t'')$, then each of the three terms in (19.3.5) gives a factor $e^{i\omega t'} e^{-i\omega t''}$. To resolve the time integrals, we note that, in the first term, t' and t'' are integrated from 0 to t separately, while in the last two terms, t' is integrated from 0 to t, but t'' is integrated from 0 to t'. This latter case can be related to the former by writing

$$\int_0^t dt' \int_0^t dt''\, e^{i\omega t'} e^{-i\omega t''} = \int_0^t dt' \int_0^{t'} dt''\, e^{i\omega(t'-t'')} + \int_0^t dt'' \int_0^{t''} dt'\, e^{i\omega(t'-t'')}. \qquad (19.3.14)$$

By a switch of variables $t' \to t''$, $t'' \to t'$, the second integral becomes the complex conjugate of the first. If we assume that there is a nearly continuous sum over states in $\hat{V}$, which leads to the rest of the integrand being nearly constant near $\omega = 0$, then the imaginary terms will vanish since they are odd in ω. The two terms on the right are therefore equal, so that we have

$$
\int_0^t dt' \int_0^{t'} dt'' \, e^{i\omega t'} e^{-i\omega t''} = \frac{1}{2} \int_0^t dt' \int_0^t dt'' \, e^{i\omega t'} e^{-i\omega t''}
$$

$$
= \frac{1}{2} 2\pi t \, \delta(\omega), \tag{19.3.15}
$$

where, in the last line, we have used formula (14.1.10) from Section 14.1. We then take t equal to some small interval dt, which gives us

$$
\frac{d}{dt} \langle \hat{\rho}_{k,k'} \rangle = \frac{2\pi}{\hbar} \langle \psi_i | (\hat{V} \hat{\rho}_{k,k'} \hat{V} - \frac{1}{2} \hat{V} \hat{V} \hat{\rho}_{k,k'} - \frac{1}{2} \hat{\rho}_{k,k'} \hat{V} \hat{V}) | \psi_i \rangle \delta(\tilde{E}), \tag{19.3.16}
$$

where $\tilde{E}$ is the sum of the changes in energy brought about (and then undone) by the interaction $\hat{V}$; the δ-function enforces energy conservation in these changes. The term in the square brackets is known as the *Lindbladian* operator for the time evolution.[1]

The zero- and first-order terms (19.3.8) and (19.3.12) both give a pure phase precession of the density matrix, since for a real-valued ("Hermitian") interaction $\hat{V}$, they give terms for the time evolution that are purely imaginary. The second-order term here can give a change of the amplitude of the density matrix elements, which means that phase coherence can either decrease or grow in time.

Decoherence. Let us take a specific case of an interaction energy which is proportional to a number of particles in state k which is one of the states sampled by the $\hat{\rho}_{k,k'}$ operator. For example, this might be the occupation number of the upper state of a two-level electron system modeled by the Bloch sphere. We take $i = j$ in the terms of $\hat{V}$, corresponding to a pure T_2 process, and assume that all the other operators in $[A_q]$ commute with $\hat{\rho}_{k,k'}$. We then have

$$
\sum_q [A_q]^\dagger [A_q] \left(\hat{N}_k \hat{a}_k^\dagger \hat{N}_k - \frac{1}{2} \hat{N}_k^2 a_k^\dagger - \frac{1}{2} a_k^\dagger \hat{N}_k^2 \right) a_{k'}. \tag{19.3.17}
$$

Taking the expectation value then gives

$$
\sum_q |A_q|^2 \left((N_k + 1) \hat{N}_k - \frac{1}{2} (N_k + 1)^2 - \frac{1}{2} N_k^2 \right) \langle a_k^\dagger a_{k'} \rangle
$$

$$
= -\frac{1}{2} \sum_q |A_q|^2 \langle \hat{\rho}_{k,k'} \rangle, \tag{19.3.18}
$$

for both fermions and bosons. The calculation for the $a_{k'}$ in $\hat{\rho}_{k,k'}$ gives the same result. Since the pair $[A_q]^\dagger [A_q]$ does not change the state, but only gives a numerical factor, we therefore have, for $k \neq k'$,

---

[1] If the interaction term $\hat{V}$ commutes with H_0, then the second-order integrals in (19.3.5) will not have any oscillating terms, and therefore will simply be proportional to t^2, corresponding to the square of the first-order term (19.3.4) in the exponential expansion. Thus, to obtain the dissipative behavior generated by the Lindbladian, discussed in the following, it is crucial that the interaction $\hat{V}$ couples the system of interest to other states.

$$\frac{d\langle\hat{\rho}_{k,k'}\rangle}{dt} = -\langle\hat{\rho}_{k,k'}\rangle\frac{2\pi}{\hbar}\sum_q |A_q|^2\delta(\tilde{E})$$

$$\equiv -\frac{\langle\hat{\rho}_{k,k'}\rangle}{T_2}. \qquad (19.3.19)$$

This gives an exponential decay of any off-diagonal correlation terms, proportional to an interaction rate in the same form as Fermi's golden rule. The same can be shown to occur for higher-order correlation terms (Snoke 2012). Therefore our assumption used earlier, in deriving the quantum Boltzmann equation and elsewhere, that the off-diagonal correlation terms are negligible, is self-consistent.

As we will see in Section 21.1, however, processes that lead to a macroscopic influx of particles into a state can lead to the reverse effect, an increase of phase coherence, for boson systems. Our assumption of strong decoherence will break down in that case, that is, in the case of superconductors, superfluids, and large classical waves such as sound and light waves.

19.4 Quantum Trajectories

The quantum trajectories method is based on the time evolution of the density matrix of a system. In Section 19.3, we derived the time evolution of a correlation function, which could be viewed as an off-diagonal element of a density matrix. Here, we will start by deriving the time evolution of the full density matrix itself. The recipe for the quantum trajectories method is given in Section 19.4.2. Much of the material in this section follows the useful review of Daley (2014).

19.4.1 Derivation of the Time-Evolution Equations for the Density Matrix

As a specific example, we suppose that we have a single two-level system that couples to a larger environment. We write the initial total wave function as

$$|\psi\rangle = (\alpha_0|1\rangle + \beta_0|2\rangle)|\text{env}\rangle, \qquad (19.4.1)$$

where $|1\rangle$ and $|2\rangle$ are the two states of interest, α and β are complex coefficients, and $|\text{env}\rangle$ is the full many-body state of the larger environment.

After a short time of interaction with the environment, we can in general write the state of the system as

$$|\psi'\rangle = \alpha|1\rangle|\text{env}_1\rangle + \beta|2\rangle|\text{env}_2\rangle, \qquad (19.4.2)$$

where the environment coupled to the two states after the interaction can be different; because the time evolution is unitary, we still have $|\alpha|^2 + |\beta|^2 = 1$. However, the environment can be assumed to rapidly equilibrate. We can write the state of the environment in equilibrium as

$$|\text{env}\rangle = \sum_n e^{i\varphi_n} \gamma_n |n\rangle, \tag{19.4.3}$$

where φ_n is some random phase factor, and γ_n is a real-valued weight factor given by the equilibrium occupation of the eigenstates n,

$$\gamma_n^2 = P(n) = \frac{1}{Z} e^{-E_n/k_B T}, \tag{19.4.4}$$

and Z is the partition function from statistical mechanics. We therefore can write

$$|\psi'\rangle = \alpha |1\rangle \sum_n e^{i\varphi_n} \gamma_n |n\rangle + \beta |2\rangle \sum_n e^{i\tilde{\varphi}_n} \gamma_n |n\rangle, \tag{19.4.5}$$

where the phase factors $\varphi_n, \tilde{\varphi}_n$ depend on the history of the interactions.

Averaging over the environment. If we want a particular correlation function $\langle \hat{\rho}_{12} \rangle = \langle a_1^\dagger a_2 \rangle$, as defined in Section 19.2, we can write this by inserting a complete set of eigenstates $|l\rangle$ as

$$\begin{aligned}
\langle \psi' | \hat{\rho}_{12} | \psi' \rangle &= \sum_l \langle \psi' | l \rangle \langle l | \hat{\rho}_{12} | \psi' \rangle \\
&= \sum_l \langle l | \hat{\rho}_{12} | \psi' \rangle \langle \psi' | l \rangle \\
&\equiv \text{Tr } \hat{\rho}_{12} | \psi' \rangle \langle \psi' |.
\end{aligned} \tag{19.4.6}$$

This is known as taking the trace of the system; that is, summing over all the diagonal elements of the density matrix.

Suppose now that we do a partial sum only over all possible eigenstates $|n\rangle$ of the environment. We write

$$\begin{aligned}
\sum_n \langle n | \hat{\rho}_{12} | \psi' \rangle \langle \psi' | n \rangle &= \sum_{n,n',n''} \langle n | \hat{\rho}_{12} \gamma_{n'} | n' \rangle (\alpha e^{i\varphi_{n'}} |1\rangle + \beta e^{i\tilde{\varphi}_{n'}} |2\rangle) \\
&\quad \times (\alpha^* e^{-i\varphi_{n''}} \langle 1| + \beta^* e^{-i\tilde{\varphi}_{n''}} \langle 2|) \gamma_{n''} \langle n'' | n \rangle \\
&= \hat{\rho}_{12} \sum_n P(n) \left(\alpha e^{i\varphi_n} |1\rangle + \beta e^{i\tilde{\varphi}_n} |2\rangle \right) ((\alpha e^{i\varphi_n})^* \langle 1| + (\beta e^{i\tilde{\varphi}_n})^* \langle 2|), \\
&= \hat{\rho}_{12} \sum_n P(n) \left(\alpha_n |1\rangle + \beta |2\rangle \right) (\alpha_n^* \langle 1| + \beta_n^* \langle 2|).
\end{aligned} \tag{19.4.7}$$

Note that the sum in the last line has the same form as the mixed state (19.1.7), that is, a sum of pure states with $|\alpha_n|^2 + |\beta_n|^2 = 1$, weighted by the probability $P(n)$. Although there has been unitary time evolution, the coupling to the large, random environment makes the system essentially a statistical mixture. Tracing over the environment therefore reduces the problem from a sum over the vast number of possible environmental states to just a mixed density matrix for the two states of interest. If the phases φ_n and $\tilde{\varphi}_n$ are completely randomized, then the sum for the off-diagonal terms will average to zero, and the density matrix will have only diagonal elements.

Taking the trace over states $|1\rangle$ and $|2\rangle$ in (19.4.7) will give us the correlation function for these two states. We can therefore generally write for these two states the *reduced density matrix*

$$\hat{\rho}_r = \sum_n P(n)\left(\alpha_n|1\rangle + \beta|2\rangle\right)(\alpha_n^*\langle 1| + \beta_n^*\langle 2|)\,, \tag{19.4.8}$$

where the sum is over all the environmental eigenstates.

Lindbladian for the density matrix operator. We now write down the time evolution for a density matrix operator. We adopt the interaction representation, introduced in Section 11.4, with $|\psi_t\rangle = e^{-iH_0t/\hbar}|\psi(t)\rangle$. As in previous sections, we evolve the wave function forward in time a short amount, and then determine the rate of change to get the time derivative. We write

$$\begin{aligned}
d\hat{\rho} &= |\psi_t\rangle\langle\psi_t| - |\psi_0\rangle\langle\psi_0| \\
&= e^{-iH_0t/\hbar}|\psi(t)\rangle\langle\psi(t)|e^{iH_0t/\hbar} - \hat{\rho},
\end{aligned} \tag{19.4.9}$$

for some small time step dt. We then use the expansion (14.1.5), which gives us, to second order in $\hat{V}$,

$$\begin{aligned}
d\hat{\rho} &= e^{-iH_0t/\hbar}|\psi(t)\rangle\langle\psi(t)|e^{iH_0t/\hbar} - \hat{\rho} \\
&= e^{-iH_0t/\hbar}\left(1 + \frac{1}{i\hbar}\int_0^t dt'\,\hat{V}(t') + \frac{1}{(i\hbar)^2}\int_0^t dt'\int_0^{t'} dt''\,\hat{V}(t')\hat{V}(t'') + \cdots\right)\hat{\rho} \\
&\quad \times \left(1 - \frac{1}{i\hbar}\int_0^t dt'\,\hat{V}(t') + \frac{1}{(i\hbar)^2}\int_0^t dt'\int_0^{t'} dt''\,\hat{V}(t')\hat{V}(t'') + \cdots\right)e^{iH_0t/\hbar} - \hat{\rho}.
\end{aligned} \tag{19.4.10}$$

For the lowest-order term, we take the short-time approximation $e^{-iH_0t/\hbar} \simeq 1 - (i/\hbar)H_0$, which gives us

$$d\hat{\rho} = -\frac{i}{\hbar}[H_0, \hat{\rho}]t, \tag{19.4.11}$$

that is, for $t = dt$,

$$\frac{d\hat{\rho}}{dt} = -\frac{i}{\hbar}[H_0, \hat{\rho}], \tag{19.4.12}$$

which is the same as the Liouville equation (19.1.5) derived previously, in the absence of interactions.

The term which is first-order in $\hat{V}$ is

$$d\hat{\rho} = \frac{1}{i\hbar}e^{-iH_0t/\hbar}\int_0^t dt'\,[\hat{V}(t'), \hat{\rho}]e^{iH_0t/\hbar}. \tag{19.4.13}$$

Taking the same short-time approximation for each exponential phase factor, we find that the leading-order term, proportional to t, gives us

$$\frac{d\hat{\rho}}{dt} = -\frac{i}{\hbar}[\hat{V}, \hat{\rho}]. \tag{19.4.14}$$

Finally, the second-order term is

$$d\hat{\rho} = e^{-iH_0t/\hbar} \left(\frac{1}{(i\hbar)^2} \int_0^t dt' \int_0^{t'} dt'' \left(\hat{V}(t')\hat{V}(t'')\hat{\rho} + \hat{\rho}\,\hat{V}(t')\hat{V}(t'') \right) \right.$$

$$\left. - \frac{1}{(i\hbar)^2} \int_0^t dt' \int_0^t dt'' \, \hat{V}(t')\hat{\rho}\,\hat{V}(t'') \right) e^{iH_0t/\hbar}. \qquad (19.4.15)$$

This term will give a time evolution analogous to (19.3.16), used in Section 19.3.

Interaction with the environment. We now choose a specific interaction, for the two states of interest, in the Lindbladian (19.4.15). We write

$$\hat{V} = \sum_q A_q \left(a_1^\dagger a_2 a_q + a_2^\dagger a_1 a_q^\dagger \right). \qquad (19.4.16)$$

We can think of $a_q^\dagger$ as giving the emission of a photon (or phonon) which escapes into the environment.

We perform the partial trace over the environmental states $|n\rangle$, and insert a sum over a complete set of states, which gives us for the first of the three terms in (19.4.15)

$$\sum_{n,m} \frac{1}{(i\hbar)^2} \int_0^t dt' \int_0^{t'} dt'' \langle n|\hat{V}(t')\hat{V}(t'')|m\rangle\langle m|\hat{\rho}|n\rangle$$

$$\simeq \sum_n \frac{1}{(i\hbar)^2} \int_0^t dt' \int_0^{t'} dt'' \langle n|\hat{V}(t')\hat{V}(t'')|n\rangle\langle n|\hat{\rho}|n\rangle, \qquad (19.4.17)$$

where in the second line, as in Section 19.3, we have kept only second-order terms in which one term of $\hat{V}$ undoes the action of the other term, so that we are not left with higher-order correlation functions, which we assume are negligible.

Picking an explicit term of $\hat{V}$, and assuming no stimulated transitions, we then have

$$\sum_n \sum_q \frac{1}{(i\hbar)^2} \int_0^t dt' \int_0^{t'} dt'' \langle n|e^{iH_0t'/\hbar}(A_q a_2^\dagger a_1 a_q)e^{-iH_0t'/\hbar}$$

$$\times e^{iH_0t''/\hbar}(A_q^* a_1^\dagger a_2 a_q^\dagger)e^{-iH_0t''/\hbar}|n\rangle\langle n|\hat{\rho}|n\rangle$$

$$= \sum_q \frac{1}{(i\hbar)^2} \int_0^t dt' \int_0^{t'} dt'' \, e^{i\Delta E(t'-t'')/\hbar}|A_q|^2(a_2^\dagger a_1)(a_1^\dagger a_2) \sum_n \langle n|\hat{\rho}|n\rangle$$

$$= \left[\frac{2\pi t}{\hbar} \sum_q |A_q|^2\delta(\Delta E) \right] \frac{1}{2}(a_1^\dagger a_2)^\dagger(a_1^\dagger a_2)\hat{\rho}_r, \qquad (19.4.18)$$

where we have resolved the time integrals into an energy-conserving δ-function as in Section 19.3, and $\hat{\rho}_r$ is the reduced density matrix of (19.4.8).

We can then take $t = dt \to 0$, so that we finally have

$$\frac{d\hat{\rho}_r}{dt} = \left[\frac{2\pi}{\hbar} \sum_q |A_q|^2\delta(\Delta E) \right] \frac{1}{2}c_{12}^\dagger c_{12}\hat{\rho}_r, \qquad (19.4.19)$$

where we have defined $c_{12} = a_1^\dagger a_2$ for the transition between states 1 and 2. The term in the square brackets is just the same as what we would have gotten from using Fermi's golden rule for the transition rate from 2 to 1. In other words, we can view this process as summing over all possible photon emission processes with photon momentum q.

It is easy to show that the second term in (19.4.15) gives the same result but with the different ordering $\hat{\rho}_r c_{12}^\dagger c_{12}$. The last of the three terms is

$$- \sum_{n,n',n''} \frac{1}{(i\hbar)^2} \int_0^t dt' \int_0^t dt'' \, \langle n|\hat{V}(t')|n'\rangle \hat{\rho} \langle n''|\hat{V}(t'')|n\rangle. \qquad (19.4.20)$$

For the same photon emission process, we pick the terms from the interaction Hamiltonian that give

$$- \sum_{n,n',n''} \sum_q \frac{1}{(i\hbar)^2} \int_0^t dt' \int_0^t dt'' \, |A_q|^2 \langle n|e^{iH_0 t'/\hbar} a_1^\dagger a_2 a_q^\dagger e^{-iH_0 t'/\hbar}|n'\rangle$$
$$\times \hat{\rho} \, \langle n''|e^{iH_0 t''/\hbar} a_2^\dagger a_1 a_q e^{-iH_0 t''/\hbar}|n\rangle. \qquad (19.4.21)$$

In this case, the environmental state $|n\rangle$ must have one more photon than the state $|n'\rangle$ or $|n''\rangle$. Following the same logic as for the first two terms gives the result for this term,

$$\frac{d\hat{\rho}_r}{dt} = -\left[\frac{2\pi}{\hbar} \sum_q |A_q|^2 \delta(\Delta E) \right] c_{12} \hat{\rho}_r c_{12}^\dagger. \qquad (19.4.22)$$

The final total result, adding all the terms up to second order, is then

$$\frac{d\hat{\rho}_r}{dt} = -\frac{i}{\hbar}[H_0 + \hat{V}, \hat{\rho}] + \sum_m \frac{1}{\tau_m} \left(\frac{1}{2} c_m^\dagger c_m \hat{\rho}_r + \frac{1}{2} \hat{\rho}_r c_m^\dagger c_m - c_m \hat{\rho}_r c_m^\dagger \right), \qquad (19.4.23)$$

where c_m can refer to the transition between states 1 and 2 in either direction, and τ_m is the Fermi's golden rule transition time calculated for each process. This is another version of the master equation, for a reduced density matrix.

Effective time evolution equation. Result (19.4.23) can be further simplified by writing the Schrödinger equation in the general form

$$\frac{\partial}{\partial t}|\psi_t\rangle = \frac{1}{i\hbar}(\text{Re } H - i\text{Im } H)|\psi_t\rangle. \qquad (19.4.24)$$

The imaginary term will lead to nonunitary time evolution; namely, the amplitude of the wave function will decay. This makes sense if we assume that we have focused our attention on just part of the total wave function of the system, and that part loses amplitude while other parts gain.

The time evolution of the density matrix is then

$$\frac{\partial}{\partial t}|\psi_t\rangle\langle\psi_t| = \left(\frac{\partial}{\partial t}|\psi_t\rangle \right) \langle\psi_t| + |\psi_t\rangle \left(\frac{\partial}{\partial t}\langle\psi_t| \right)$$
$$= \frac{1}{i\hbar}(\text{Re } H - i\text{Im } H)\hat{\rho} - \frac{1}{i\hbar}\hat{\rho}(\text{Re } H + i\text{Im } H). \qquad (19.4.25)$$

We can therefore rewrite (19.4.23) as

$$\frac{d\hat{\tilde{\rho}}_r}{dt} = -\frac{i}{\hbar}(H_{\text{eff}} \, \hat{\tilde{\rho}}_r - \hat{\tilde{\rho}}_r H_{\text{eff}}^\dagger) + \sum_m \frac{1}{\tau_m} c_m \hat{\tilde{\rho}}_r \, c_m^\dagger, \qquad (19.4.26)$$

with

$$H_{\text{eff}} = H_0 + \hat{V} - \frac{i}{2}\sum_m \frac{1}{\tau_m} c_m^\dagger c_m. \qquad (19.4.27)$$

19.4.2 The Quantum Trajectories Recipe

The time evolution of the density matrix given by (19.4.25) allows us to define the process for computing quantum trajectories. Since we are only concerned about the density matrix of the two states of interest, we write the wave function $|\psi(t)\rangle = \alpha(t)|1\rangle + \beta(t)|2\rangle$. The quantum trajectories method is then the following:

- Evolve the state forward in time using the effective Hamiltonian of (19.4.24), using the Schrödinger equation for a very small time interval dt:

$$|\psi(t+dt)\rangle = (1 - iH_{\text{eff}}dt)|\psi(t)\rangle. \qquad (19.4.28)$$

- Compute the change of weight of this state dp, according to

$$\begin{aligned}\langle\psi(t+dt)|\psi(d+dt)\rangle &= \langle\psi(t)|(1 + iH_{\text{eff}}^\dagger dt)(1 - iH_{\text{eff}}dt)|\psi(t)\rangle \\ &\approx 1 - i\langle\psi(t)|(H_{\text{eff}} - H_{\text{eff}}^\dagger)|\psi(t)\rangle \\ &\equiv 1 - dp. \end{aligned} \qquad (19.4.29)$$

The change dp will be nonzero because the Hamiltonian has a part that corresponds to decay. From our definition of H_{eff}, we have

$$dp = \sum_m dp_m \equiv \sum_m \frac{dt}{2\tau_m}\langle\psi(t)|c_m^\dagger c_m|\psi(t)\rangle, \qquad (19.4.30)$$

where the processes m are all possible transitions including loss to the environment, with the associated rates $1/\tau_m$.

- Pick the next state randomly according to the following rule: With probability $1 - dp$, renormalize the state to have its original weight:

$$|\psi(t+dt)\rangle \rightarrow \frac{|\psi(t+dt)\rangle}{\sqrt{1-dp}}; \qquad (19.4.31)$$

and with probability dp, apply the action of one particular interaction operator c_m:

$$|\psi(t+dt)\rangle \rightarrow \frac{c_m|\psi(t)\rangle}{\sqrt{dp_m(\tau_m/dt)}}, \qquad (19.4.32)$$

where a specific choice of m is chosen out of all the possibilities with probability dp_m/dp.

To see that this recipe gives the same average evolution as (19.4.26), we write the reduced density matrix

$$\hat{\rho}_{12}(t) = |\psi(t)\rangle\langle\psi(t)|. \qquad (19.4.33)$$

The statistical average of this density matrix after a short time dt will then be given by the sum of each of the two possible outcomes, multiplied by the probability of each:

$$\overline{\rho(t+dt)} = (1-dp)\frac{|\psi(t+dt)\rangle}{\sqrt{1-dp}}\frac{\langle\psi(t+dt)|}{\sqrt{1-dp}}$$

$$+dp\sum_m\left(\frac{dp_m}{dp}\right)\frac{c_m|\psi(t)\rangle}{\sqrt{dp_m(\tau_m/dt)}}\frac{\langle\psi(t)|c_m^\dagger}{\sqrt{dp_m(\tau_m/dt)}}. \qquad (19.4.34)$$

Substituting (19.4.28) for the wave function after the time interval dt, and canceling the various factors in the numerators and denominators, we have

$$\overline{\rho_{12}(t+dt)} = \hat{\rho}_{12}(t) - idt(H_{\text{eff}}\hat{\rho}_{12}(t) - \hat{\rho}(t)H_{\text{eff}}^\dagger) + \frac{dt}{\tau_m}\sum_m c_m\hat{\rho}_{12}c_m^\dagger, \qquad (19.4.35)$$

which gives just the same evolution as Equation (19.4.26).

The action of the operator c_m on the state $|\psi(t)\rangle$ in the last step of this recipe is a type of collapse. For example, if the state is $|\psi\rangle = \alpha|1\rangle+\beta|2\rangle$, where $|1\rangle$ and $|2\rangle$ are single-particle fermion states, then the action of c_m is

$$c_m|\psi\rangle = a_1^\dagger a_2|\psi\rangle = \beta|1\rangle. \qquad (19.4.36)$$

This corresponds to a definite transition from a superposition to state 1.

As discussed in Section 6.4, the quantum trajectories method shows that having a system in which spontaneous collapses of the wave function occur randomly in time in a large system is far from an unheard-of hypothesis; it leads to tractable calculations that agree with real experimental results. Since all the detectors we have involve fermions (namely, electrons), fermion collapse into energy eigenstates would be sufficient for detection results.

Note that nonlocal collapse is automatically accounted for in the quantum trajectories method. If a two-particle state is $|\psi\rangle = \alpha|1\rangle|1\rangle + \beta|2\rangle|2\rangle$, where the two states in each product are for electrons in different detectors, then a *single* operator c_m acting on one of the electrons will force a collapse into either $|1\rangle|1\rangle$ or $|2\rangle|2\rangle$.

References

A. Daley, "Quantum trajectories and open many-body quantum systems," *Advances in Physics* **63**, 77 (2014).

D. W. Snoke, G.-Q. Liu, and S. M. Girvin, "The basis of the second law of thermodynamics in quantum field theory," *Annals of Physics* **327**, 1825 (2012).

D. W. Snoke, "Mathematical formalism for nonlocal spontaneous collapse in quantum field theory," *Foundations of Physics* **53**, 34 (2023).

B. Tamir and E. Cohen, "Introduction to weak measurements and weak values," *Quanta* **2**, 7 (2013).

20

Proposed Model for Spontaneous Collapse of Fermion States

One could simply posit that the quantum trajectories process described in Section 19.4, which is normally taken as a computational method, actually occurs in reality: with a probability proportional to the dephasing rate, project each system onto one of its eigenstates via a definite transition process. One can go further, however, to write down equations that give a physical mechanism for this. This chapter reviews recent work in Snoke 2021, Snoke 2022, and Snoke and Maienshein 2023 which flesh out such an approach. As discussed in Section 6.1, this requires explicitly nonunitary terms.

When relativity is taken into account, we must specify the rest frame in which to define the eigenstates. Section 20.5 discusses how we might go about choosing this.

20.1 Hypothesis of Nonunitary Behavior of Quantum Fields

One can create a model for spontaneous collapse using the Bloch sphere model summarized in Section 14.3, which has three equations that describe the motion a vector pointing to a location on the sphere. This model treats only a nonunitary term in fermionic states, not bosonic states. In this model, the top of the Bloch sphere corresponds to the fermion state $N = 1$, that is, "occupied by a fermion," and the bottom of the sphere corresponds to $N = 0$, that is, "unoccupied." As discussed in Section 13.1, these two states need not be associated with billiard-ball-like particles, but can be seen as two resonances of the underlying field.

The Bloch equations for a two-state system in the rotating frame are

$$\frac{\partial U_1}{\partial t} = \tilde{\omega} U_2$$

$$\frac{\partial U_2}{\partial t} = -\tilde{\omega} U_1 - \omega_R U_3$$

$$\frac{\partial U_3}{\partial t} = \omega_R U_2, \tag{20.1.1}$$

where we have assumed that there are no external irreversible processes described by T_1 and T_2; instead, an irreversible process will be represented by an explicit sequence of random phase shifts. There is no external coherent driving term, and therefore the detuning $\tilde{\omega}$ would normally be zero, but the effect of random phase shifts due to the environment is accounted for by a fluctuating value of $\tilde{\omega}$ centered around $\tilde{\omega} = 0$.

Although there is no driving field, an effective Rabi frequency ω_R is introduced here, which gives the nonlinearity hypothesized in this model. The nonlinearity is introduced by giving the Rabi frequency a dependence on the values of U_1 and U_2 which depends on the rate of change of rotation speed in the $U_1 - U_2$ plane:

$$\omega_R = \alpha |U_\perp| f(\ddot{\phi}), \tag{20.1.2}$$

where α is a constant parameter which sets the timescale, $|U_\perp| = \sqrt{U_1^2 + U_2^2}$, and $f(\ddot{\phi})$ is some function of the acceleration of the angle ϕ in the U_1-U_2 plane. The factor $|U_\perp|$ makes the points $U_3 = \pm 1$ act as "attractors," because ω_R vanishes there. The dependence on $\ddot{\phi}$ makes the system responsive only to time-dependent fluctuations, which are characteristic of an inhomogeneous system. This implies that real "lumpiness" of the system is required, for example a collection of localized atoms, which have a natural size, as discussed in Sections 3.1 and 9.5.

The value of ϕ is assumed to be subject to random accelerations. For the numerical simulations of Snoke 2021, the time evolution was broken into a series of time steps dt, and within each time step, the value of f was picked with a probability given by the Lorentz distribution,

$$P(f) = \frac{1}{\pi} \frac{\gamma}{f^2 + \gamma^2}, \tag{20.1.3}$$

where γ characterizes the width of the peak. Figure 20.1 shows typical results of a numerical model for random motion of the Bloch vector in the limit of $\gamma \ll 1$; nonlinearity and random noise cause a superposition of two waves to pop into one or the other of the two states. In these plots, the two distinct states of the fermion are assigned the numbers 1 and -1, and numbers between these two limits correspond to superpositions with more or less weight of the two individual states. As discussed in Section 2.4, and presented

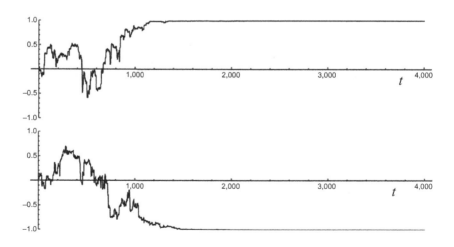

Figure 20.1   Examples of random walks that end up in one of two states (corresponding to 1 or -1) due to noise acting on nonlinearity. The horizontal axis gives the elapsed time t. From Snoke 2021.

mathematically in Section 13.1, the two allowed states of fermion fields, which are typically thought of as "a particle existing" and "a particle not existing," are just two different resonances of the field. This model gives a random probability of popping into one state or the other in agreement with the Born rule when there is strong decoherence (large fluctuation of $\ddot{\phi}$), while it leaves the system in an unchanging superposition when there is no environmental decoherence.

Verification that this system reproduces Born's rule was done by running the numerical solution many times for the same initial conditions but different random values of f, for different starting points given by $U_3 = \cos\theta_0$, $U_2 = \sin\theta_0$, and $U_1 = 0$. By Born's rule, the probability of a measurement of ± 1 should be equal to

$$P(\pm 1) = \frac{1 \pm U_3(0)}{2}. \tag{20.1.4}$$

The statistics of the $t \to \infty$ value of U_3 of many numerical solutions for each initial condition were found to agree with this rule for a sufficiently large number of random phase accelerations.

The Lorentzian probability of phase accelerations given in (20.1.3) can be justified in various ways; the simplest is simply to take note that an exponentially decaying correlation function of fluctuations in time corresponds to a Lorentzian power spectrum of the fluctuations. But the result found here is not sensitive to the exact form of $P(f)$; other peaked functions such as a Gaussian distribution or stretched Lorentzian were also used in the numerical model, with the same results. It can be proven mathematically (Snoke and Maienshein 2023) that any random walk with steps that are small compared to the total range of U_3 will give the same result of the Born rule.

20.2 Action on Superposition States

The full many-particle state of the system can be written as $|\psi\rangle$ in the form (18.2.8), as a superposition of Fock states. The outcomes of the random walks discussed in Section 20.1 amount to applying the following rule: Within a time dt, with probability dt/τ where τ is the decoherence time for state k, perform a jump on fermionic eigenstate n according to:

- With probability $\langle \hat{N}_n \rangle = \langle \psi | a_n^\dagger a_n | \psi \rangle$, apply the operator $\hat{N}_n$ to the many-particle state $|\psi\rangle$, to force $N_n = 1$, and normalize the overall amplitude of the resulting state so that $\langle \psi | \psi \rangle = 1$.
- With probability $1 - \langle \hat{N}_n \rangle$, apply the projection operator $(1 - \hat{N}_n)$ to force $N_n = 0$, and normalize the resulting many-particle state.

Consider the case of an electron with two available states, $|1\rangle$ and $|2\rangle$, as in Section 19.4.2. If the electron is initially in state $|2\rangle$, we can write this as $a_2^\dagger |0\rangle$, where $|0\rangle$ is the vacuum state. After an interaction with the environment that allows for photon emission, the state becomes

$$|\psi\rangle = \left(\alpha + \sum_m \beta_m a_m^\dagger a_1^\dagger a_2 \right) a_2^\dagger |0\rangle = \alpha a_2^\dagger |0\rangle + \sum_m \beta_m a_m^\dagger a_1^\dagger |0\rangle, \tag{20.2.1}$$

where β_m is a complex amplitude factor for each process of emission of a photon into some state m, and α is the complex amplitude factor for *not* emitting a photon. If the ground state jumps to having $N_1 = 1$, we apply $\hat{N}_1$ to the whole state, which gives

$$|\psi'\rangle = \sum_m \beta_m a_m^\dagger a_1^\dagger |0\rangle, \tag{20.2.2}$$

which must then be normalized to give $\langle\psi'|\psi'\rangle = 1$. The state now definitely has a fermion in the state $|1\rangle$, and a superposition of all possible emitted photon states m. Energy conservation is not violated, because the photons have the energy that was lost by the electron.

In the future evolution of the system, these photons may encounter other fermionic states, leading many fermion eigenstates (e.g., the electron states of many single atoms) to be in superposition states. At that point, strong decoherence of any one of them can lead to a jump that collapses the whole state to a photon hitting just one atom.

In the quantum trajectories approach of Section 19.4, a jump of an electron corresponded to an average over all photon emission processes, after which the photons were lost in the unchanging, equilibrium environment. Here, we could in principle keep track of the exiting photon superposition, but the effect on the original electron is the same.

As a second example, consider the case of a radioactive nucleus decaying, leading to a cloud chamber track like one of those shown in Figure 3.1. The radiative process that corresponds to particle emission initially creates a symmetric, spherical wave emanating from the nucleus. This wave propagates at first with no decoherence. When it enters the cloud chamber, it encounters a large number of fermions (electrons) in the atoms of the cloud chamber, and puts the system into a superposition of all of these atoms being excited. These "hot" electrons have strong decoherence due to their interaction with their environment due to collisions. Therefore, the mechanism of Section 20.1 gives an instability for the electrons to pop into a nonsuperposed state. When one of the excited electrons commits to an excited state, it projects all the other excited electrons in the superposition into their nonexcited states, and the energetic particle from the nucleus is projected into its state of having excited just that one electron, via the number-operator recipe given at the beginning of this section. This is another example of a nonlocal correlation, because all the electrons in the superposition are projected out by the jump of any one of them.

20.3 Implementation in Quantum Field Theory

In Section 20.1, we considered a simple Bloch sphere model for the superposition between an occupied and unoccupied fermion state. Here we show how to introduce this action explicitly into many-body field theory.

In general, the state in which a fermion could be either present or absent in a single-particle state can always be written as

$$|\psi_{\text{tot}}\rangle = \alpha|\psi_0\rangle|0\rangle + \beta|\psi_1\rangle|1\rangle, \tag{20.3.1}$$

where $|0\rangle$ and $|1\rangle$ are the unoccupied and occupied fermion states of the localized state n of interest, α and β are complex c-numbers normalized by $\sqrt{|\alpha|^2 + |\beta|^2} = 1$, and $|\psi_0\rangle$ and $|\psi_1\rangle$ represent the many-body state of the rest of the system, which are orthogonal and normalized. The states $|\psi_0\rangle$ and $|\psi_1\rangle$ will always be orthogonal because they have different total numbers of fermions, and total fermion number is conserved. For example, suppose we have a many-body state which is a sum of three Fock states,

$$\frac{1}{\sqrt{3}}(|0, 0, 1\rangle + |0, 1, 0\rangle + |1, 0, 0\rangle), \tag{20.3.2}$$

in which a single fermion can be in one of three single-particle states. This can be written as

$$\frac{1}{\sqrt{3}}|0, 0\rangle|1\rangle + \sqrt{\frac{2}{3}}\left(\frac{1}{\sqrt{2}}(|0, 1\rangle + |1, 0\rangle)\right)|0\rangle. \tag{20.3.3}$$

In this case, $\alpha = \sqrt{2/3}$ and $\beta = \sqrt{1/3}$.

The vertical component of a vector undergoing Rabi oscillations on a Bloch sphere representing superpositions of the occupied and unoccupied state can be written as

$$U_3 = |\beta|^2 - |\alpha|^2 = \cos \omega_R t. \tag{20.3.4}$$

The constraint of normalization means that $|\alpha|^2 + |\beta|^2 = 1$, which implies

$$2|\beta|^2 - 1 = \cos \omega_R t$$
$$|\beta|^2 = \frac{1}{2}(\cos \omega_R t + 1) = \cos^2 \frac{\omega_R t}{2}. \tag{20.3.5}$$

The time derivative is then

$$\frac{\partial}{\partial t}|\beta|^2 = 2|\beta|\frac{\partial|\beta|}{\partial t} = -\omega_R \cos \frac{\omega_R t}{2} \sin \frac{\omega_R t}{2} = -\omega_R|\beta|\sqrt{1 - |\beta|^2}$$
$$\frac{\partial|\beta|}{\partial t} = -\omega_R\sqrt{1 - |\beta|^2} = -\omega_R|\alpha|. \tag{20.3.6}$$

The constraint of normalization implies

$$\frac{\partial|\alpha|^2}{\partial t} + \frac{\partial|\beta|^2}{\partial t} = 0,$$
$$|\alpha|\frac{\partial|\alpha|}{\partial t} + |\beta|\frac{\partial|\beta|}{\partial t} = 0, \tag{20.3.7}$$

and therefore

$$\frac{\partial|\alpha|}{\partial t} = -\frac{|\beta|}{|\alpha|}\frac{\partial|\beta|}{\partial t}$$
$$= \omega_R\frac{|\beta|}{|\alpha|}\sqrt{1 - |\beta|^2}$$
$$= \omega_R\frac{|\beta|}{\sqrt{1 - |\beta|^2}}\sqrt{1 - |\beta|^2} = \omega_R|\beta|. \tag{20.3.8}$$

The magnitude of β is extracted from (20.3.1) by

$$|\beta|^2 = \langle\psi_{\text{tot}}|\hat{N}_n|\psi_{\text{tot}}\rangle \equiv \langle\hat{N}_n\rangle. \tag{20.3.9}$$

Therefore, Rabi oscillations between the two states of (20.3.1) will be given by

$$
\frac{\partial}{\partial t}|\psi_{\text{tot}}\rangle = \frac{\partial \alpha}{\partial t}|\psi_0\rangle|0\rangle + \frac{\partial \beta}{\partial t}|\psi_1\rangle|1\rangle
$$

$$
= \omega_R \frac{|\beta|}{|\alpha|} a_n a_n^\dagger |\psi_{\text{tot}}\rangle - \omega_R \frac{|\alpha|}{|\beta|} a_n^\dagger a_n |\psi_{\text{tot}}\rangle. \tag{20.3.10}
$$

In other words, Rabi oscillations between these two states can be induced by adding a term to the Hamiltonian given by

$$
H_R = i\hbar\omega_R \left(\frac{\sqrt{\langle \hat{N}_n \rangle}}{\sqrt{1 - \langle \hat{N}_n \rangle}}(1 - \hat{N}_n) - \frac{\sqrt{1 - \langle \hat{N}_n \rangle}}{\sqrt{\langle \hat{N}_n \rangle}}\hat{N}_n \right). \tag{20.3.11}
$$

As shown in Section 20.1, the states with pure $\langle \hat{N}_n \rangle = 0$ and $\langle \hat{N}_n \rangle = 1$ can be made into attractors by multiplying the rate of Rabi precession by the value $\langle \hat{N}_n \rangle - \langle \hat{N}_n \rangle^{\frac{1}{2}}$, which is proportional to the magnitude of the horizontal component $U_\perp = \sqrt{1 - U_3^2}$ in the Bloch sphere model, since $U_3 = 2\langle \hat{N}_n \rangle - 1$. This term is equal to zero for both $\langle \hat{N}_n \rangle = 0$ and $\langle \hat{N}_n \rangle = 1$, with a maximum at $\langle \hat{N}_n \rangle = \frac{1}{2}$. Multiplying (20.3.11) by this, and summing over all states, we have for the full nonunitary term

$$
H_{\text{nonunitary}} = \sum_n i\hbar\omega_{R,n} \left(\langle \hat{N}_n \rangle(1 - \hat{N}_n) - (1 - \langle \hat{N}_n \rangle)\hat{N}_n \right)
$$

$$
= \sum_n i\hbar\omega_{R,n} \left(\langle \hat{N}_n \rangle - \hat{N}_n \right), \tag{20.3.12}
$$

where $\omega_{R,n}$ depends on the fluctuations of the frequency ω_n, which in Section 20.1 we connected to a function $f(\ddot{\phi}_n)$, which had random fluctuations due to time-dependent inhomogeneities. Thus we have, in the end, an appealingly simple term to add to the Schrödinger equation. Although it is nonunitary, its time average is unitary, if time average of the fluctuations is zero.

20.4 Comparison to Weak Measurement Theory

The action of the nonunitary operator (20.3.12) resembles an early proposal by Gisin (1984). Later, Gisin and others argued that this form of nonlinear term in the Schrödinger equation, and other, similar nonlinear terms intrinsically imply the experimental prediction of superluminal communication (Gisin and Rigo 1995). Their argument was based on analysis of the density matrix; in terms of the Bloch vector for a two-state system discussed in Section 20.1, it was argued that a Bloch vector with nonunit length would increase its length as it became a pure state, and that this change could be detected at a spacelike distance by observation of other states entangled with this state. However, in the process proposed in Sections 20.1–20.3, there is always only a rotation of pure states on the Bloch sphere (albeit these states may be very complicated, involving the whole environment), so that the Bloch vector for the full wave function (20.3.1) always has exactly unit length.

Therefore, this operator always gives the same projections as standard measurements in quantum mechanics. This can be seen by comparing this model to the predictions of *weak measurement theory* and *continuous measurement theory* (for reviews, see Oreshkov, 2005; Jacobs and Steck, 2006; Clerk, 2010)

Consider the following scenario of weak measurement, following the model of Tamir and Cohen (2013). At time $t = 0$, the initial state is

$$|\psi\rangle = |\phi\rangle|X = 0\rangle, \tag{20.4.1}$$

where $|X\rangle$ is the state of an external detector with a center-of-mass value at X (e.g., the position of a needle in a meter), and $|\phi\rangle = \alpha|-1\rangle + \beta|1\rangle$ is the internal state that is in a superposition. The external state in general has position uncertainty in a Gaussian,

$$|X\rangle = \int dx \, e^{-(x-X)^2/2\sigma^2}|x\rangle, \tag{20.4.2}$$

where $|x_0\rangle$ is a state of the external detector at exactly x_0.

At $t = 0$, a weak interaction of the system with the detector is turned on briefly, of the form

$$\hat{V} = gS_z P, \tag{20.4.3}$$

where $S_z = \frac{\hbar}{2}\sigma_z$ is the standard spin operator (σ_z is defined in (13.2.5)), P is the momentum operator of the external center of mass X, and g is some small number. After a short time dt, the state of the system is

$$|\psi'\rangle = |\psi\rangle + gdt \int dx|x\rangle \left(-\alpha\frac{\partial}{\partial x}e^{-x^2/2\sigma^2}|-1\rangle + \beta\frac{\partial}{\partial x}e^{-x^2/2\sigma^2}|1\rangle\right)$$

$$= |\psi\rangle + gdt \int dx|x\rangle \left(\alpha\frac{x}{\sigma^2}e^{-x^2/2\sigma^2}|-1\rangle - \beta\frac{x}{\sigma^2}e^{-x^2/2\sigma^2}|1\rangle\right)$$

$$= \int dx|x\rangle e^{-x^2/2\sigma^2} \left(\alpha\left(1 + \frac{xgdt}{\sigma^2}\right)|-1\rangle + \beta\left(1 - \frac{xgdt}{\sigma^2}\right)|1\rangle\right)$$

$$\simeq \int dx|x\rangle \left(\alpha e^{-(x-gdt)^2/2\sigma^2}|-1\rangle + \beta \, e^{-(x+gdt)^2/2\sigma^2}|1\rangle\right). \tag{20.4.4}$$

The last line is commonly used in discussing this scenario, but we will stick with the third line.

We now do a strong measurement of the external detector to collapse it to a definite value $x = x_0$. Defining $df = xgdt/\sigma^2 \ll 1$, the state of the whole system is then

$$|\psi''\rangle = \frac{1}{\sqrt{|\alpha|^2(1+df)^2 + |\beta|^2(1-df)^2}} (\alpha(1+df)|-1\rangle + \beta(1-df)|1\rangle) \, |x_0\rangle$$

$$\simeq \frac{1}{\sqrt{1 + 2|\alpha|^2 df - 2|\beta|^2 df}} (\alpha(1+df)|-1\rangle + \beta(1-df)|1\rangle) \, |x_0\rangle$$

$$\simeq \left(1 - |\alpha|^2 df + |\beta|^2 df\right) (\alpha(1+df)|-1\rangle + \beta(1-df)|1\rangle) \, |x_0\rangle. \tag{20.4.5}$$

The change in time is then

$$d|\psi\rangle = |\psi''\rangle - |\psi\rangle = gdtx_0\left(\alpha(1 - |\alpha|^2 + |\beta|^2)|-1\rangle - \beta(1 + |\alpha|^2 - |\beta|^2)|1\rangle\right)|x_0\rangle$$
$$= gdtx_0\left((|\beta|^2 - |\alpha|^2)(\alpha|-1\rangle + \beta|1\rangle) - (-\alpha|-1\rangle + \beta|1\rangle)\right)|x_0\rangle$$
$$= gdtx_0(\langle S_z\rangle - S_z)|\phi\rangle|x_0\rangle. \tag{20.4.6}$$

We have thus obtained a nonunitary operator exactly of the same form as (20.3.12), because the S_z operator acts just like the $\hat{N}_n$ operator on two states; the nonunitarity arises in this case from the fact that measurement is an intrinsically nonunitary process.

If the variance σ is large, then x_0 is equally probable to be positive or negative. Therefore, it we apply the same process of weak measurement multiple times, each time allowing the position of the needle in the detector to gain uncertainty due to normal wave spreading, and then doing a measurement of the classical center of mass of the needle, we get a random walk that is exactly equivalent to the results discussed in Section 20.1. This is another type of "quantum trajectory."

Since we know that the experiment described here is physically possible, we can view this theory as a template for a spontaneous collapse model, taking the "measurements" as representing some intrinsic nonunitary effect in the physical world that does not require human observers. In the latter case, we can simply adopt the perspective that the operator (20.3.12) is fundamental. In that case, all "strong" measurements are simply the $t \rightarrow$ limit of the random walk prescribed by repeated application of this operator. (As discussed in Section 20.1, we know that these random walks give the Born rule for the average of the outcomes.) Since the prescription of weak measurement theory is based entirely on standard quantum mechanical measurement theory, which does not allow superluminal communication (though it does allow nonlocal correlations), we know that the action of the operator (20.3.12), applied with random fluctuations of direction, also *cannot* give any predictions that allow superluminal communication. For more discussion of the possibility of superluminal communication, see Appendix A of Snoke and Maienshein 2023.

20.5 Relativistic Considerations

If we considered just one rest frame of the system, then the nonunitary operator (20.3.12) solves the measurement problem: it gives us nonlocal, spontaneous collapse in agreement with the Born rule. However, when we take into account special relativity and look at the system in different moving frames, we are confronted with questions of *narrativity*, that is, of how to have a consistent story of "what really happened" that different observers can agree on.

As discussed in Section 6.6, when two events are correlated nonlocally, it means that either one of them could be viewed as occurring first, in some reference frame. This doesn't create a problem for the final outcome, because the joint probability of both events is the same in every rest frame. To see this, suppose that we send out a light wave from a single source such that the wave intensity on one detector is 75% and is 25% on a second, space-like separated detector. In one rest frame, we view the first detector as triggering with probability of 75% to detect a particle, thereby collapsing the other detector to not detect a particle. In a different rest frame, the second detector triggers first, to *not* detect a particle (i.e., definitely have a fermion in the number state 0), with 75% probability, thereby collapsing the other detector to definitely detect a particle. The outcome is the same in both cases, with the same probability.

This scenario is problematic, however, if we try to construct a single, consistent story that applies in all frames of reference. If we take the approach that the cause of the collapse comes from local fluctuations, and the collapse then acts nonlocally at one point in time on the entire wave function of the system, we seem to have a contradiction about the "fact of the matter." In a rest frame in which the first event was the detector registering a count, we would say that local fluctuations there caused the collapse, which then enforced that the second detector would not register a count. However, in the rest frame in which the other detector encounters the wave first, we would say that local fluctuations *there* led to a collapse into the $N = 0$ state, which then caused the other detector to have the $N = 1$ state. The problem arises because we envision collapse to happen on a horizontal slice in the world diagram in the frame of reference under consideration. Since the time axis is defined differently for different reference frames, the horizontal time slices will be different in different frames of reference.

Natural rest frames. One way to solve this problem is to define a "natural" rest frame, and have all nonlocal actions occur on equal time slices in this rest frame. An obvious choice is to adopt the rest frame of the cosmic microwave background, which is the rest frame of the center of mass of the universe (see, e.g., Consoli, 2018; Aurich, 2021).

The time slices of this rest frame will be tilted in other frames, which means that events in the natural rest frame could affect events earlier in time in some other frames of reference. This backward-in-time collapse in some rest frames will not cause grandfather paradoxes, however, because it only affects spacelike-separated points. Therefore, none of those points can causally connect forward in time back to the source of the collapse.

Figure 20.2 illustrates how we might think about a detector-caused collapse in this case. In this space–time diagram, we see the world lines of the detectors and wave packets of the experiment of Figure 16.1, in which two entangled wave packets propagate in opposite directions. For convenience, we take the rest frame of the detectors to be the same as the natural rest frame (Frame 1). In this frame, the equal-time slices are horizontal, as indicated by the horizontal dashed line. If we adopt a moving frame of reference (Frame 2), then the equal-time slices are parallel to the tilted dashed line in Figure 20.2, and the detector on the left encounters the wave packet first. In Frame 2, the detector on the right encounters its wave packet first. However, we can say that the collapse was caused by local effects at the detector on the left in the "natural" rest frame. In this case there is a "fact of the matter" that one of the detectors randomly underwent a local collapse and caused the other to act consistently with it, even if that controls the action of the other detector that was earlier in time. This allows us to have a single, consistent narrative that applies across all frames of reference.

There would still remain the question, however, of what happens if both detection events occurred at exactly the same time in the natural rest frame. This is where the picture of

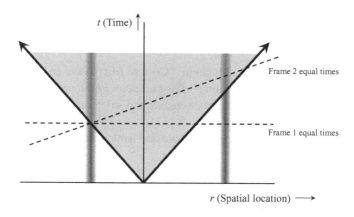

Figure 20.2

World lines of the experiment of Figure 16.1. The vertical, fuzzy gray bars represent the world lines of the detectors in the natural rest frame; the solid black arrows give the world lines of the outgoing wave packets, and the gray area represents the causally connected region of space time in which these two packets can be entangled. In this rest frame (Frame 1), a wave packet hits the detector on the left before the other wave packet hits the detector on the right. In a different frame of reference (Frame 2), this event occurs after the detector on the right encounters the wave packet, since the equal-time slices are parallel to the tilted dashed line.

random walk introduced in Section 20.1 helps. Instead of viewing a collapse event as a single, all-or-nothing process at one point in time, we can instead view each detector as giving a sequence in time of collapse "kicks" which push the state of the system closer to or further from one outcome or the other. This is precisely the scenario presented in Section 20.1, that is, a random walk between collapse outcomes due to a large series of small kicks from environmental fluctuations.

Nonlocal collapse and the transactional interpretation. The scenario considered in this section of allowing spacelike action has some similarities to the transactional interpretation of Cramer and Kastner, discussed in Section 16.4. Those authors argue that the equations of physics allow advanced waves, that is, waves traveling backward in time, which would be restricted to the reverse light cone pointing downward, unlike the mechanism here which places the effects of collapse in spacelike-separated regions. These authors have also argued that the existing equations of quantum field theory have intrinsic nonunitary aspects which can give instability and collapse; as discussed in Section 16.4, although it is often introduced without adequate explanation in textbooks, this imaginary term does not imply that the full, normal quantum field theory is nonunitary. As discussed in Section 6.1, in any spontaneous collapse model, an explicit nonunitary term must be added to the standard equations of quantum mechanics.

References

R. Aurich and D. Reinhardt, "Determining our peculiar velocity from the aberration in the cosmic microwave background," *Monthly Notices of the Royal Astronomical Society* **506**, 3259 (2021).

A. Clerk, M. Devoret, S. Girvin, F. Marquardt, and R. Schoelkopf, "Introduction to quantum noise, measurement, and amplification," *Reviews of Modern Physics* **82**, 1155 (2010).

M. Consoli and A. Pluchino. "Cosmic Microwave background and the issue of a fundamental preferred frame," *The European Physical Journal Plus* **133**, 295 (2018).

N. Gisin, "Quantum measurements and stochastic processes," *Physical Review Letters* **52**, 1657 (1984).

N. Gisin and M. Rigo, "Relevant and irrelevant nonlinear Schrödinger equations," *Journal of Physics A* **28**, 7375 (1995).

K. Jacobs and D. A. Steck, "A straightforward introduction to continuous quantum measurement," *Contemporary Physics* **47**, 279 (2006).

O. Oreshkov and T. A. Bru, "Weak measurements are universal," *Physical Review Letters* **95**, 110409 (2005).

D. W. Snoke, "A model of spontaneous collapse with energy conservation," *Foundations of Physics* **51**, 100 (2021).

D. W. Snoke and D. Maienshein, "Experimental predictions for norm-conserving spontaneous collapse," *Entropy* **25**, 1489 (2023).

B. Tamir and E. Cohen, "Introduction to weak measurements and weak values," *Quanta* **2**, 7 (2013).

21

Spontaneous Coherence: Lasers, Superfluids, and Superconductors

In the previous chapters, we have discussed decoherence, or dephasing, which is the process by which phase information is lost when there are many degrees of freedom in a system. However, there is a special case when the opposite occurs: when many particles spontaneously acquire a collective coherence, which can also be called *enphasing*. As discussed in Section 8.5, this occurs in lasers, superconductors, and superfluids, which have important technological applications.

21.1 Spontaneous Coherence

The rate of onset of phase coherence in a homogeneous condensate can be found by a simple extension of the second-order rate equations for phase already presented in Section 19.3. As in that section, we write, for correlations of any state with the ground state,

$$\frac{d}{dt}\langle\hat{\rho}_{k,0}\rangle^{(0)} = \frac{i}{\hbar}\langle\psi_i|[H_0, \hat{\rho}_{k,0}]|\psi_i\rangle \tag{21.1.1}$$

$$\frac{d}{dt}\langle\hat{\rho}_{k,0}\rangle^{(1)} = \frac{i}{\hbar}\langle\psi_i|[\hat{V}, \hat{\rho}_{k,0}]|\psi_i\rangle, \tag{21.1.2}$$

and

$$\frac{d}{dt}\langle\hat{\rho}_{k,0}\rangle^{(2)} = \frac{2\pi}{\hbar}\langle\psi_i|\left(\hat{V}\hat{\rho}_{k,0}\hat{V} - \frac{1}{2}\hat{V}\hat{V}\hat{\rho}_{k,0} - \frac{1}{2}\hat{\rho}_{k,0}\hat{V}\hat{V}\right)|\psi_i\rangle\delta(\tilde{E}), \tag{21.1.3}$$

for the zero-, first-, and second-order contributions to the evolution, respectively.

Instead of assuming that there is a coupling to an external system, we instead use a interaction for momentum-conserving scattering between two particles, of the form used in Section 18.2.2,

$$\hat{V} = \sum_{k_1,k_2,k_3} \frac{U}{2V}a^\dagger_{k_4}a^\dagger_{k_3}a_{k_2}a_{k_1}, \tag{21.1.4}$$

where $\vec{k}_1 + \vec{k}_2 = \vec{k}_2 + \vec{k}_4$. Using this, we need to compute all of the operator terms in the evolution equations. We find for the zero-order term,

$$\frac{d}{dt}\langle\hat{\rho}_{k,0}\rangle^{(0)} = \frac{i}{\hbar}\langle\psi_i|\sum_{k_1}\hbar\omega_{k_1}(\pm a^\dagger_k a_0\delta_{k_1,0} + a^\dagger_k a_0\delta_{k,k_1})|\psi_i\rangle$$

$$= i(\omega_k - \omega_0)\langle a^\dagger_k a_0\rangle. \tag{21.1.5}$$

This term gives us a circular precession in the complex plane for a nonzero value of $\hat{\rho}_{k,0}$.

For the first-order term, we resolve the commutator as

$$[a_{k_4}^\dagger a_{k_3}^\dagger a_{k_2} a_{k_1}, a_k^\dagger a_0] = \mp a_k^\dagger a_{k_3}^\dagger a_{k_2} a_{k_1} \delta_{k_4,0} - a_k^\dagger a_{k_4}^\dagger a_{k_2} a_{k_1} \delta_{k_3,0}$$
$$\pm a_{k_4}^\dagger a_{k_3}^\dagger a_{k_1} a_0 \delta_{k,k_2} + a_{k_4}^\dagger a_{k_3}^\dagger a_{k_2} a_0 \delta_{k,k_1}. \qquad (21.1.6)$$

In keeping with our hierarchical approach, we assume that fourth-order correlation terms are negligible, and keep only second-order terms. Furthermore, we assume that only correlation functions with a_0 are relevant. Substitution into (21.1.2) then gives us

$$\frac{d}{dt}\langle \hat{\rho}_{k,0} \rangle^{(1)} = 0, \qquad (21.1.7)$$

when U is a constant. More generally, this term gives a frequency correction to the rotation of the phase in the complex plane, proportional to the difference between the mean-field shifts of the two states involved in the correlation function.

We now move on to the second-order term. We use (21.1.6) to resolve the operator term in (21.1.3),

$$\hat{V}\rho_{k,0}\hat{V} - \frac{1}{2}\hat{V}\hat{V}\hat{\rho}_{k,0} - \frac{1}{2}\hat{\rho}_{k,0}\hat{V}\hat{V} = -\frac{1}{2}\hat{V}[\hat{V},\hat{\rho}_{k,0}] + \frac{1}{2}[\hat{V},\hat{\rho}_{k,0}]\hat{V}, \qquad (21.1.8)$$

which on substitution gives us for the first term on the right side,

$$-\frac{1}{8}\sum_{k_1',k_2',k_3'}\sum_{k_1,k_2,k_3} a_{k_4'}^\dagger a_{k_3'}^\dagger a_{k_2'} a_{k_1'} \left(\mp a_k^\dagger a_{k_3}^\dagger a_{k_2} a_{k_1} \delta_{k_4,0} - a_k^\dagger a_{k_4}^\dagger a_{k_2} a_{k_1} \delta_{k_3,0} \right.$$
$$\left. \pm a_{k_4}^\dagger a_{k_3}^\dagger a_{k_1} a_0 \delta_{k,k_2} + a_{k_4}^\dagger a_{k_3}^\dagger a_{k_2} a_0 \delta_{k,k_1} \right), \qquad (21.1.9)$$

and a similar result for the second term.

As with the lower-order terms, we pick only terms in the second $\hat{V}$ operator that reset the particles changed by the first operator, because otherwise we would have higher-order correlation functions. The calculation then gives[1]

$$\frac{d}{dt}\langle \hat{\rho}_{k,0} \rangle^{(2)} = \langle a_k^\dagger a_0 \rangle \frac{2\pi}{\hbar}\frac{1}{4V^2}\sum_{k_1,k_2}(U_D \pm U_E)^2$$
$$\times \left[\pm N_{k_1}N_{k_2}(1 \pm N_{k_1+k_2}) - N_{k_1+k_2}(1 \pm N_{k_1})(1 \pm N_{k_2})\delta(\tilde{E}) \right]$$
$$+ \langle a_k^\dagger a_0 \rangle \frac{2\pi}{\hbar}\frac{1}{4V^2}\sum_{k_1,k_3}(U_D \pm U_E)^2$$
$$\times \left[\pm N_{k_3}N_{k+k_1-k_3}(1 \pm N_{k_1}) - N_{k_1}(1 \pm N_{k_3})(1 \pm N_{k+k_1-k_3})\delta(\tilde{E}) \right], \qquad (21.1.10)$$

where U_D and U_E are the direct and exchange k-dependent interaction terms introduced in Section 18.2.2.

The terms in this equation are proportional to the same initial and final-states factors used in the quantum Boltzmann equation, for in-flow and out-flow in the limit when the

---

[1] This result implies there is a missing factor of $1/2$ in the related result in Snoke 2012.

states of interest have $N \simeq (1 + N)$. The first term in the first sum gives the in-flow into the ground state, while the second term gives the out-flow from the ground state, and the second sum has the same two terms for the in-flow and out-flow from state k.

This is crucial result for thinking about decoherence. For fermions, both terms are negative, and therefore any coherence in a fermion system is always lost over time. For bosons, at low density the second term dominates over the first term, because it is proportional to N while the first term is proportional to N^2, and therefore dephasing will also occur. Both of these scenarios give the same decoherence as deduced in Section 19.3.

For bosons at high density, however, the in-flow term can dominate over the out-flow term, which is to say that in-flow can dominate over out-flow, for a state with high amplitude. When both the ground state and the state k have high occupation and therefore high in-flow, the overall sign of the integral will be positive, which means that the equation has the form

$$\frac{d}{dt}\langle a_k^\dagger a_0 \rangle = \frac{1}{\tau}\langle a_k^\dagger a_0 \rangle, \tag{21.1.11}$$

which gives exponential growth of the coherent amplitude.

When there is a region of k-space near $k = 0$ with net influx, coherence will be amplified, with exponential growth of the amplitude of the phase. Therefore, if there is any small nonzero correlation $\langle a_k^\dagger a_0 \rangle$, it will rapidly grow to become macroscopic. The growth of the coherent amplitude will cease in equilibrium. This justifies the assumption often made in the literature that states near to $k = 0$ in a Bose–Einstein condensate are coherent and can be treated as classical waves governed by the Gross–Pitaevskii equation.[2]

The question remains how to get an initial nonzero value of $\langle a_k^\dagger a_0 \rangle$ which can be amplified. This could be created by elastic scattering of the form (14.2.22) given in Section 14.2.3, but this requires spatial inhomogeneity, that is, asymmetry in space. On the other hand, one can imagine that, even in a spatially homogeneous universe, there is some small non-unitary term which acts on bosonic states of the type hypothesized for fermionic states in Chapter 20. This would imply that not only spontaneous fermionic collapse to particle-like behavior occurs, but also bosonic collapse to wavelike behavior. Note that the method of calculation leading to Equation (21.1.10) applies to any boson system, including ones without number conservation, such as phonons with a Planck distribution in equilibrium. Whenever occupation numbers exceed unity, there can be amplification of coherent phase fluctuations. It can be argued that this is the fundamental reason why macroscopic systems such as water waves are always in definite-amplitude states instead of Fock states.

The time evolution that leads to onset of phase coherence in condensates has been an active topic of study over the past two decades (see, e.g., Kagan 1995; Stoof 1995). The results of this section apply to a homogeneous gas in the thermodynamic limit. Berloff and Svistunov (2002) showed that after there is phase coherence locally in a condensate, in what may be called "coarse grains" that are nearly homogeneous, the system will eventually obtain long-range coherence on a timescale that is proportional to the size of the system.

---

[2] For the definition and derivation of the Gross–Pitaevskii equation, see, for example, Snoke 2020, Section 11.3.

21.2 Why Phase Coherence Leads to "Super" Behavior

A system with a Bose–Einstein condensate is often called *superfluid*, because it has special properties. These properties arise from the coherent wave nature of the condensate.

To discuss macroscopic condensates, we adopt the *coarse graining* approximation, which takes the phase of a condensate as nearly constant over a small region (a "grain") but allows for it to vary slowly over long distances. Writing the coherent wave function of the condensate as $\psi = \sqrt{n(\vec{x})}e^{i\theta(\vec{x})}$, the proper quantum mechanical operator for the number current density is the symmetrized operator (Cohen-Tannoudji 1977)

$$\vec{g} = \text{Re}\left(\frac{1}{m}\psi^*(-i\hbar\nabla\psi)\right) = \frac{i\hbar}{2m}\left(\psi\nabla\psi^* - \psi^*\nabla\psi\right). \tag{21.2.1}$$

Using the definition of the wave function, this becomes

$$\begin{aligned}
\vec{g} &= \frac{i\hbar}{2m}\left(\sqrt{n_0}e^{i\theta}(e^{-i\theta}\nabla(\sqrt{n_0}) - i\sqrt{n_0}e^{-i\theta}\nabla\theta)\right.\\
&\quad \left. -\sqrt{n_0}e^{-i\theta}(e^{i\theta}\nabla(\sqrt{n_0}) + i\sqrt{n_0}e^{i\theta}\nabla\theta)\right)\\
&= \frac{\hbar}{m}n_0\nabla\theta.
\end{aligned} \tag{21.2.2}$$

In other words, the current is equal to the gradient of the phase. We see here that the phase gradient is a macroscopic observable, in contrast to the common notion that phase is only a mathematical construct. Phase plays the same role in relation to the superfluid current as electric potential plays in relation to electric field. Although the absolute value of the phase cannot be measured, relative changes of the phase are macroscopically observable.

Several theorems of vector calculus allows us to write down some relations based on this result. First, if the density n_0 is constant, which is the case if the temperature is constant, then we know that

$$\nabla \times \vec{g} = \frac{\hbar}{m}n_0\nabla \times (\nabla\theta) = 0 \tag{21.2.3}$$

since the curl of a gradient is zero. This implies that, if the condensate density is the same everywhere, then the liquid is irrotational, since by the Stoke's theorem, for any path that does not contain a pole,

$$\int_A (\nabla \times \vec{g}) \cdot d\vec{A} = \oint_L \vec{g} \cdot d\vec{l} = 0, \tag{21.2.4}$$

where L is a closed loop and A is the area enclosed. More generally, if we allow for the possibility of a pole inside the loop, then we have

$$\begin{aligned}
\oint_L \vec{g} \cdot d\vec{l} &= \frac{\hbar}{m}n_0 \oint_L \nabla\theta \cdot d\vec{l}\\
&= \frac{\hbar}{m}n_0(\theta_2 - \theta_1),
\end{aligned} \tag{21.2.5}$$

where θ_1 and θ_2 are the beginning and ending values of θ in going around the loop. If θ is single-valued, then we must have

$$\theta_2 - \theta_1 = 2\pi N, \tag{21.2.6}$$

where N is an integer. The flow in any closed loop is therefore quantized,

$$\oint_L \vec{g} \cdot \vec{dl} = N \left(\frac{h}{m} n_0 \right). \tag{21.2.7}$$

We identify the number N as the number of *vortices*, that is, poles in the superfluid current.

The Navier–Stokes equation governing the flow of an incompressible fluid is (Chaikin 2000)

$$\rho \frac{\partial \vec{v}}{\partial t} + \rho \vec{v} \cdot \nabla \vec{v} = -\nabla P + \eta \nabla^2 \vec{v}, \tag{21.2.8}$$

where ρ is the mass density, $\vec{v}$ is the fluid velocity, P is the pressure, and η is the viscosity. Since the current $\vec{g}$ is the density times the average velocity, the velocity of the superfluid is $\vec{g}/n_0$, which implies

$$\vec{v} = \frac{\vec{p}}{m} = \frac{\hbar \nabla \theta}{m}. \tag{21.2.9}$$

Therefore, assuming that n_0 is constant, $\vec{\nabla} \times \vec{g} = 0$ implies $\vec{\nabla} \times \vec{v} = 0$.

We now use the vector identity

$$\nabla^2 \vec{v} = \vec{\nabla}(\nabla \cdot \vec{v}) - \vec{\nabla} \times (\vec{\nabla} \times \vec{v}), \tag{21.2.10}$$

and substitute it into the Navier–Stokes equation to obtain

$$\frac{\partial \vec{v}}{\partial t} + \vec{v} \cdot \nabla \vec{v} = -\frac{1}{\rho} \vec{\nabla} P + \eta(\vec{\nabla}(\nabla \cdot \vec{v}) - \vec{\nabla} \times (\vec{\nabla} \times \vec{v})). \tag{21.2.11}$$

Since $\vec{\nabla} \times \vec{v} = 0$, and in the incompressible case $\nabla \cdot \vec{v} = 0$, the last two terms drop out, and we are left with

$$\frac{\partial \vec{v}}{\partial t} + \vec{v} \cdot \nabla \vec{v} = -\frac{1}{\rho} \vec{\nabla} P. \tag{21.2.12}$$

This is the fluid equation for a liquid with no viscosity. The condensate is therefore a "super"-fluid.

This analysis does not mean that superfluids (and superconductors, which are the same thing, as shown in Section 21.3) have no dissipation whatsoever. As shown by several recent works (e.g., McDonald 2016, and references therein), condensates still have friction-type drag and dissipation (cooling off) due to interaction with a separate, incoherent system. This has the effect of giving some decoherence to the otherwise coherent condensate, but not necessarily so much that all phase coherence is lost.

21.3 Superconductors and Superfluids are the Same Thing

The most common type of superconductor is known as a BCS (Bardeen–Cooper–Schreiffer) superconductor, after the three physicists who published the theory for the effect in the later 1950s (Bardeen 1957; for comprehensive review of the theory, see Lifshitz 1980; Leggett 2006). The effects of superconductivity were known much earlier, from the time of the first experiments with cryogenic refrigerators that could go to nearly absolute zero temperature (Kammerlingh Onnes 1911). The theory of superconductivity continues to be a major subject of solid-state physics, especially with the discovery of high-temperature superconductivity.

Before Bardeen, Cooper, and Schreiffer published their theory, Schafroth (1954) proposed that superconductivity could be explained as Bose–Einstein condensation (BEC) of electron pairs. When the BCS theory was written down, it did not immediately appear that it was equivalent to BEC. The BCS wave function is written as

$$|\Psi\rangle = \prod_{\vec{k}} \left(u_k + v_k b^{\dagger}_{\downarrow,-\vec{k}} b^{\dagger}_{\uparrow,\vec{k}} \right) |0\rangle, \tag{21.3.1}$$

where $b^{\dagger}_{\downarrow,-\vec{k}}$ and $b^{\dagger}_{\uparrow,\vec{k}}$ are the creation and destruction operators for electrons with spin down and spin up, respectively. The complex number v_k gives the probability amplitude for finding a pair occupied, and u_k the probability amplitude for finding it unoccupied, with the normalization condition

$$u_k^2 + v_k^2 = 1. \tag{21.3.2}$$

This appears to depend only on fermionic properties.

A long history of the theory of *BEC–BCS crossover* has shown that the BCS theory and standard BEC are two limits of the same theory.[3]

We start by writing the *Cooper pair* creation operator with center-of-mass momentum equal to zero as

$$c^{\dagger}_0 = \sum_{\vec{k}} \phi(\vec{k}) \, b^{\dagger}_{\downarrow,-\vec{k}} b^{\dagger}_{\uparrow,\vec{k}} . \tag{21.3.3}$$

We assume that the wave function $\phi(\vec{k})$ is the Fourier transform of the real-space wave function for the orbital motion of the two fermions around each other. This is the same form of wave function as found, for example, for the motion of an electron paired to a proton in a hydrogen atom. (See Section 15.1.3 for the derivation of the Cooper pairing interaction; for more details see Snoke 2020, Section 11.7.)

Using these pair operators, we can create a coherent state with the same properties as a coherent state of pure bosons, discussed in Section 12.5. A coherent state is a state with definite phase; that is, definite complex amplitude, which is the essence of BEC, as we have

---

[3] Very early, Blatt (1964) showed that a BEC state is equivalent to a coherent state of bosons, and the derivation here follows his work. Later work was done by Keldysh, Noziéres, and Randeria, and coworkers; see Keldysh (1995) and Randeria (1995).

seen. We insert the operator (21.3.3) into the formula for the coherent state (12.5.5), which gives us

$$|\alpha\rangle = e^{-|\alpha|^2/2} \sum_N \frac{(\alpha c_0^\dagger)^N}{N!} |0\rangle, \qquad (21.3.4)$$

where $\alpha = \sqrt{N_0} e^{i\theta}$ is the complex amplitude of the coherent wave, also called the *order parameter* of the condensate, since the phase transition to condensation involves this value becoming nonzero.

The equivalence of (21.3.1) with (21.3.4) can be seen by rewriting (21.3.1) as

$$|\Psi\rangle = \prod_{\vec{k}'} u_{k'} \prod_{\vec{k}} \left(1 + \alpha\phi(\vec{k}) b_{\downarrow,-\vec{k}}^\dagger b_{\uparrow,\vec{k}}^\dagger \right) |0\rangle, \qquad (21.3.5)$$

where $\alpha\phi(\vec{k}) = v_k/u_k$, assuming u_k is never strictly equal to zero for any $\vec{k}$. In the expansion of the product over $\vec{k}$, the presence of the 1 means that we have every possible number of pairs of operators. The expansion gives us

$$\prod_{\vec{k}} \left(1 + \alpha\phi(\vec{k}) b_{\downarrow,-\vec{k}}^\dagger b_{\uparrow,\vec{k}}^\dagger \right)$$

$$= 1 + \alpha \sum_{\vec{k}} \phi(\vec{k}) b_{\downarrow,-\vec{k}}^\dagger b_{\uparrow,\vec{k}}^\dagger + \frac{1}{2} \left(\alpha \sum_k \phi(\vec{k}) b_{\downarrow,-\vec{k}}^\dagger b_{\uparrow,\vec{k}}^\dagger \right)^2 + \cdots$$

$$= \sum_{N=0}^\infty \frac{(\alpha c_0^\dagger)^N}{N!} |0\rangle. \qquad (21.3.6)$$

The higher-order terms are divided by $N!$ because there are $N!$ terms that have a sum over $\vec{k}$ to the Nth power, but these are all the same after changing the order of the operators. In the products of the sums in the expansion of (21.3.6), there are terms with more than one creation operator for the same state, which do not appear in the product (21.3.5), but these terms are zero because of Pauli exclusion.

The prefactor $C \equiv \prod u_k$ that appears in (21.3.5) is just a normalization constant, determined by the condition

$$\langle\Psi|\Psi\rangle = 1 \qquad (21.3.7)$$

$$= \langle 0| \sum_N \frac{1}{N!} \left(\alpha^* \sum_k \phi^*(\vec{k}) b_{\uparrow,\vec{k}} b_{\downarrow,-\vec{k}} \right)^N |C|^2 \sum_{N'} \frac{1}{N'!} \left(\alpha \sum_k \phi(\vec{k}) b_{\downarrow,-\vec{k}}^\dagger b_{\uparrow,\vec{k}}^\dagger \right)^{N'} |0\rangle.$$

There are $N!$ nonzero terms in each Nth order term in the sum. The normalization condition becomes

$$\langle\Psi|\Psi\rangle = |C|^2 \langle 0| \sum_N \left(1 + |\alpha|^2 \sum_{\vec{k}} |\phi(\vec{k})|^2 + \frac{1}{2} \left(|\alpha|^2 \sum_{\vec{k}} |\phi(\vec{k})|^2 \right)^2 + \cdots \right) |0\rangle$$

$$= |C|^2 e^{|\alpha|^2} = 1, \qquad (21.3.8)$$

where we have used the normalization of the wave function $\sum |\phi(\vec{k})|^2 = 1$. This implies

$$|\Psi\rangle = e^{-|\alpha|^2/2} \sum_N \frac{(\alpha c_0^\dagger)^N}{N!} |0\rangle, \qquad (21.3.9)$$

which is exactly the coherent state (21.3.4) we wrote for the condensate. The value of α is determined by

$$\prod_{\vec{k}} u_k = \prod_{\vec{k}} \sqrt{1 - v_k^2}$$

$$= \prod_{\vec{k}} \left(1 - \frac{1}{2}v_k^2 + \frac{1}{8}v_k^4 + \cdots\right)$$

$$\simeq 1 - \frac{1}{2}\sum_{\vec{k}} v_k^2 + \frac{1}{2}\left(\frac{1}{2}\sum_{\vec{k}} v_k^2\right)^2 + \cdots$$

$$= \exp\left(-\frac{1}{2}\sum_{\vec{k}} v_k^2\right) = \exp(-N_0/2), \qquad (21.3.10)$$

where we use the normalization condition $\sum v_k^2 = N_0$, the total probability in all states equals the number of pairs. This implies $|\alpha| = \sqrt{N_0}$, as we assumed in writing (21.3.4).

The operators $c_{\vec{k}}^\dagger, c_{\vec{k}}$ do not exactly obey the boson commutation relations at high density (Combescot 2005). Even in the high density limit, however, the operator c_0 always has exactly the same action on the condensate state as the a_0 operator has for the boson condensate, namely

$$c_0|\Psi\rangle = \left(\sum_{\vec{k}} \phi^*(\vec{k}) b_{\uparrow,\vec{k}} b_{\downarrow,-\vec{k}}\right) e^{-|\alpha|^2/2} \left(1 + \alpha \sum_{\vec{k}'} \phi(\vec{k}') b_{\downarrow,-\vec{k}'}^\dagger b_{\uparrow,\vec{k}'}^\dagger\right.$$

$$\left. + \frac{1}{2}\alpha^2 \sum_{\vec{k}',\vec{k}''} \phi(\vec{k}')\phi(\vec{k}'') b_{\downarrow,-\vec{k}'}^\dagger b_{\uparrow,\vec{k}'}^\dagger b_{\downarrow,-\vec{k}''}^\dagger b_{\uparrow,\vec{k}''}^\dagger + \cdots\right)|0\rangle$$

$$= e^{-|\alpha|^2/2} \left(\alpha \sum_{\vec{k}} |\phi(\vec{k})|^2 + \alpha^2 \sum_{\vec{k}} |\phi(\vec{k})|^2 \sum_{\vec{k}'} \phi(\vec{k}') b_{\downarrow,-\vec{k}'}^\dagger b_{\uparrow,\vec{k}'}^\dagger + \cdots\right)|0\rangle$$

$$= \alpha|\Psi\rangle, \qquad (21.3.11)$$

where we have again used the normalization condition $\sum |\phi(\vec{k})|^2 = 1$. In both this case and in the weakly interacting condensate, the operator acts like a simple complex number, not changing the state of the condensate.

It is often not appreciated that the BCS state (21.3.1) is equivalent to a coherent state. Figure 21.1 shows the crossover between the two limits of BEC and BCS. At low density, the occupation number of the fermion states is just proportional to the Fourier transform of the pair wave function. If we keep increasing the number of pairs in the ground state, however, the probability of finding a single particle in a given $\vec{k}$-state cannot keep increasing

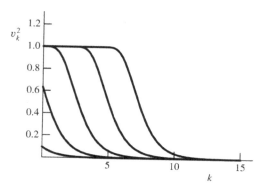

Figure 21.1   The probability of finding a fermion as a function of k, for various pair densities.

linearly, because the occupation of any fermion state cannot exceed unity. Therefore, when $N > V/a^3$, the ground state cannot be simply the product of pair states (21.3.4). The function $\phi(\vec{k})$ must be altered, to keep a maximum of one fermion per $\vec{k}$-state. As the electron density increases, the pair wave function evolves to keep the total probability per state less than or equal to unity. At high density, the pair wave function will resemble a Fermi sea, with all states filled below a density-dependent Fermi energy.[4]

What are the fermions doing? The picture in Figure 21.1 is useful for addressing an important question about superconductors, and also for all known Bose–Einstein condensates, which are also made of composite bosons – for example, two electrons, two protons, and two neutrons in a helium atom, which is an even number of fermions, with a total integer spin. How can fermions act like bosons? Underlying all the equations, they are still fermions, after all.

As seen in Figure 21.1, the fermions, even at high density, never have an occupation number of any one state of greater than 1. Therefore, every fermion is doing something different.

What we have essentially done is to separate the motion of the pairs into two different degrees of freedom: the center of mass of each pair, and the motion of the fermions relative to that center of mass. The center-of-mass motion acts like a boson excitation, as we have seen, but the total quantum state of each fermion is that center-of-mass motion *plus* its motion relative to the center of mass. Therefore, each fermion has a different state.

In relation to the spontaneous collapse model of Section 20, the center-of-mass degree of freedom has very low dephasing (and, as we have seen in Section 21.1, *enphasing* away from equilibrium), and therefore is unaffected by the spontaneous collapse mechanism. The relative motion degrees of freedom, however, will tend to collapse into eigenstates with a definite number. In this case, the proper eigenstates are not plane-wave k-states, but are the orbitals of the pairs, analogous to the orbital states of hydrogen atoms. The lowest orbital of each pair is a distinguishable fermion state, with exactly one fermion in each state.

---

[4] The form of the BCS wave function at high density has been worked out using variational theory; see, for example, Snoke 2020, Section 11.8.

21.4 Lasers Also Involve Spontaneous Symmetry Breaking

Lasers fundamentally exist because of the same property of bosons invoked for Bose–Einstein condensates and superconductors, namely the spontaneous phase coherence of bosons at high amplitude. One difference, however, is that that they do not normally involve a thermodynamic phase transition with a critical temperature. Instead, a different externally controlled parameter drives the phase transition to lasing, namely the power of a pump of incoherent photons into the system. In this section, we derive an amplification equation that gives spontaneous symmetry breaking for lasers, completely analogous to the result of Section 21.1. Under certain conditions, though, it is also possible to have photons form a Bose–Einstein condensate via a thermodynamic transition.[5]

The coupling of the dipole moments in a medium and the electromagnetic field is given by Maxwell's wave equation.[6] We assume here that the electromagnetic wave has the simple form $E(\vec{k}, t) = E_0 e^{i(\vec{k}\cdot\vec{x} - \omega t)}$, which allows us to resolve the spatial second derivative that appears in Maxwell's wave equation as

$$\nabla^2 E = -k^2 E = -(\omega^2/c^2)E. \tag{21.4.1}$$

For an isotropic and spatially homogeneous system, Maxwell's wave equation is then

$$-\omega^2 E = \frac{\partial^2 E}{\partial t^2} + \frac{1}{\varepsilon_0}\frac{\partial^2 P}{\partial t^2}. \tag{21.4.2}$$

We now drop consideration of the spatial dependence of the electromagnetic field and write simply $E = E_0(t)e^{-i\omega t}$.

The polarization of the medium P that appears in this equation can be connected to quantum mechanical states of an ensemble of two-level oscillators using the methods developed in Section 14.3. The standard model for a laser is a set of isolated two-level oscillators that can be pumped into their excited states by an incoherent external source (e.g., an electrical current, or an incoherent light), as illustrated in Figure 21.2. In the case when the coherent driving field is resonant with the energy gap between the two levels, the equations for this system are

$$\frac{\partial U_1'}{\partial t} = -\frac{U_1'}{T_2}$$

$$\frac{\partial U_2'}{\partial t} = -\frac{U_2'}{T_2} - \omega_R U_3'$$

$$\frac{\partial U_3'}{\partial t} = -\frac{U_3' + 1}{T_1} + \omega_R U_2', \tag{21.4.3}$$

---

[5] *Polaritons* are photons strongly coupled to electronic excitations; in an optical cavity they have an effective mass, and undergo Bose–Einstein condensation (see, e.g., Kasprzak 2006, Balili 2007; Liu 2015). Photons with an effective mass but with very weak interactions have also been shown to condense by means of absorption and reemission by a dye (Klaers 2010). For reviews of these effects, see articles in Proukakis 2017.

[6] For a derivation of Maxwell's wave equation, see, for example, Snoke 2020, Section 3.5.

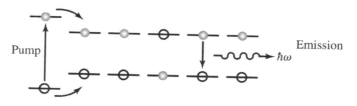

Figure 21.2   Basic picture of a solid-state laser.

where U_1' and U_2' are the real and imaginary parts of the off-diagonal term of the density matrix in the rotating frame, $U_3' = \rho_{cc} - \rho_{vv}$ is the population inversion, and $\omega_R = e|\langle x \rangle|E_0/\hbar$ is the Rabi frequency.

We now add one extra term, which represents an incoherent pump that acts to promote electrons from the lower to the upper state, as in the process shown in Figure 21.2. This process will saturate due to Pauli exclusion when the upper level is fully occupied, that is, when $U_3' = 1$. We write

$$\frac{\partial U_1'}{\partial t} = -\frac{U_1'}{T}$$

$$\frac{\partial U_2'}{\partial t} = -\frac{U_2'}{T} - \omega_R U_3'$$

$$\frac{\partial U_3'}{\partial t} = -\frac{U_3' + 1}{T} + \omega_R U_2' + G(1 - U_3'), \qquad (21.4.4)$$

where G is the rate of generating excited electrons in the upper state. Here we have also set $T_2 = T_1 \equiv T$, for simplicity.

The polarization in the medium is proportional to the off-diagonal density matrix element, which is given by the Bloch vector. We write

$$P(t) = \frac{qN}{V} \langle c|x|v \rangle (U_1'(t) + iU_2'(t))e^{-i\omega t}. \qquad (21.4.5)$$

To find a self-consistent solution for the electric field amplitude $E_0(t)$, we need to use this polarization in the Maxwell wave equation (21.4.2). Because the field amplitude E_0 appears in the Rabi frequency ω_R, the self-consistent solution can in general be a complicated function of time.

To simplify the solution, we assume that the timescale for change of the amplitude E_0 is long compared to all other timescales (this is known as the slowly varying envelope approximation). Taking the rate of change of E_0 as slow compared to the relaxation rate $1/T$ and the pumping rate G allows us to find the Bloch vector $\vec{U}'$ as the steady-state solution of the equations (21.4.4) for a given value of E_0. Setting all time derivatives to zero yields

$$U_1' = 0, \quad U_2' = -\omega_R T \frac{GT - 1}{GT + 1 + \omega_R^2 T^2}, \quad U_3' = \frac{GT - 1}{GT + 1 + \omega_R^2 T^2}. \qquad (21.4.6)$$

When the Rabi frequency is small, for example, when the electric field is not too strong, we can rewrite the formula for U_2' as

$$U_2' = -\omega_R T \left(\frac{GT-1}{GT+1} \right) \frac{1}{1 + \omega_R^2 T^2/(GT+1)}$$

$$\approx -\omega_R T \left(\frac{GT-1}{GT+1} \right) \left(1 - \frac{\omega_R^2 T^2}{GT+1} \right). \tag{21.4.7}$$

Using the definition of the Rabi frequency gives us the polarization of the medium

$$P(t) = -i \frac{e^2 \langle x \rangle^2 N}{\hbar V} E_0(t) T \left(\frac{GT-1}{GT+1} \right) \left(1 - e^2 \langle x \rangle^2 E_0^2(t) \frac{T^2}{GT+1} \right) e^{-i\omega t}, \tag{21.4.8}$$

which can be written simply as

$$P(t) = -i(AE_0(t) - BE_0^3(t))e^{-i\omega t}, \tag{21.4.9}$$

where A and B are constants that depend on the properties of the medium and the incoherent pumping rate G. The constant A can be either positive or negative depending on whether the generation rate G is larger than the relaxation rate $1/T$.

Substituting this into the Maxwell's wave equation (21.4.2), we obtain

$$-\omega^2 E_0(t)e^{-i\omega t} = \left(1 - \frac{iA}{\varepsilon_0} \right) \frac{\partial^2}{\partial t^2} E_0(t)e^{-i\omega t} + \frac{iB}{\varepsilon_0} \frac{\partial^2}{\partial t^2} E_0^3(t)e^{-i\omega t}. \tag{21.4.10}$$

Since we assume that the rate of change of E_0 is slow, such that $(1/E_0)\partial E_0/\partial t \ll \omega$, we can approximate the time derivatives as

$$\frac{\partial^2}{\partial t^2} E_0(t)e^{-i\omega t} \approx \left(-2i\omega \frac{\partial E_0}{\partial t} - \omega^2 E_0 \right) e^{-i\omega t}$$

$$\frac{\partial^2}{\partial t^2} E_0^3(t)e^{-i\omega t} \approx -\omega^2 E_0^3 e^{-i\omega t}, \tag{21.4.11}$$

so that (21.4.10) becomes

$$0 = \left(-2i\omega \frac{\partial E_0}{\partial t} \right) + \left(\frac{iA}{\varepsilon_0} \right) \left(-2i\omega \frac{\partial E_0}{\partial t} - \omega^2 E_0 \right) - \frac{iB}{\varepsilon_0} \left(-\omega^2 E_0^3 \right). \tag{21.4.12}$$

For typical doping densities in laser media, $A/\varepsilon_0 \ll 1$. We can therefore drop the term in which A/ε_0 multiplies the time derivative of E_0 as small compared to the first term, and rewrite this equation as

$$\frac{\partial E_0}{\partial t} = \frac{\omega}{2\varepsilon_0}(AE_0 - BE_0^3). \tag{21.4.13}$$

If A is negative, in other words, if the relaxation exceeds the excitation, then any coherent electric field amplitude will decay exponentially. If A is positive, in other words, if there is net *gain* in the system, then a small coherent electric field will grow exponentially. It will continue to grow until it reaches a steady state value given by the solution of

$$- AE_0 + BE_0^3 = 0. \tag{21.4.14}$$

Just as with the condensate considered in Section 21.1, any fluctuation that gives rise to a small coherent field will be amplified until a macroscopic coherent field is established.

Note that this treatment of lasing did not require any discussion of mirrors. All that is needed is for the gain to outweigh the loss for one state. Using mirrors to reflect photons back through the medium is one way to decrease the loss for one state. We have, of course, introduced many simplifications in this model of lasing, but the effect of spontaneous onset of macroscopic optical coherence is a general one.

References

R. Balili, V. Hartwell, D. W. Snoke, L. N. Pfeiffer, and K. West, "Bose–Einstein condensation of microcavity polaritons in a trap," *Science* **316**, 1007 (2007).

J. Bardeen, L. Cooper, and J. R. Schriffer, "Theory of superconductivity," *Physical Review* **108**, 1175 (1957).

N. G. Berloff and B. V. Svistunov, "Scenario of strongly nonequilibrated Bose–Einstein condensation," *Physical Review A* **66**, 013603 (2002).

J. M. Blatt, *Theory of Superconductivity*, (Academic Press, 1964).

P. M. Chaikin and T. C. Lubensky, *Principles of Condensed Matter Physics*, (Cambridge University Press, 2000), p. 449.

C. Cohen-Tannoudji, B. Diu, and F. Laloë, *Quantum Mechanics* (Wiley, 1977), Section 3.D.1.c.

M. Combescot and O. Betbeder-Matibet, "How composite bosons really interact," *European Physical Journal B* **48**, 469 (2005).

Yu. Kagan, "Kinetics of Bose–Einstein condensate formation in an interacting Bose gas," in *Bose–Einstein Condensation*, A. Griffin, D. W. Snoke, and S. Stringari, eds., (Cambridge University Press, 1995).

H. Kammerlingh Onnes, "Further experiments with liquid helium. C. On the change of electric resistance of pure metals at very low temperatures etc. IV. The resistance of pure mercury at helium temperatures," *Proceedings of the Section of Sciences* **13**, 1274 (1911).

J. Kasprzak, M. Richard, S. Kundermann, et al., "Bose–Einstein condensation of exciton-polaritons," *Nature* **443**, 409 (2006).

L. Keldysh, "Macroscopic coherent states of excitons in semiconductors," in *Bose–Einstein Condensation*, A. Griffin, D. W. Snoke, and S. Stringari, eds., (Cambridge University Press, 1995).

K. Klaers, J. Schmitt, F. Vewinger, and M. Weitz, "Bose–Einstein condensation of photons in an optical microcavity," *Nature* **468**, 545 (2010).

A. Leggett, *Quantum Liquids*, (Oxford University Press, 2006).

E. M. Lifshitz and L. P. Pitaevskii, *Statistical Physics: Part 2*, (Pergamon Press, 1980).

G.-Q. Liu, D. W. Snoke, A. J. Daley, L. N. Pfeiffer, and K. West, "A new type of half-quantum circulation in a macroscopic polariton spinor ring condensate," *Proceedings of the National Academy of Sciences (USA)* **112**, 2676 (2015).

R. G. McDonald and A. S. Bradley, "Brownian motion of a matter-wave bright soliton moving through a thermal cloud of distinct atoms," *Physical Review A* **93**, 063604 (2016).

N. P. Proukakis, *Universal Themes of Bose–Einstein Condensation*, N. Proukakis, D. W. Snoke, and P. B. Littlewood, eds., (Cambridge University Press, 2017).

Randeria 1995. M. Randeria, "Crossover from BCS theory to Bose–Einstein condensation," in *Bose–Einstein Condensation*, A. Griffin, D. W. Snoke, and S. Stringari, eds., (Cambridge University Press, 1995).

M. R. Schafroth, "Theory of superconductivity," *Physical Review* **96**, 1442 (1954).

D. W. Snoke, G.-Q. Liu, and S. M. Girvin, "The basis of the second law of thermodynamics in quantum field theory," *Annals of Physics* **327**, 1825 (2012).

D. W. Snoke, *Solid State Physics: Essential Concepts*, 2nd ed., (Cambridge University Press, 2020).

H. T. C. Stoof, "Condensate formation in a Bose gas," in *Bose–Einstein Condensation*, A. Griffin, D.W. Snoke, and S. Stringari, eds., (Cambridge University Press, 1995).

Appendix A **Summary of quantum interpretations**

This appendix gives a shorthand summary of different interpretations of quantum mechanics, including some not discussed elsewhere in this book.

1. **Local hidden variables**

   Summary: Each particle has some hidden local property that can make it act differently from others under identical external circumstances.
   Proponents: Albert Einstein
   Appeal: Agrees with many physicists' intuition.
   Negatives: Disagrees with experiments.
   Discussion: Sections 4.3 and 16.1

2. **Copenhagen ("knowledge waves")**

   Summary: The waves of quantum mechanics are not "real"; only the particles are. When a measurement occurs, the wave function changes nonlocally to correspond to updated human knowledge.
   Proponents: Niels Bohr, most modern textbooks.
   Appeal: Seems to solve the nonlocal causality problem; agrees with experiments.
   Negatives: No intrinsic justification of the Born rule; still has causality paradoxes; the modern understanding of wave functions makes them seem very "real."
   Discussion: Sections 4.4, 6.1, 6.6, and 7.1

3. **Many-worlds**

   Summary: At every quantum event, the entire universe splits into billions of ever-decreasing fractions, representing all possible options.
   Proponents: Henry Everitt, John Wheeler, and David Wallace
   Appeal: Treats the wave functions as real, in agreement with physicists' intuition.
   Negatives: Like Copenhagen, it still has causality paradoxes and no intrinsic justification of the Born probability rule; it requires linearity of the fields to a fantastic degree, going against most scientists' intuition.
   Discussion: Sections 5.1, 6.6, and 16.2

4. Pilot waves

Summary: Both waves and particles exist as separate entities; particle motion is influenced by the waves.

Proponents: David Bohm

Appeal: Treats the wave functions as real, in agreement with physicists' intuition. The math works out for some cases.

Negatives: The particles of this hypothesis cannot be equated with the particles of quantum field theory; there are sometimes bizarre implications for particle motion; there is no version for relativistic or massless particles.

Discussion: Sections 5.2, and 16.3

5. Revised Copenhagen

Summary: Any macroscopic detector plays the same role as human consciousness in the Copenhagen interpretation. Standard laws of quantum mechanics will eventually find a way to account for this.

Proponents: The default view of many physicists.

Appeal: Removes the special role for human brains that seems counterintuitive; agrees with Copenhagen otherwise.

Negatives: Incompatible with existing unitary laws of quantum mechanics.

Discussion: Section 6.1

6. Spontaneous localization

Summary: On a random basis, particle wave functions collapse to localized values. A universal noise source, which may be related to gravity, causes this.

Proponents: Ghirardi, Rimini, Weber (GRW), Diósi, and Penrose

Appeal: Provides an explicit mechanism for collapse; the math works out in some cases.

Negatives: Violates energy conservation; some versions have been shown to disagree with experiments.

Discussion: Sections 5.4 and 6.5

7. Transactionalism

Summary: Waves traveling backwards in time give the nonlocality of quantum mechanical correlations; spontaneous jumps are triggered by the interaction of forward and backward waves.

Proponents: John Cramer, Carver Mead, and Ruth Kastner

Appeal: Explicitly addresses the nonlocality of quantum mechanics; the math works out for some simple cases.

Negatives: The math that gives spontaneous jumps has not been worked out for the general case of a three-dimensional space.

Discussion: Sections 5.4 and 16.4

8. Spontaneous fermionic collapse

Summary: Local environmental fluctuations lead to spontaneous collapse of fermion states due to a nonunitary term in the Schrödinger equation.

Proponents: Presented in this book.

Appeal: Conserves energy; collapse is connected to decoherence as expected; works with many-particle quantum field theory.

Negatives: Seems to require reference to a universal reference frame; no direct experimental verification or falsification yet.

Discussion: Sections 6.4 and 6.5, and Chapter 20

9. Positivism

Summary: Don't ask questions, just do the calculations.

Proponents: Rudolph Carnap, and Kurt Gödel; the default view of many physicists.

Appeal: Easily dismisses philosophical enigmas.

Negatives: Science has historically progressed by people asking "what is really going on?" and not just calculating.

Discussion: Sections 4.4 and 5.3

10. Consistent histories

Summary: A novel mathematical method is used to construct a past history for each particle after a set of measurements is done.

Proponents: Robert Griffiths

Appeal: Allows one to maintain the picture of particles as real entities.

Negatives: The additional mathematical methods are not favored by the underlying quantum field theory; sometimes very counterintuitive implications for particle behavior.

Discussion: Section 5.3

11. Quantum logic

Summary: The regular laws of logic must be dropped, and we must accept that things that seem irrational by normal logic are valid in a new quantum logic.

Proponents: John von Neumann

Appeal: We know that computers can be programmed to obey different "logics" (sets of rules), so why not allow a new logic?

Negatives: Human logic is not fundamentally defined by a set of rules arbitrarily relating 0 and 1 values; it is defined by what we can conceive without self-contradiction.

Discussion: Section 5.3

12. Quantum Bayesianism (qubism)

Summary: A variation of positivism, with no commitment to the reality of quantum entities; only the results of certain types of measurements.
Proponents: David Mermin
Appeal: Like positivism, it is a "shut up and calculate" approach.
Negatives: Like positivism, it leaves many physicists feeling it takes an easy way out of answering hard questions.

13. Relationalism

Summary: Essentially another variation of positivism, but with much in common with the many-worlds approach. Fields are real and evolve deterministically, and there is no special role for observers or knowledge, but only results of measurements in relation to us are relevant.
Proponents: Carlo Rovelli
Appeal: A fully field-theoretical approach.
Negatives: No specific mechanism for random collapse; like positivism, it dismisses some questions as invalid.

Index